Data Engineering

Data Engineering

Fuzzy Mathematics in Systems Theory and Data Analysis

Olaf Wolkenhauer

A Wiley-Interscience Publication
JOHN WILEY & SONS, INC.
New York / Chichester / Weinheim / Brisbane / Singapore / Toronto

This text is printed on acid-free paper. ♾

Published simultaneously in Canada.

For ordering and customer service, call 1-800-CALL-WILEY.

Library of Congress Cataloging in Publication Data is available.

ISBN 0-471-41656-8

10 9 8 7 6 5 4 3 2 1

Contents

Preface

> We see an ever-increasing move toward inter and trans-disciplinary attacks upon problems in the real world [..]. The system scientist has a central role to play in this new order, and that role is to first of all understand ways and means of how to encode the natural world into "good" formal structures.
>
> —John L. Casti (1992)

Contemporary systems which have become increasingly complex, constitute generally problems of an interdisciplinary nature. Such systems are usually endowed with various types of uncertainties in the system parameters, system structure, and in the environment in which such systems operate. With an increase in complexity of the systems considered, with either very small or very large amounts of data available, the representation of uncertainty in a mathematical model becomes of vital importance. Fuzzy mathematical concepts such as fuzzy sets, fuzzy logic and similarity relations represent one of the most influential currents in engineering and operational research[1] and is expected to play an increasingly important role in the representation of uncertainty, systems modelling and data analysis.

[1]Operational research (or Operations Research) is the application of scientific principles to business management, providing a quantitative basis for complex decisions.

Intended as a guide to system modelling, system identification and decision making (*i.e.* classification, prediction and control), various on-the-surface unconnected methodologies are presented. To cope with uncertainties, several methods based on probability theory, fuzzy mathematics and possibility theory are introduced. Presupposing little familiarity with system theory, the book endeavors to steer the reader through a number of concepts interconnected by fuzzy mathematics. For most of the theory presented in this book, I cannot claim that the given material is novel or that its arguments have great originality. Indeed, much of what I say draws on the insight of prominent researchers in the various areas addressed. However, in drawing together separately developed concepts, restating the case from a systems engineering[2] perspective, I hope to show the continuing vitality and attractiveness of fuzzy mathematics in face of new challenges appearing in science and engineering.

Textbooks on fuzzy systems, system theory, time-series and data analysis traditionally describe a theoretical framework or particular methodology and then apply these concepts to problems. I believe that such a strategy is not optimal, nor does it seem adequate to deal with the current challenges in science and engineering. Researchers in system and control theory have, over the last few decades, "zoomed in" on certain aspects of the theory, refined their mathematical tools to tremendous depths, and at the same time established various schools of thought. The subsequent accumulation of a vast amount of theoretical material within any particular area has condemned young researchers to specialize in a particular technique whereby they tend to lose the 'big picture'. Such overspecialization in training often makes it more difficult to *choose* an adequate framework within which to work. I contend that starting from the problem at hand, with the available information and an understanding of the uncertainty involved, we require knowledge of more than one methodology and how the different theoretical frameworks can be related. We shall therefore start in this book with system models and the information available to build them. Uncertainty is considered by formalizing the intuitive concept of 'expectation', from which we subsequently derive quantitative measures of uncertainty (statistics, fuzzy measures). Fuzzy sets and fuzzy relations will emerge 'naturally' from considering the application of theories (models) to observations (data). To relate formal models to sampled data, we have to generalize set-operations (as the basis for comparisons) and transitivity (as the basis for reasoning).

In the spirit of John Casti's view quoted above, control engineers should be prepared to move 'up-scale' and consider large-scale systems. Leaps in the

[2]Systems engineering is a generic term used to describe the branch of engineering which is interdisciplinary in nature and is concerned with the integration and interfacing of differing techniques (technologies) to analyze (develop) complex systems.

technology for data acquisition, storage and processing make it not necessarily feasible but necessary to deal with complex or large-scale phenomena[3]. As the complexity of systems increases and the nature and the quality of the data varies, the formalization and quantification of uncertainty becomes increasingly important. It is the aim of fuzzy mathematics and possibility theory to be precise about uncertainty by:

▷ combining quantitative data and qualitative information;

▷ integrating functional and rule-based models;

▷ complementing statistical with fuzzy mathematical concepts.

This book introduces fuzzy systems sympathetically explaining its philosophical implications and practical applications. There are four main themes or arguments running through the book:

1. Reformulating systems theory to take account of fuzzy systems and possibility theory.

2. Introducing *data engineering* as the discipline, which for a given set of complex, uncertain data, extracts information by detecting pattern, and thereby turns information into the evidence used in decision making (classification, prediction and control).

3. The quest for a methodology which enables us to combine quantitative formal analysis with qualitative context-dependent expert knowledge.

4. That when solving a real-world problem, (matching observations with a model) an empirical approach (implying heuristics)[4] is perfectly acceptable. Certainty is a myth; there is no single 'correct' methodology.

My philosophy or motivation for this book is similar to what I believe has been the primary interest of people like N. Wiener (Cybernetics), M. Mesarovic (Abstract Systems Theory), H. Haken (Synergetics), R. Rosen (Anticipatory Systems) and J. Casti (Complexification). We are primarily interested in the overall properties of systems and the character of the models used to capture the behavior of these systems or processes. An important aspect of our approach includes understanding the constraints of our enquiries.

As for terminology, I am reluctant to use the terms 'system theory' or 'control theory' as they may suggest abstraction or unrelatedness to the practise of solving 'real-world' problems. On the other hand, we may follow the

[3]If we define a large-scale system as one described in 'fine-grain' terms, complexity is only relative to the scale employed.

[4]The term *empirical* means based or acting on observation or experiment, not on theory; deriving knowledge from experience. The term *heuristic* means allowing or assisting to discover, proceeding to a solution by trial and error.

advice of mathematician David Hilbert, who once commented that "there is nothing more practical than a good theory". The terms 'systems engineering' and 'control engineering' have subsequently been used to emphasize the more applied side of modelling, estimation, filtering and control. However, this still reflects some kind of separation. Somewhat surprisingly, probability theory and statistics are often mistaken as synonymous whereas they could not be more 'apart' conceptually – for any real-world problem, probabilistic models are an abstraction through which we generalize. Whereas some in many cases modelling, as a generalization, has been the main purpose, we call 'data engineering' the practice of matching data with models, or more generally with the art of *turning data into information*. Data engineering is then related to possibility theory with the latter providing the conceptual framework for the former and thus constitutes the 'toolbox' which allows *reasoning about data in the presence of uncertainty*.

There have been many approaches to modelling, identification, control, forecasting and decision making[5], and the literature is full of detailed expositions of these individual techniques which usually lend themselves to specific, mostly well understood problems in science and engineering. While over the last few decades researchers have focused on refining these approaches and with respect to specific problems, I am convinced that current challenges – for example, in business process analysis, financial forecasting, and molecular biology – require us to "zoom out" to consider principles, paradigms and methodologies from a global or more abstract point of view. As we shall see, various methodologies, often perceived as distinct or competitive, are in fact closely related (*i.e.* complementary). Having knowledge of formal relationships and semantic differences we are now in the position to choose the appropriate representation for a given problem.

For example, in molecular biology, sequences of nucleotides (describing genomes) and amino acids (describing proteins) form the basic "information units" from which scientists try to extract *patterns*. Such patterns hint at relationships between often numerous variables. These relationships may be time-dependent, they may be 'logical' in nature (truth relations), or they simply state some unqualified form of coexistence or association (mapping). Reasonably accurate formal and possibly dynamic models may not be realistic yet but one can safely assume that no one single modelling paradigm will be sufficient to capture all the aspects of any given system. Instead it will be important to understand the principle ideas of formal mathematical modelling, the encoding of time and the formulation of states and relationships.

[5]As we shall see, the purpose of a model is, if initially for understanding hidden relationships, ultimately for decision making. Forecasting, prediction, control, classification and prioritization are only examples of decision making for which, assuming a minimum degree of rationality, we require a model.

A specific approach is then selected among well-understood principles such as finite automaton, black-box model, rule-based system, and so forth. The choice will not just depend upon the nature of these complex relationships but also upon how we perceive and we wish to use them. For this reason, we shall not be at all concerned with the intrinsic aspects of models. We will not consider, for instance, approximation and forecasting accuracy but will instead focus on the global properties, the relationships and similarities between different approaches. Hopefully, the reader will be surprised how closely related, or similar, many concepts usually considered as distinct actually are. Apart from a 'natural' and desirable specialization in order to study any particular approach in greater detail, to gain a better understanding of it, there is an unpleasant tendency in systems and control theory to emphasize differences, usually by pointing out advantages (w.r.t. particular properties), "to rubbish" other ideas in order to get your own approach considered as 'new' (publishable). This unnecessary obsession with separation, is partly fueled by the way funding is provided for research – which requires the researcher to speculate about (fairly specific) applications and to claim advantages over other concepts. This entails that, in order to be 'successful', one has to focus on one particular approach (knowledge and time are limited) and constantly point out differences rather than seeking synergy or integration (which requires a broader knowledge base, more time, and does not lead immediately to applications.).

In modern systems theory and its applications as described above, the physical structure of the system under consideration is of secondary interest. Instead, a model is intended to appropriately represent the *behavior* of the system under consideration. We therefore require methodologies and paradigms which provide *interpretability* in addition to accuracy. Accuracy is thereby understood not as the ability to provide a numerical value which minimizes (for example) the mean-square-error, but rather as the ability to be precise and honest about uncertainty. The forecast of a conventional time-series model, based on various simplifying assumptions, is not necessarily more accurate than the forecast of a rule-based fuzzy system generating a possibility distribution. A fuzzy set may be more precise than a real number!

In this book, we are concerned with the fundamental aspects of data analysis, system modelling and uncertainty calculi. It is organized as follows: Key concepts in modelling, identification and clustering, including their fuzzy mathematical extension, are reviewed. Standard paradigms to modelling and identification are discussed with the conclusion that a particular mathematical framework is a matter of preference with any experimentally verifiable approach being valid. The combination of classical state-space approaches and fuzzy systems, suggests a new paradigm for a propositional calculus of systems. Subsets of the state-space are identified with logical propositions. The motivation behind the proposed methodology is to put together two streams of ideas: a) fuzzy clustering inducing fuzzy equivalence classes with respect to generalized equivalence relations and b) the question of finding characteristics of non-transitive systems using propositional logical systems.

Typing this book, clustered into clearly separated sections forming an ordered sequence, reflected the struggle of my mind with its content – we tend to impose a linear ordering (subsequent sections) on issues where we know a web-like structure is more realistic. We also tend to *reduce* problems into smaller (linear) subproblems (chapter, section, subsection), such that the whole is represented as the sum of its parts. Generating such hierarchical structures, free of interconnections suggests dichotomies where, in fact, the whole is more than the sum of its parts. Unable to overcome the two-dimensional world of paper, screens and the linear order of words, I have avoided clearly separated chapters and encourage the reader to skip forward and backward as he or she goes along reading the book in the usual way. However, this is not intended as an excuse for neglect to structure in the layout of this book. Although I concede that the spread of arguments across various sections may at first appear a little confusing (and perhaps even daunting to some), I believe this structure to be a necessary evil and that it will eventually prove very rewarding. I fully accept that it is the responsibility of the author to control the process of forming thoughts into words, and then transcribing them onto paper as text, to be recreated in someone else's mind. It is astonishing how precisely we can communicate experiences verbally but are unable to formalize them. Instead of providing a mathematical concept in textual form it sometimes appears more convenient to focus just on the math. For example, instead of describing the context in which data are generated in order to understand what it means if they are correlated, we 'simply' establish a functional relationship where semantics seems of little importance... that is, until we try to learn something from data which we did *not* know already! The fact that the mathematical world is embodied in percepts while existing in independence from them, is often used as an excuse to "let the equations speak". It is true that mathematical truth is likewise manifest in, but independent of, any material embodiment and is thus outside the perceptual categories of space, time and causality. However, I find this only initially satisfying since we quickly realize that we *use* mathematics to describe aspects of the phenomenal world and that syntactic rules cannot be independently discussed from semantics (*i.e.* the meaning and interpretation of the context which the mathematics is intended to describe). Therefore by suggesting formal equivalence or similarity we may forget that the mathematics is about "something", that is, that the propositions express percepts or qualities. A fact or datum by itself is essentially meaningless; it is only the interpretation assigned to it that has significance. On the other hand we are informally free to interpret these propositions in any way that we want. It therefore seems to me that in order to make the struggle for completeness, consistency and clarity bearable, we need to turn confusion into a wonder for diversity and complexity.

Although the material of this book has been taught to students studying control theory, I hope it is of interest across the engineering disciplines, mathematics, and the sciences. This book is dedicated to those whose minds cross boundaries.

O. WOLKENHAUER

Manchester, April 2001

Acknowledgments

Special thanks go to Professor Peter Wellstead for providing constant encouragement and valuable advice. During the writing of the book, my research was supported by the Engineering & Physical Sciences Research Council (EPSRC) under grant GR/L 95151. Also the support of the British Council, provided to collaborate with the Control Laboratory, Ruhr-University Bochum in Germany as part of ARC Project No.934, is gratefully acknowledged. Janos Abonyi, Kung Jui Lee, Javier Nuñez-Garcia and Victor Collazo contributed simulations and plots and Serkan Impram helped at the final stages of the manuscript preparation. Between November 1999 and June 2000, I followed an invitation by Robert Babuška to work at the TU Delft in the Netherlands. The visit was made possible by the generous support of the ITS faculty. I am grateful to Allan Muir for many inspiring discussions and useful comments. Finally, the manuscript was prepared using MIKTEX, a freeware installation of LATEX 2ε, made available by Christian Schenk. His efforts have indirectly had a considerable influence on my work.

O. W.

Introduction

Scientific theories deal with concepts - not reality. Formula and theories are so formulated as to correspond in some 'useful' way to the real world. The quest for precision is analogous to the quest for certainty and both - precision and certainty are impossible to attain.

—Karl Popper

Control theory evolved within systems theory and from cybernetics by focussing on closed-loop (feedback) control of engineering systems. Mathematical control theory has made considerable progress not only with respect to linear time-invariant systems but also in the analysis of nonlinear systems. As a result of the progress, researchers have become specialists in usually a fairly narrow view of how to model and control a system. Current challenges in science and engineering seem to require a different breed of researchers. I believe it is again time to take a broader look at (dynamic) systems and students studying control theory should be in a good position to play a useful role in Science and Engineering: For a control engineer, 'Model-Predictive Control' is what in General Systems Theory is described by *anticipatory systems*; 'Optimal Control' describes *goal seeking systems* and possibly most general, any control system is, in fact, successive decision making in the context of *policy making*.

Within engineering, models have long been used to predict the temporal evolution of the attributes of a physical system. The successful application of

Newton's laws in describing physical systems by means of differential equations has lead to a focus on simulation and the transfer of these principles to other areas such as biology, social sciences, operations research, economics and so forth. We now realize that owing to *complexity*, and although 'in principle' possible, classical modelling is not a practical way forward. With complexity I mean in this context the problem of multi-variable, multi-level interconnected – 'complex' systems for which formal models if applicable (significant) to a larger class of processes are not specific (precise) enough for a particular problem and if accurate for a particular problem they are usually not generally applicable.

So if models, describing the physical structure, of a system are not a complete solution, what else can be done to improve our chances to obtain law-like formal models for some aspects of the real world? In molecular biology, complexity defeats physical modelling of even comparatively simple processes. Similarly, in operations research there is considerable interest in models of business processes but one goes hardly further than using simple flowcharts. Both domains, however, are currently accumulating vast amounts of data stored in databases creating new business in the areas of *data mining* and *data warehousing*.

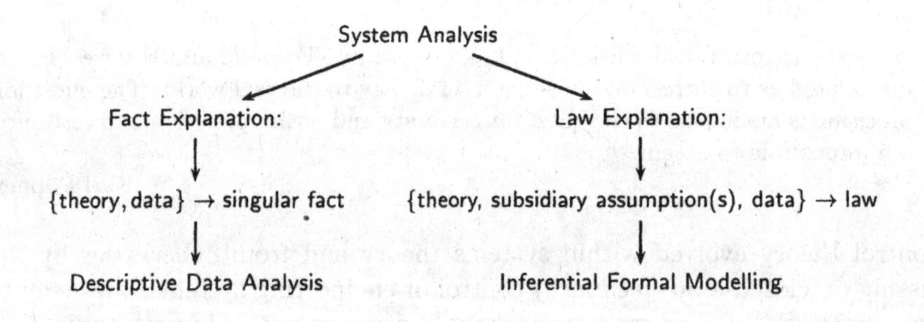

Fig. I.1 Descriptive vs. law-like explanation in system analysis [Bun98].

Figure I.1 outlines the two main modes of system analysis: *Fact Explanation*, that is, the descriptive analysis of empirical data (e.g., density and parameter estimation) and *Law Explanation*, that is, using context-dependent theoretical knowledge to describe functional relationships between variables representing the system under consideration. The purpose of the latter is usually to explain mechanisms (relationships between variables) of the system, assuming the model describes a part of the real-world accurately – "as it is". I contend that progress could be made if we instead focus on models that describe a system – "as observed". Focusing on observations and context-dependent knowledge requires us to go beyond conventional statistics and systems theory, to combine different paradigms in a pragmatic fashion and to generalize uncertainty techniques. Fuzzy mathematics, possibility theory and data engineering are the vehicles I promote in this book as a tentative way forward.

It is my view that *a priori* knowledge is vital for the analysis of complex systems and should therefore be integrated into systems analysis. This statement may, at first, seem nonsense as decades have been spent with the objective to abolish the need for *a priori* knowledge in multi-variate analysis. I believe that further advances in systems engineering will depend on the success to integrate *all kinds* of knowledge into the tools used for the analysis of complex systems. This book drafts a formal framework in which such a synergy of data and knowledge engineering could take place. The philosophy of such a program is outlined in Figure I.2. It is assumed that data sets alone would not provide sufficient information for decision making in the presence of uncertainty (*i.e.* forecasting, control and classification). The dilemma that even the most sophisticated mathematical techniques for identification and estimation still require substantial knowledge and understanding of the process under consideration, is reflected in Norbert Wiener's complaint (From 'Cybernetics: or Control and Communication in the Animal and the Machines', 1961):

> *"I may remark parenthetically that the modern apparatus of the theory of small samples, once it goes beyond the determination of its own specially defined parameters and becomes a method for positive statistical inference in new cases, does not inspire me with any confidence unless it is applied by a statistician by whom the main elements of the dynamics of the situation are either explicitly known or implicitly felt."*

In this book we shall present various methodologies which, in combination, should help us to improve our ability to turn numerical data into knowledge. If only a relatively small number of numerical data are available, we are forced to make assumptions about regions of the data space we have no numerical evidence for. This form of *extrapolation*, can often be guided by *rule-based* expert or operator experience. More recently, the area of 'Data Mining' has attempted to extract new knowledge from large volumes of multi-variate data. In these applications the amount of data available are often very large and we are required to find a form of *abstraction* to replace samples by some kind of *distribution* capturing properties of the underlying variable(s). We should note that in both cases the data available usually do not satisfy statistical assumptions of repeated trials under fixed conditions and therefore distributions used in abstraction and extrapolation become (fuzzy) restrictions of the underlying (sub)spaces. Despite all the progress made in data analysis, we must admit that even today we have to have a good idea of what we are looking for in searching data and that a good knowledge of the context in which data are generated is indispensable. More general, I shall claim that there are fundamental limits in analyzing systems. The success in modelling complex systems depends therefore very much on our ability to formalize and combine the various distinct forms of uncertainty relevant to systems engineering (*i.e.* randomness, fuzziness, vagueness, ambiguity, and imprecision).

So far the two branches of quantitative analysis (e.g., time-series analysis) and qualitative knowledge processing (e.g., expert systems) have been developed separately. I believe that the search for pattern in data with the aim to make decisions would benefit from the integration of context-dependent knowledge into quantitative multi-variate analysis. Economic forecasting is an example, illustrating the fact that though there may well be cycles which can be extracted from data sets, (rule-based) knowledge of the political context provides valuable information on trends as well[6].

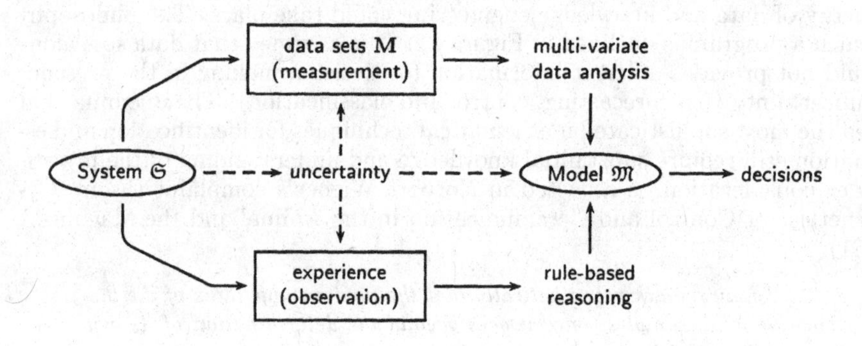

Fig. I.2 The combination of quantitative and qualitative analysis.

I.1 OVERVIEW OF THE REMAINING CHAPTERS

We start by introducing the notion of a (dynamic) system and the representation of such systems for purposes of analysis. Next, various techniques of coping with uncertainty are addressed (including least-squares criterion, maximum likelihood estimation, stochastic processes and so forth). Then, system identification (viewed as learning from data) is considered, utilizing a probabilistic approach (estimates of kernel densities and function approximation). This is followed by discussions of specifying propositions as subsets in the data space; system identification in the context of fuzzy system formulation and as random-set modelling; fuzzy inference engines; fuzzy classification; fuzzy control; and fuzzy mathematics. At the end, several appendices are provided.

[6]In stock market forecasting, the proposed methodology corresponds to a fusion of *technical analysis* and *fundamental analysis* (as the two main broad groups of approaches to predicting changes in security investments). Technical analysis is based on the premise that there are identifiable trends in the past data. Technical analysts, often called 'chartists', search data for turning points, business cycles and trends. Fundamental analysis uses information from annual reports, financial and business news and so forth. to estimate the 'intrinsic value' of a given security.

In Section 1, I discuss classical modelling as the process which specifies a set of functions. Learning from data is subsequently described as the inference of some dependency - choosing an 'appropriate' function from a given set of functions. Section 3 outlines the classical parametric approach to system identification: We try to identify a functional dependency, linear in its parameters, which is a *reasonable approximation* to the desired function. The motivation for this philosophy comes from the Weierstrass theorem, according to which any continuous function can be approximated on an interval by polynomials.

A critical discussion of the classical and fuzzy systems approaches, suggests a very different philosophy of modelling: a propositional calculus of (dynamic) systems, based on discrete sampling of a compact $(r+m)$-dimensional manifold Ξ embedded in a vector space $\mathbb{R}^{r+m}$.[7] Then, sampled data induce clusters, hence partitions, in the multi-variate data space. The approach is motivated by two developments: The logic of quantum mechanics as described in the 1930s by Birkhoff and Von Neumann and fuzzy clustering in a multi-variate data space.

Birkhoff and Von Neumann showed that for a mathematical description of a physical system one can expect to find a calculus of propositions which is formally indistinguishable from the calculus of linear (function) subspaces. Fuzzy clustering, on the other hand has been successful in pattern recognition due to the introduction of fuzzy partition spaces. I combine both ideas by describing the underlying mathematical problem as the description of a generalized equivalence relation by the set of its clusters in the same way as equivalence relations are uniquely determined by their sets of equivalence classes. Fuzzy clustering is introduced in Section 4.

In Section 7, I argue that identification of system models from data will inevitably induce similarity relations, that is, uncertainty is certain! I then set out to explore the algebraic structure of the data-induced quotient with respect to similarity relations, in order to describe a (fuzzy) propositional calculus for dynamic systems. The advantage of such an approach is that quantitative time-series analysis can easily be merged with a qualitative analysis of the context in which reasoning takes place. Such combined strategy should improve our predictions or at least increase confidence in forecasts about the evolution of a system. The structure of reasoning can be summarized as follows:

[7] I use the term manifold in general to describe a collection of objects of a set. For some dynamical systems, the data space turns out not to be a vector space but is instead some type of topological space. For example, hypersurfaces in $\mathbb{R}^{r+m}$ or affine subsets of a vector space are manifolds.

Systems → modelling → identification → regression → clustering → uncertainty, non-transitivity → equivalence relations → quotient → categories → algebras → (fuzzy) logics → approximate reasoning ⇒ synergy of quantitative and qualitative analysis.

Combining system identification and rule-based systems (Section 5 and Section 7.2), we can achieve a synergy of quantitative models (using product space clustering) and qualitative information (using approximate reasoning). Fuzzy models which can be constructed from fuzzy clustering are described in Section 5.

I will demonstrate how probability theory, statistics, are *complemented* by fuzzy mathematics and possibility theory in the field which we refer to as *Data Engineering*. A good example for this view is the following chain of formal relationships which we show to be equivalent or at least closely related:

The least-squares principle (Sec.2.1) → Fourier series (Sec.2.1.2) → kernel density estimation (Sec.3.2) → universal, basis function approximation (Sec.3.3) → fuzzy rule-based systems (Sec.5).

Virtually all the material is derived from, or directly related to, two fundamental concepts: The formulation of a system using some dependent variable(s), denoted $\mathbf{y}$, as a function of the independent variable(s) $\mathbf{x}$:

$$\boxed{\mathbf{y} = f(\mathbf{x}; \cdot)}$$

Secondly, the expectation operator forms the basis for describing (*measuring*) (un)certainty in systems and data:

$$\boxed{E[h(\cdot)] = \int g(\cdot) \circ h(\cdot)}$$

The text is self-contained to the extent that I have tried to explain briefly all concepts involved with secondary material presented in the appendices. The interested reader will not, however, be able to avoid some decision making for further reading. Sections in this book are closely connected and some themes run across more than one section. In an attempt to keep the flow of the main arguments, examples often contain extensions and new material not covered in the preceding sections. A summary of how a system can be modelled, using various paradigms introduced in this book, is given in Section 12.

I.2 SUMMARY OF KEY CONCEPTS AND IDEAS

Section 1: System Analysis

☐ *Scientific theories deal with concepts, not with reality.*

- ☐ *System theory used mathematical concepts to describe aspects of the 'real-world'.*
- ☐ *A formal model is a graph, i.e. a subset of a product space formed by variables characterizing a system or process.*
- ☐ *An observable is some characteristic of a system which can, in principle, be measured.*
- ☐ *A state is a specification of a system or process at a specific instant.*
- ☐ *A dynamic system or process is a system in which the state changes with time.*
- ☐ *Differential equations are a common way to encode dynamics.*
- ☐ *There are many alternative and equally valid ways to represent a system...*

Section 2: Uncertainty Techniques

- ☐ *The expectation operator is a generic concept to summarize information in an underlying universe of discourse.*
- ☐ *Averaging information leads to probability measures and statistics.*
- ☐ *Aggregating information leads to fuzzy measures and possibility measures in particular.*
- ☐ *Matching data with a (parametric) model, a criterion for how well the data are fitted is required.*
- ☐ *The least-squares criterion provides optimal parameter estimates for linear models.*
- ☐ *A geometric (vector) representation of the regression problem shows that the optimal solution implies orthogonality.*
- ☐ *The Fourier series is an example for function approximation using the orthogonality principle.*
- ☐ *The least-squares principle does not require a statistical framework to make sense.*
- ☐ *Maximum likelihood estimation is a statistical framework for parameter estimation.*
- ☐ *Stochastic processes are a probabilistic framework to study time-series.*
- ☐ *The Kalman-Bucy filter is a good example how a probabilistic framework, orthogonality and the expectation operator can be used to develop a new concept to model data.*

Section 3: Learning from Data: System Identification

- ☐ *The identification of a model is an approximation of the function which relates independent (e.g., input-) and dependent (e.g., output-) variables.*
- ☐ *Linear parametric regression, employing the least-squares principle, is an efficient tool to identify parameters from data - to learn linear functional relationships.*
- ☐ *In a probabilistic framework data are assumed to be distributed according to some unknown probability density function.*

- ☐ *Statistical learning can be seen as a generalization of density estimation.*
- ☐ *Like the Fourier series, Kernel density estimation provides another example of the approximation of an unknown function by means of so-called basis functions.*

Section 4: Propositions as Subsets of the Data Space

- ☐ *A more general concept to represent data sampled from a system is that of a data space.*
- ☐ *System properties and behavior are reflected by clusters of data.*
- ☐ *Clusters may be interpreted as linear submodels of an overall nonlinear system.*
- ☐ *Clusters may also be interpreted as if-then rules relating properties of the variables that form the data space.*
- ☐ *Fuzzy clustering provides least-squares solutions to identify clusters, to partition the data space into clusters or classes.*
- ☐ *Fuzzy boundaries between clusters are differentiable functions and hence are computationally attractive.*
- ☐ *For many real-world problems a fuzzy partitioning of the underlying space is more realistic than 'hard clustering'.*

Section 5: Fuzzy Systems and Identification

- ☐ *Fuzzy clustering provides an effective way to identify fuzzy rule-based models from data.*
- ☐ *Various fuzzy model structures exist and can be distinguished in terms of their simplicity, interpretability and suitability in diverse problems such as classification, prediction and control.*
- ☐ *Fuzzy systems are equivalent to the basis function expansion model.*
- ☐ *Fuzzy clustering does not make assumptions about the randomness but can be related to regression analysis.*
- ☐ *Fuzzy c-regression models yield simultaneous estimates for the parameters of c-regression models.*

Section 6: Random-Set Modelling and Identification

- ☐ *Random sets can be viewed multi-valued maps or random variables.*
- ☐ *For very small data sets, local uncertainty models (random subsets) can be used to generalize information in the data space.*
- ☐ *The estimation of coverage functions of random sets yields possibility distributions.*
- ☐ *While probabilities describe whether or not an events occurs, possibility describes the degree of confidence or feasibility to which some condition exists.*
- ☐ *A random-set approach to identification leads to qualitative predictions.*
- ☐ *Qualitative predictions are fuzzy restrictions and may therefore provide a mechanism to combine quantitative data analysis with rule-based systems describing qualitative expert knowledge.*

Section 7: Certain Uncertainty

- ☐ *Scientific investigation relies on two principal concepts: comparing and reasoning.*
- ☐ *Mathematical formulations of distance and transitivity are at the core of the modelling problem.*
- ☐ *The Poincaré paradox describes the indistinguishability of individual elements in non-mathematical continua and hence proves that uncertainty is certain.*
- ☐ *Taking account of uncertainty leads to similarity (fuzzy) relations.*
- ☐ *Fuzzy concepts therefore occur 'naturally' from fundamental analysis.*
- ☐ *Fuzzy relations motivate approximate reasoning.*
- ☐ *Approximate reasoning is a concept to capture qualitative (context-dependent) expert knowledge.*

Section 8: Fuzzy Inference Engines

- ☐ *There are various distinct ways to realize a fuzzy rule-based system, distinguished by the way rules are combined and the inference engine employed.*
- ☐ *Fuzzy systems are nonlinear mappings.*
- ☐ *Fuzzy systems are universal function approximators.*

Section 9: Fuzzy Classification

- ☐ *Fuzzy clustering groups unlabelled data into a fixed number of classes and hence can be used to design classifiers.*
- ☐ *Specific fuzzy classifiers can be shown to be formally equivalent to optimal statistical classifiers.*
- ☐ *If-then rule-based fuzzy classifiers provide an intuitive framework to interpret data.*

Section 10: Fuzzy Control

- ☐ *Fuzzy rule-based systems can also be used to devise control laws.*
- ☐ *Fuzzy control can be particularly useful if no linear parametric model of the process under control is available.*
- ☐ *Fuzzy control is not 'model-free' as a good understanding of the process dynamics may be required.*
- ☐ *Fuzzy control lacks of design methodologies.*
- ☐ *Fuzzy controllers are easy to understand and simple to implement.*

Section 11: Fuzzy Mathematics

- ☐ *Fuzzy concepts are 'natural' generalizations of conventional mathematical concepts.*
- ☐ *Probability and possibility are complementary.*
- ☐ *Possibility theory is the attempt to be precise about uncertainty, to related statistical objects with rule-based and fuzzy concepts.*

I.3 SYMBOLS AND NOTATION

The notation used generally follows the usual conventions in the literature. Matrices are denoted by bold capital letters while, $\mathbf{x}$ can have different meanings: small bold letters describe random or fuzzy variables, vectors and combinations of them. $\mathbf{x}$ is also used to denote the regressor vector and $\mathbf{x}_j$ describes a vector of measured values for the variables in the regressor vector. Indices are generally used with equivalent meaning; j is primarily used to index measurements $\mathbf{m}_j$ while i is used to index rules, regressors, and so forth. t denotes (continuous-) time, t_k or k discrete-time. Dependence on time is usually described by using brackets $\mathbf{x}(k)$. X denotes a space which may also be of dimension greater than one. For example, $X = X_1 \times \cdots \times X_r$. If subspaces are identified with the real line, $\mathbb{R}$, we may write $X = \mathbb{R}^r$ to denote the r-dimensional Euclidean space. Calligraphic letters such as $\mathcal{B}$ are used to represent sets of sets. In few cases the same symbol is used in different contexts. For example, σ is used to denote the standard deviation of a random-variable; $\sigma_{\mathbf{x},\mathbf{y}}$ denotes the covariance between $\mathbf{x}$ and $\mathbf{y}$, while σ_Ω denotes a σ-algebra on Ω.

$:$	"for which" or "given".
$\exists$	"there exists".
$\forall$	"for all".
$\doteq$	"defined".
$\approx$	"approximately".
$\Leftrightarrow$	"if and only if" (iff).
$\therefore$	"therefore".
$\mapsto$	"maps to".
$\Rightarrow$	"implies", material implication.
$\rightarrow$	"maps from to", general mapping.
$\mathfrak{S}$	system.
$\mathfrak{M}$	formal model.
ξ	observable.
$T = \{T_t\}$	dynamic or flow.
$I(x)$	trajectory.
$O(x)$	orbit.
Ξ	manifold, data space.
$\mathbf{o}$	observation in Ξ.
$\mathcal{C} = \{C_j\}$	set of subsets in Ξ
X	state-space, regressor space, set of abstract states.
U	input space.
Y	output space.

$\leq$ less or equal.
$\preceq$ partial ordering.
$\subseteq$ subsethood.
$\in$ elementhood.
$\cup$ union.
$\cap$ intersection.
$\vee$ disjunction.
$\wedge$ conjunction.
$\neg$ negation.
^{c} complement.
T, S triangular (t-)norm and t-conorm.
$\#\{\cdot\}$ cardinality, size, count.
$\circ$ compositional operator.
$*$ binary operation.
$\hat{}$ estimate.
* optimal value.
$'$ any other value.
$\dot{}$ derivative.

$\{\cdot\}$ general set.
$\langle\cdot\rangle$ sequence.
$\mathbb{R}$ set of real numbers, real line.
$[\cdot]$ vector.
$\vec{Y}$ vector space.
^{T} transposed.
$\|\cdot\|$ vector norm.
$d(\cdot,\cdot)$ distance, metric.
L_1 distance $\sum_i |x_i - y_i|$ between $\mathbf{x}$ and $\mathbf{y}$.
L_2 Euclidean norm.
$|\cdot|$ absolute value.
$|\mathbf{A}|$ determinant of matrix $\mathbf{A}$.
ln logarithm to base e.
log logarithm to base 2.
∇ gradient.
$\mathcal{P}$ power class, set of (crisp) sets.
$\mathcal{F}$ set of fuzzy sets.
$\mathcal{B}$ Borel algebra.
$\mathcal{T}$ Topology.
$M = (L, *, \preceq)$ cl-monoid.

s	Laplace operator.
z	z-transform operator.
$\mathbf{j}$	complex number, imaginary part.
ϑ	damping factor.
Δ	unit delay.
$\mathbf{\Phi}$	state-transition matrix.
$\mathbf{F}$	system matrix.
$\mathbf{G}$	gain matrix.
$\mathbf{x}$	regressor vector, random variable.
$\mathbf{x}(k)$	sequence of random variables.
$\mathbf{x}_j$	measurement of $\mathbf{x}$.
$\mathbf{m}_j \in \mathbf{M}$	measurement vector, data matrix.
$\mathbf{Y}$	output matrix (vector).
$\mathbf{X}$	regressor matrix.
$\mathbf{W}$	weighting matrix.
e	prediction error.
a_i, b_i	ARX, ARMA, and TS-model parameters.
τ	lag.
$\theta, \boldsymbol{\theta}$	parameter (vector).
$E[\cdot]$	expectation operator.
ζ	crisp set indicator function.
μ	fuzzy set membership function, fuzzy restriction.
Pr	probability measure.
p	probability distribution, density function.
Π	possibility measure.
π	possibility distribution.
F	cumulative distribution function.
g	fuzzy measure.
$\omega \in \Omega$	elementary outcomes.
σ	(co)variance.
η	mean.
ρ	correlation.
R^2	multiple correlation coefficient.
$N(\eta, \sigma)$	normal or Gaussian distribution.
$L(\cdot), R(\cdot)$	loss, risk functional.
$K(\cdot)$	kernel function.
$\phi(\cdot)$	basis function.
ℓ	likelihood function.
$Q(\cdot)$	risk function.
$\mathcal{L} = \log \ell$	log-likelihood function.

n	general limit.
m	number of outputs.
r	number of regressor variables.
d	number of data.
l	number of parameters.
n_u, n_y	NARX system order.
n_R	number of rules.
n_p	number of projected subsets.
n_n	number of nearest neighbors.
n_d	order of MA model.
i, j, k	indices.
δ	threshold tolerance.
ε	noise, disturbance.

$f(\cdot)$	regression function, hypersurface.
F	graph.
$\widetilde{F}$	fuzzy graph.
$\tilde{f}$	fuzzy mapping.
ψ	classifier.
$N : \Xi \to \Xi/E$	natural map.
Γ	multi- or set-valued mapping.
$⨍$	Sugeno or fuzzy integral.
R^α	level-set, α-cut.
$\alpha \in L$	co-domain for fuzzy restrictions.
$ran[\cdot]$	range.
R	rule, relation, set.
E	equivalence relation.
$\widetilde{E}$	similarity relation.
$[x]$	equivalence class.
X/E	quotient (set), factor set.
$\mathcal{E}_{\widetilde{E}}(\alpha)$	partition induced by E_α.

$\beta(\mathbf{x})$	rule fulfillment.
$\mathbf{C} = [\mathbf{c}^{(1)}, .., \mathbf{c}^{(c)}]$	cluster prototypes.
$\lambda, \mathbf{\Phi}$	eigenvalue, eigenvector.
$\mathbf{U}$	partition matrix $u_{ij} \in \mathbf{U}$.
J	objective or cost function.
l	loop counter.
c	number of clusters.
$w, \mathbf{w}$	weighting, weights.
M_{hc}, M_{fc}	hard, fuzzy partition space.
$\mathbf{F}^{(i)}$	fuzzy covariance matrix.
V_{cd}	set of real $c \times d$ matrices.
$\mathbf{I}$	identity matrix.
$\mathbf{T}$	transformation matrix.
$\mathbf{R}$	correlation matrix.
$\mathbf{\Upsilon}$	matrix with σ_i as diagonal elements.
$\mathbf{S}$	autocorrelation matrix.
$\mathbf{\Sigma}$	covariance matrix.
η	mean vector.
V_{dd}	set of $d \times d$ positive definite matrices.
M_0	set of typical points.

1

System Analysis

- ☐ *Scientific theories deal with concepts, not with reality.*
- ☐ *Systems theory uses mathematical concepts to describe aspects of the 'real-world'.*
- ☐ *A formal model is a graph, i.e. a subset of a product space formed by variables characterizing a system or process.*
- ☐ *An observable is some characteristic of a system which can, in principle, be measured.*
- ☐ *A state is a specification of a system or process at a specific instant.*
- ☐ *A dynamic system or process is a system in which the state changes with time.*
- ☐ *Differential equations are a common way to encode dynamics.*
- ☐ *There are many alternative and equally valid ways to represent a system...*

So far as the laws of mathematics refer to reality, they are not certain. And so far as they are certain, they do not refer to reality.

—Albert Einstein

System theory uses mathematical concepts to describe physical, natural or social systems in order to gain an understanding of the processes involved. It is important to realize that [Pap91]: "scientific theories deal with concepts, not with reality. All theoretical results are derived from certain axioms by deductive logic. In physical sciences, the theories are so formulated as to

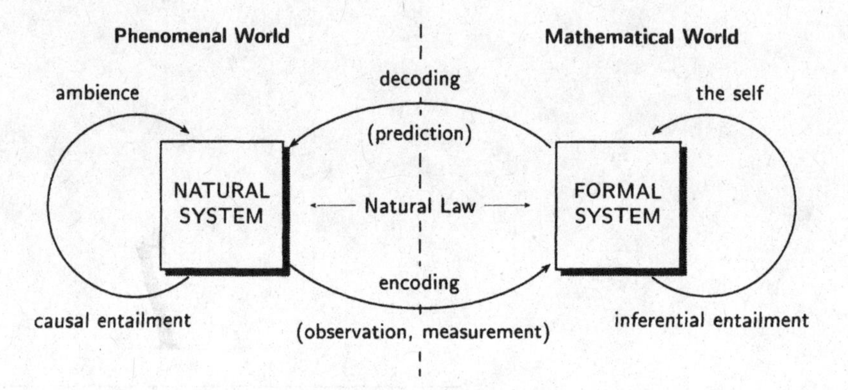

Fig. 1.1 The modelling relation between a natural system $\mathfrak{S}$ and a formal system $\mathfrak{M}$. If the modelling relation brings both systems into congruence by suitable modes of encoding and decoding, it describes a *Natural Law*. In this case $\mathfrak{M}$ is a *model* of $\mathfrak{S}$, that is, $\mathfrak{S}$ is a realization of $\mathfrak{M}$. (Picture adopted from R. Rosen [Cas92]).

correspond in some useful sense to the real world, whatever that may mean. However, this correspondence is approximate, and the physical justification of all theoretical conclusions is based on some from of inductive reasoning." This means that the model under consideration is a formal model, it does not pretend to model reality adequately and hence model assumptions are in a sense arbitrary, that is, the model builder can freely decide which model characteristics he chooses. Hereafter, a *system* $\mathfrak{S}$ is understood, very generally, as a part of the "external world" we wish to describe by means of a *formal model* $\mathfrak{M}$. The situation is illustrated in Figure 1.1.

Let us first consider a simple example. Assume we are observing three variables (attributes) denoted by x_1, x_2 and x_3 taking values from the sets $X_1 = \{1,2,3,4\}$, $X_2 = \{1,2,3,4\}$, and $X_3 = \{2,8,18,32\}$, respectively. An experiment generates the following measurements:

$$\mathbf{m}_1 = (1,1,1),\ \mathbf{m}_2 = (2,2,8),\ \mathbf{m}_3 = (3,3,18),\ \mathbf{m}_4 = (4,4,32).$$

The process or system under consideration is therefore represented by the collection of all pairs

$$\mathbf{M} = \{\mathbf{m}_j\} \qquad j = 1,\ldots,d\,. \tag{1.1}$$

Though $\mathbf{M}$ can be considered as a model, representing the system under consideration, a more general and constructive description would be desirable. From the observations we may consider the following equation as a law that describes the generation of the observations above:

$$x_3^2 = x_1^2 + x_2^2\,. \tag{1.2}$$

If variables x_1, x_2 take their values in the sets X_1 and X_2, equation (1.2) gives the values for x_3. We may say that x_3 is a function of x_1 and x_2, denoted $x_3 = f(x_1, x_2)$. An equivalent alternative to the functional description (1.2) is to describe the model by a set of ordered pairs, generalizing (1.1):

$$F = \left\{ (x_1, x_2, x_3) \ : \ x_3 = f(x_1, x_2) \right\} .$$

This simple example captures most aspects of the problem we will consider hereafter. To arrive at a formal model which represents (is a model of) the phenomenon under consideration it is necessary to:

▷ Select variables (attributes) to be observed.

▷ Specify the range of values observed variables can take.

▷ Sample or measure data (ordered pairs) in an experiment.

▷ Identify a formal model $F \subset X_1 \times \cdots \times X_n$.

▷ Give an encoding (constructive formulation) $f(\cdot)$ representing the system under consideration. For example, the relation, rule, correspondence or mapping $f(\cdot)$ can be a regression model, differential equations, many-valued logical functions, a probability density and so forth.

Observed variables may, of course, be time functions over the time interval $\{t \colon t_1 \leq t \leq t_2\}$ of experimentation. Typically one variable will be of particular interest and is therefore considered as dependent on the other variables. We will denote this situation by describing the model as the composite of two spaces X (referring to independent variables), U (referring to inputs) and Y (outputs or dependent variables), that is, $F \subset U \times Y$. Since in general, F is a proper relation, for any given input there might be many outputs. This violates the classical view on causality and motivates the introduction of the concept of *state*. Once a formal mathematical model is obtained we can:

▷ Study the properties of the system (model) using deductive methods or simulation.

▷ Interpret the meaning of the derived properties in the context of the experimental evidence.

▷ Using the model, forecasts or classifications can be used in decision making, specifically control and prioritization.

Figure 1.2 outlines the process of system analysis in the way it is considered in subsequent sections.

What follows is a more formal treatment of the aspects discussed above. We first introduce the basic notions of *state* and *observable* as discussed by R.

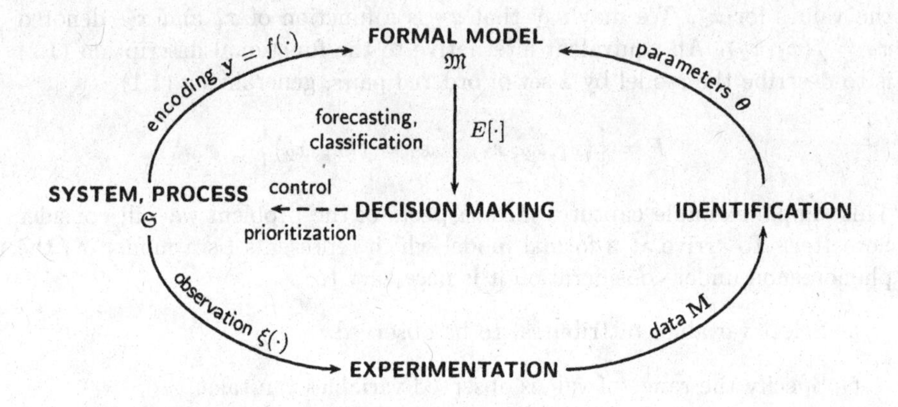

Fig. 1.2 System Analysis.

Rosen[1]. Instead of trying to model the internal physical structure of a system, we describe formalism to represent the process of 'recognition', 'measurement', 'discrimination' and 'classification' - related to observed quantities. It is the dynamical behavior of these 'observables' that provides the basis for learning about the original system $\mathfrak{S}$. In some sense, we therefore assume that either, as suggested above, we cannot know principally the exact inner structure of $\mathfrak{S}$, or complexity suggests a more pragmatic approach to modelling.

Intuitively, an *observable* of a system is some characteristic of a system which can, in principle, be measured, and a *state* is a specification of what our system is like at a specific instant of time. The basic philosophy on which this view rests is summarized by the following two propositions:

1. Observable events can be represented by the evaluation of factors on abstract states.

2. An observable can be regarded as a mapping from states to real numbers.

Assume the system under consideration exists in a (finite, infinite or uncountable) set of distinct, *abstract states* $X = \{x_1, x_2, \ldots\}$, for example, $X = \{x_1 = \text{"OFF"}, x_2 = \text{"ON"}\}$. We may not be able to determine which of these states $\mathfrak{S}$ is in. This uncertainty will depend on various factors such

[1] R. Rosen's work established a considerable advance to classical modelling and simulation in which physical models, successfully employed in engineering, were less successfully applied to science. The reasons why Rosen's ideas, which, though a major advance in theory, have not had the impact they deserve, may be seen in their abstract nature. A good summary of Rosen's work can be found in [Cas92].

as the nature of the information available, the quality of the model, assumptions made and so on. Observing the system we assume that the system is characterized by a set of observables (maps) ξ, which associate a number or point with each state $x \in X$ of $\mathfrak{S}$. Hence, a state is not an intrinsic property of the system itself but rather a mathematical construct, depending upon the way we analyze and model $\mathfrak{S}$. We therefore assume, that the system $\mathfrak{S}$ under consideration, consists of an abstract state-space and a finite set of observables

$$\xi_j : X \to \mathbb{R} \qquad j = 1, 2, \ldots, n+m+l \ .$$

In principle, we would require an infinite number of observables in order to observe the system $\mathfrak{S}$ completely. In practice, we however assume the system is sufficiently described by a finite set of observables:

$$\mathfrak{S} = (X, \xi_1, \xi_2, \ldots, \xi_{n+m+l}) \ .$$

1.1 UNCERTAINTY

To compare observables with respect to the information they convey, let X be a set of states, and consider the observable

$$\xi \ : \ X \to \mathbb{R} \ .$$

The *encoding* of an observable as a mapping $\xi\colon X \to \mathbb{R}$ from a set X of abstract states into the real numbers $\mathbb{R}$ implies that an observation represents a description, or *encoding*, of that state. In most experimental contexts we will find that an observable ξ on X induces an *equivalence relation*

$$E_\xi(x, x') = 1 \quad \text{if and only if} \quad \xi(x) = \xi(x')$$

and hence *equivalence classes* $[x]_\xi$ for which elements in X are indistinguishable w.r.t. ξ:

$$[x]_\xi = \{x' \ : \ \xi(x') = \xi(x)\} \ .$$

The set of equivalence classes on X is called *quotient set* and is denoted by X/E_ξ. Therefore, what we actually observe is usually not X but the set of *reduced states* X/E_ξ. Equivalence relations will play an important role throughout what follows. We shall return to the problem of system uncertainty and equivalence relations in Section 7.

1.2 THE ART OF MODELLING: LINKAGE

The relation E_ξ specifies the extent to which the elements of X can be resolved by the observable ξ. In other words, E_ξ specifies these distinct elements of X which result in the same observation. Thus, as far as ξ is concerned, the replacement of any state x by another state in the same equivalence class is *not an observable event*. In fact, using only one observable ξ, what we actually observe is not the state set X, but rather the quotient set X/E_ξ. For us, then, this quotient set would *be* the set of states of our system. Now, consider two observables ξ and ξ' and let Z be the set of $E_{\xi'}$ equivalence classes that intersect $[x]_\xi$. Intuitively, the observables provide two independent descriptions of the same system and we are interested in how these descriptions are to be compared. Using Rosen's terminology, we say

1. ξ' is ***totally linked*** to ξ at $[x]_\xi$ if Z is a single ξ' equivalence class (E_ξ implies $E_{\xi'}$).
2. ξ' is ***partially linked*** to ξ at $[x]_\xi$ if Z is more than one ξ' equivalence class but not all of the quotient set $X/R_{\xi'}$.
3. ξ' is ***unlinked*** to ξ at $[x]_\xi$ if $Z = X/E_{\xi'}$.

We shall discuss three cases for which two factors are 'unlinked', 'linked' and 'partially linked'. We assumed that an observable takes its values in $\mathbb{R}$. Let us for the sake of simplicity in notation, denote the range of values by $\bar{X}$ such that an observable is a mapping from X to $\bar{X}$:

$$\begin{aligned} \xi\colon X &\rightarrow \bar{X} \\ x &\mapsto \bar{x}\,. \end{aligned}$$

We first consider the illustration in Figure 1.3 defining two factors ξ and ξ' which partition X in different ways.

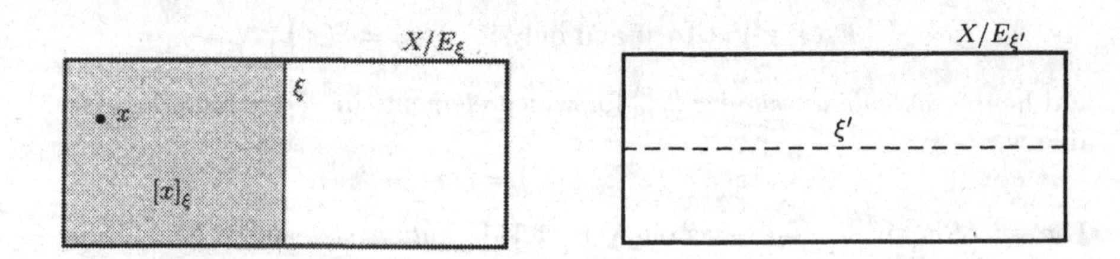

Fig. 1.3 Example of two totally unlinked factors ξ and ξ'. The grey area on the left is the equivalence class $[x]_\xi$ generated by ξ on X.

The concept of linkage between factors ξ and ξ' becomes plausible by assuming a given $[x]_\xi$ in X/E_ξ and subsequently to discuss which ξ'-equivalence classes intersect with $[x]_\xi$. From Figure 1.3, we find that factor ξ' splits the classes of E_ξ such that ξ' can distinguish between values, indistinguishable via ξ. We say that the greater the extent of the splitting of $[x]_\xi$ by ξ', the more unlinked ξ' is to ξ at $[x]_\xi$. We find that

▷ The whole of $X/E_{\xi'}$, *i.e.* both ξ'-classes intersect with $[x]_\xi$: ξ' is said to be *unlinked* to ξ at $[x]_\xi$.

▷ ξ' is unlinked to ξ at each $[x]_\xi$; every E_ξ-class intersects every $E_{\xi'}$-class and conversely: ξ' is said to be *totally unlinked* to ξ.

Having fixed some value $\bar{x}$ in $\xi(X)$, $\xi'(x)$ is *not* arbitrary in $\xi'(X)$; the coordinates $f(x)$, $g(x)$ of $x \in X$ are not independently variable in X/E_ξ, $X/E_{\xi'}$, respectively.

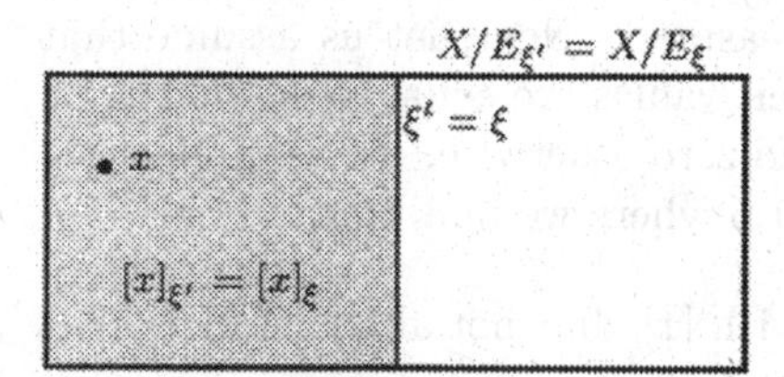

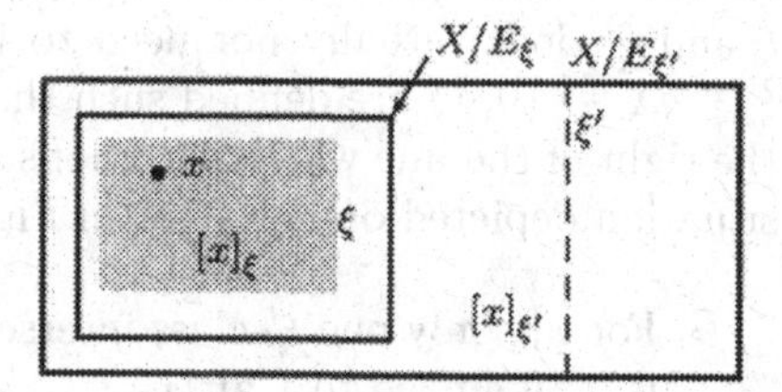

Fig. 1.4 Two examples of two totally linked factors ξ and ξ' such that E_ξ refines $E_{\xi'}$.

Figure 1.4 illustrates the second extreme: total linkage. We make the following observations:

▷ Only a single ξ'-class intersects with $[x]_\xi$: ξ' is said to be *linked* to ξ at $[x]_\xi$.

▷ Since ξ' is linked to ξ at each $[x]_\xi$, every class of E_ξ intersects exactly one class of $E_{\xi'}$, namely, the one which contains it: ξ' is said to be *totally linked* to ξ.

If ξ' and ξ are totally linked, E_ξ is said to *refine* $E_{\xi'}$, ξ' does not split the classes of E_ξ and no new information is obtained from an additional factor ξ'. The coordinates $\xi(x)$ and $\xi'(x)$ are *independently variable* in X/E_ξ, $X/E_{\xi'}$ respectively. That is, having fixed some value $\bar{x}$ in $\xi(X)$ we may find a value in X such that $\xi(x) = \bar{x}$ and $\xi'(x)$ is arbitrary in $\xi'(X)$.

In general, let E_ξ, $E_{\xi'}$ be equivalence relations on a set X. E_ξ is said to be a refinement of $E_{\xi'}$ if $E_\xi(x_1, x_2)$ implies $E_{\xi'}(x_1, x_2)$. In terms of equivalence class, this means that every E_ξ-equivalence class is a subset of some $E_{\xi'}$-equivalence class or in other words, E_ξ refining $E_{\xi'}$ means that elements of the partition from $E_{\xi'}$ are further partitioned by E_ξ and blocks of the $E_{\xi'}$ partition can be obtained from the set-theoretic union of E_ξ-blocks. If E_ξ is a refinement of $E_{\xi'}$, then there is a unique mapping

$$\begin{aligned} h: \quad X/E_\xi &\to X/E_{\xi'} \\ [x]_\xi &\mapsto h([x]_\xi) = [x]_{\xi'} \end{aligned} \tag{1.3}$$

which makes the following diagram commute:

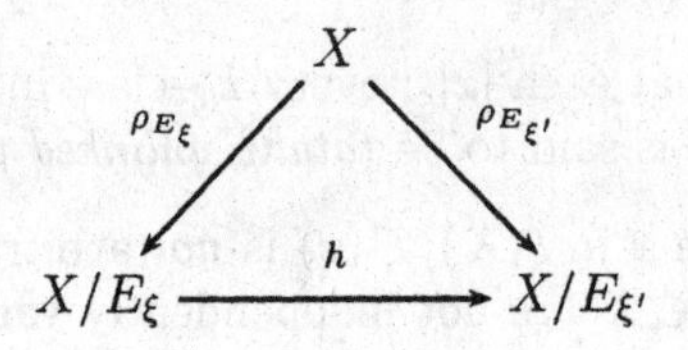

Thus the value of ξ' on x in X is completely determined by the value of ξ on that x through the relation $\xi'(x) = h(\xi(x))$. That is, ξ' is a function of ξ and ξ' does actually not need to be measured. Next, let us assume that $\xi, \xi' \colon X \to \{0, 1\}$ are defined such that their values are equal to one if x is on the right of the line which partitions X and zero otherwise. We then have the situation depicted on the right in Figure 1.5 where we find that:

▷ For x_1, only one ξ'-class intersects with $[x]_\xi$ but not all of $X/E_{\xi'}$. That is, ξ' is linked to ξ at $[x]_\xi$.

▷ For x_2, both ξ'-classes intersect with $[x]_\xi$ and hence ξ' is unlinked to ξ at $[x]_\xi$.

We also note that the linkage relationship between ξ and ξ' is not symmetric, that is, the linkage of ξ' to ξ at $[x]_\xi$ can be different from the linkage of ξ to ξ'.

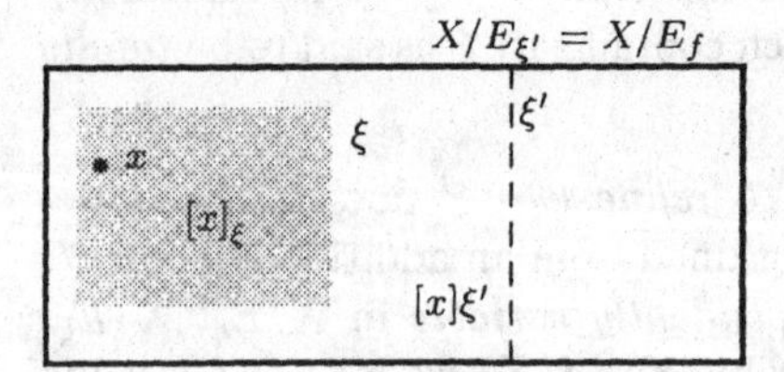

Fig. 1.5 Two examples of partial linkage between factors.

Before concluding this subsection, we look at another illustration of linkage. From Figure 1.6, we have the following equivalence classes for ξ and ξ' from which we find that ξ and ξ' are totally unlinked.

$$\begin{array}{lll}
[x_1]_\xi = \{x_1, x_2\} & [x_1]_{\xi'} = \{x_1, x_3\} & X/E_\xi = \{\{x_1, x_2\}, \{x_3, x_4\}\} \\
[x_2]_\xi = \{x_1, x_2\} & [x_2]_{\xi'} = \{x_2, x_4\} & X/E_{\xi'} = \{\{x_1, x_3\}, \{x_2, x_4\}\} \\
[x_3]_\xi = \{x_3, x_4\} & [x_3]_{\xi'} = \{x_1, x_3\} & \\
[x_4]_\xi = \{x_3, x_4\} & [x_4]_{\xi'} = \{x_2, x_4\} & X/E_{\xi\xi'} = \{\{x_1\}, \{x_2\}, \{x_3\}, \{x_4\}\}
\end{array}$$

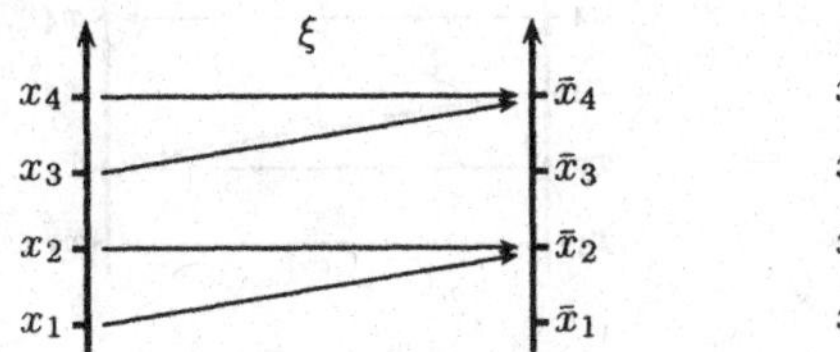

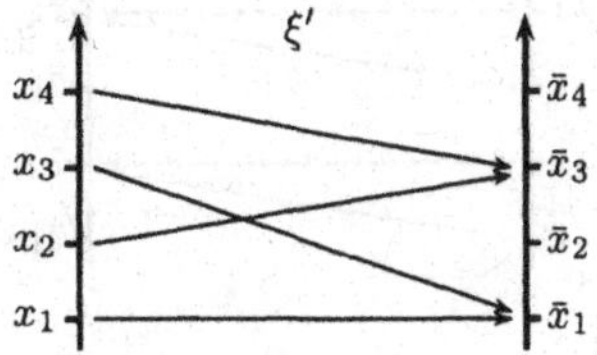

Fig. 1.6 Example of two totally unlinked factors ξ and ξ'.

In Figure 1.7, we find an example of total linkage. The equivalence classes and quotient sets are as follows:

$$
\begin{array}{lll}
[x_1]_\xi = \{x_1, x_2\} & [x_1]_{\xi'} = \{x_1\} & X/E_\xi = \{\{x_4\}, \{x_3\}, \{x_1, x_2\}\} \\
[x_2]_\xi = \{x_1, x_2\} & [x_2]_{\xi'} = \{x_2\} & X/E_{\xi'} = \{\{x_3, x_4\}, \{x_2\}, \{x_1\}\} \\
[x_3]_\xi = \{x_3\} & [x_3]_{\xi'} = \{x_3, x_4\} & \\
[x_4]_\xi = \{x_4\} & [x_4]_{\xi'} = \{x_3, x_4\} & X/E_{\xi\xi'} = \{\{x_1\}, \{x_2\}, \{x_3\}, \{x_4\}\}
\end{array}
$$

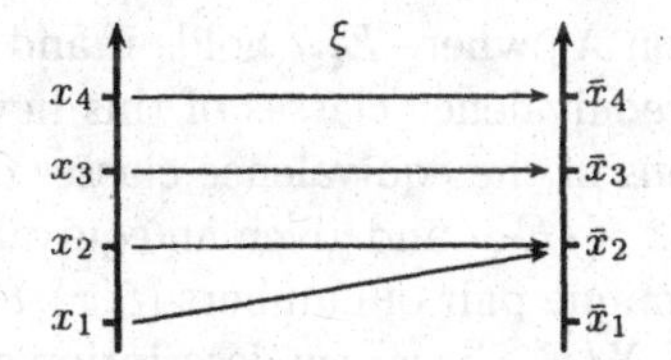

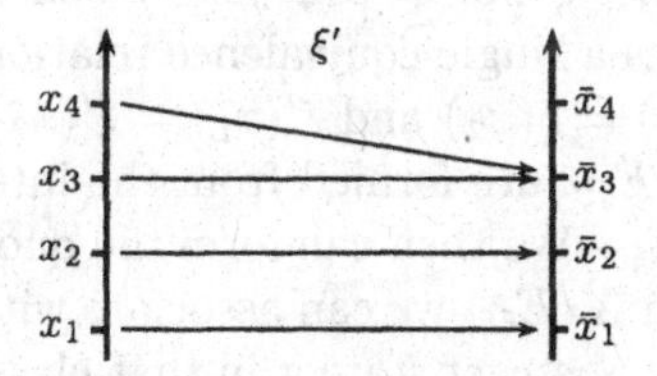

Fig. 1.7 Example of two totally linked factors ξ and ξ'.

Finally we look at an example of partial linkage, illustrated in Figure 1.8. The equivalence classes and quotient sets are:

$$
\begin{array}{lll}
[x_1]_\xi = \{x_1, x_2\} & [x_1]_{\xi'} = \{x_1\} & X/E_\xi = \{\{x_1, x_2\}, \{x_3, x_4\}\} \\
[x_2]_\xi = \{x_1, x_2\} & [x_2]_{\xi'} = \{x_2, x_3\} & X/E_{\xi'} = \{\{x_1\}, \{x_2, x_3\}, \{x_4\}\} \\
[x_3]_\xi = \{x_3, x_4\} & [x_3]_{\xi'} = \{x_2, x_3\} & \\
[x_4]_\xi = \{x_3, x_4\} & [x_4]_{\xi'} = \{x_4\} & X/E_{\xi\xi'} = \{\{x_1\}, \{x_2\}, \{x_3\}, \{x_4\}\}
\end{array}
$$

With respect to the linkage of ξ' to ξ we find that for all x in X, ξ' is *partially linked* to ξ at $[x]_\xi$ since it intersects with more than one ξ'-class but not all of $X/E_{\xi'}$. The linkage of ξ to ξ' at $[x]_{\xi'}$ is, however, different:

▷ Linkage at $[x_1]_{\xi'}$: Intersects with a single ξ-class.

▷ Unlinked at $[x_2]_{\xi'}$ and $[x_3]_{\xi'}$: Intersections with all of X/E_ξ.

▷ Linkage at $[x_4]_{\xi'}$.

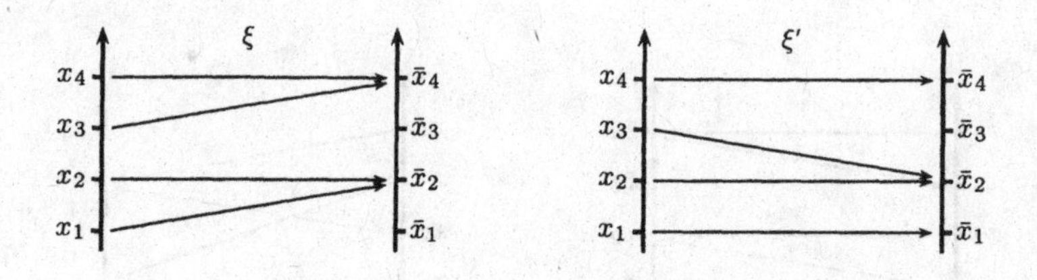

Fig. 1.8 Example of partial linkage between ξ and ξ'.

The aim of a *product space representation* is to obtain for a given $x \in X$ a unique representation in form of 'coordinates':

$$\begin{aligned} X &\rightarrow \xi(X) \times \xi'(X) \\ x &\mapsto (\xi(x), \xi'(x)) \; . \end{aligned}$$

The expectation is therefore that two observables provide a more comprehensive description of elements of X. Considering a pair of observables (ξ, ξ'), as if they describe one, we would say that two abstract states are equivalent if neither ξ nor ξ' can distinguish between them. That is, if the pair (ξ, ξ') imposes a single equivalence relation $E_{\xi\xi'}$ on X, where $E_{\xi\xi'}$ holds if and only if $\xi(x_1) = \xi(x_2)$ and $\xi'(x_1) = \xi'(x_2)$. The equivalence classes of this new relation $E_{\xi\xi'}$ are formed from the intersections of the equivalence classes of E_ξ with $E_{\xi'}$. We then can form the quotient set $X/E_{\xi\xi'}$ and given an equivalence class in $X/E_{\xi\xi'}$ we can associate with it a unique pair of numbers $(\xi(x), \xi'(x))$ for any abstract state x in that class. Thus $X/E_{\xi\xi'}$ is a new description of X, playing the same role as the the state descriptions X/E_ξ and $X/E_{\xi'}$ but finer than either of them. $X/E_{\xi\xi'}$ assumes the role of what is usually called the *phase space* or *state-space*, with ξ and ξ' playing now the role of *state variables*. The case for two observables can be generalized to an arbitrary family of observables, in which case the symbol $\prod$ is used to denote the product.

The *linkage relation* between observables ξ and ξ' can be represented geometrically in the two-dimensional 'state-space' $\rightarrow \xi(X) \times \xi'(X)$ as a curve, that is, as a relation

$$f(\xi(x), \xi'(x)) = 0$$

in which $\xi'(x)$ can be considered as the independent variable. More general relationships between observables are described in the *equation of state*

$$f_i(\xi_1, \ldots, \xi_{n+m+l}) = 0 \qquad i = 1, 2, \ldots, m \; . \tag{1.4}$$

We note that an equation of state establishes a *deterministic* relationship between observables but not necessarily conveys information about *causal* links. In [Cas92] the following example is given: Let $\mathfrak{S}$ be a closed vessel containing an ideal gas. Take X to be the positions and velocities of the molecules making up the gas, and define the three observables for properties of the gas

$$
\begin{aligned}
P(x) &= \text{pressure when in state } x, \\
V(x) &= \text{volume when in state } x, \\
T(x) &= \text{temperature when in state } x\,.
\end{aligned}
$$

Then the ideal gas law asserts the single equation of state

$$f(P,V,T) = 0 \qquad \text{specifically} \qquad f(p,v,t) = pv - t\,.$$

Observables whose values remain fixed for every state $x \in X$ are called *parameters*, $\xi_i(x) = \theta_i$. For l parameters we write $i = n+m+1, n+m+2, \ldots, n+m+l$,

$$f\left(\xi_1, \ldots, \xi_{n+m};\ \theta_1, \theta_2, \ldots, \theta_l\right) = 0\,.$$

If in addition m observables $\xi_{n+1}, \xi_{n+2}, \ldots, \xi_{n+m}$ are functions of the remaining observables $\xi_1, \xi_2, \ldots, \xi_n$, we use the notation

$$
\begin{aligned}
\mathbf{u} &\doteq [\xi_1, \xi_2, \ldots, \xi_n] \\
\mathbf{y} &\doteq [\xi_{n+1}, \xi_{n+2}, \ldots, \xi_{n+m}] \\
\boldsymbol{\theta} &\doteq [\theta_{n+m+1}, \theta_{n+m+2}, \ldots, \theta_{n+m+l}]
\end{aligned}
$$

and obtain for the equation of state,

$$f(\mathbf{u}; \boldsymbol{\theta}) = \mathbf{y}\,. \tag{1.5}$$

We may then interpret the independent observables $\mathbf{u}$, as *inputs* to the system and dependent observables $\mathbf{y}$ as the resulting *outputs*. We can think that the inputs somehow 'determine'[2] the outputs. Before considering other model structures, let us consider the fundamental question of how to decide whether two descriptions $f(\mathbf{u}; \boldsymbol{\theta})$ and $f(\mathbf{u}; \boldsymbol{\theta}')$ of $\mathfrak{S}$ are equivalent, that is, contain the same information. We note that $f(\mathbf{u}; \boldsymbol{\theta})$ is, in fact, a family of models, indexed by $\boldsymbol{\theta}$. Then the two maps

$$f_{\boldsymbol{\theta}} : U \to Y \qquad \text{and} \qquad f_{\boldsymbol{\theta}'} : U \to Y\,,$$

where U and Y are the input and output spaces, respectively, are considered to be equivalent if there exist bijections (one-to-one and onto)[3]

$$g_{\boldsymbol{\theta},\boldsymbol{\theta}'} : U \to U \qquad \text{and} \qquad h_{\boldsymbol{\theta},\boldsymbol{\theta}'} : Y \to Y\,,$$

[2] We should emphasize that causation is not mirrored by a functional representation such as $y = f(u)$. Causal connections are essentially ***asymmetrical*** but $y = f(u)$ can be inverted to yield $u = f^{-1}(y)$, so that the relative places of the 'cause', and of the 'effect' can be exchanged.

[3] An overview of different types of mappings is provided in Appendix 13.1.

which enable us to *transform* f_θ into $f_{\theta'}$, that is, the following diagram commutes:

$$\begin{array}{ccc} U & \xrightarrow{f_\theta} & Y \\ {\scriptstyle g_{\theta,\theta'}} \downarrow & & \downarrow {\scriptstyle h_{\theta,\theta'}} \\ U & \xrightarrow{f_{\theta'}} & Y \end{array}$$

This diagram is called *commutative diagramm* since the maps satisfy the following relation:

$$f_{\theta'} \circ g_{\theta,\theta'} = h_{\theta,\theta'} \circ f_\theta \; .$$

An important conclusion is that only those properties of $\mathfrak{M}$ which remain invariant under such transformation, are intrinsic properties of the system $\mathfrak{S}$. The symbol $\circ$ denotes the composition of two functions defined as follows: Given any two functions g and h, when the codomain of g is the domain of h as in

$$U \xrightarrow{\;g\;} X \xrightarrow{\;h\;} Y$$

the *composite function* $h \circ g$ is defined as the set of ordered pairs

$$\big\{(u,y) \;:\; u \in U, y \in Y, \text{ and } \exists\, x \in X \text{ with } (u,x) \in g \text{ and } (\omega, y) \in h\big\} \; . \quad (1.6)$$

Illustrating the composition with the commutative diagram

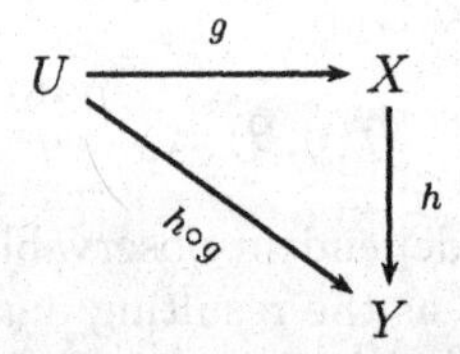

we can interpret the composition as the rule "first apply g, then apply h", $(h \circ g)(u) = h(g(u))$, formalizing the idea of two operations carried out in succession. We may therefore use the composition of mappings as a means to describe *dynamics* - changes in the 'state' of a system.

We have outlined the general framework of system analysis, (Figures 1.1 and 1.2), and are now in the position to discuss various more specific mathematical models. State equation (1.5) suggests a model of $\mathfrak{S}$ where $f(\cdot)$ is a mapping that relates the inputs $\mathbf{u}$ directly to the outputs $\mathbf{y}$ of the system without considering 'inner states':

$$\begin{aligned} f : U &\rightarrow Y \\ \mathbf{u} &\mapsto \mathbf{y} \,. \end{aligned} \quad (1.7)$$

Then any specific model describes a *graph* F of the mapping which represents system $\mathfrak{S}$:

$$F \subset U \times Y \, . \quad (1.8)$$

Modelling refers to the task of relating observations with the mapping described by the set F. Figure 1.9 illustrates the concept of a graph, where the function $f: X \to \mathbb{R}$ is defined by the set of ordered triples $(u_1, u_2, f(u_1, u_2))$ such that each triple is belonging to $\mathbb{R}^3$ forming a surface $F = \{(u_1, u_2, y) \in \mathbb{R}^3 : y = f(u_1, u_2)\}$. As we shall see shortly further below, the result of the operation f is not always a real number and one then refers to f as a *mapping*. For example, in (1.9) below, the *vector-valued function* $f: \mathbb{R}^n \to \mathbb{R}^m$ takes the vector $\mathbf{u} \in \mathbb{R}^n$ to a unique vector $f(\mathbf{u}) \in \mathbb{R}^m$.

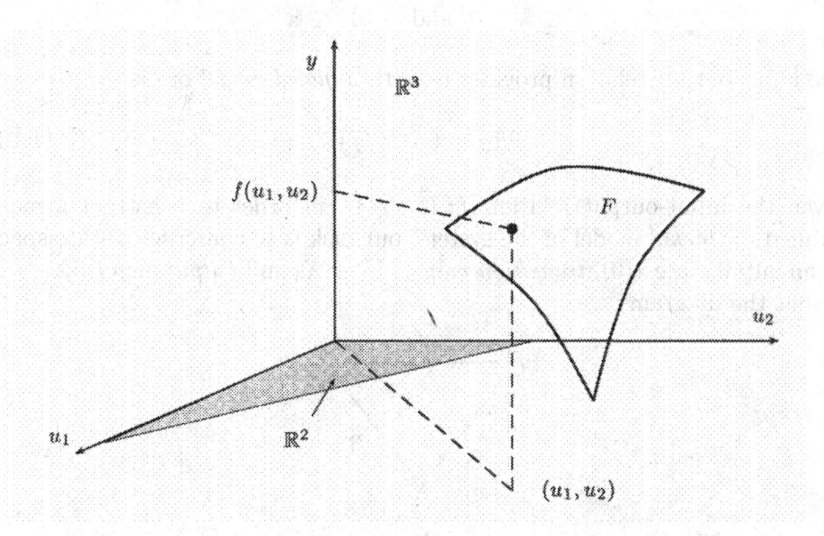

Fig. 1.9 Graph of $f: U \to \mathbb{R}$, with $U \subset \mathbb{R}^2$.

Now, let us assume we are given the input-output relation, characterized by observations of the inputs and outputs. We then wish to study the *internal* properties of the system. More specifically, let the input-output relation be given by

$$\begin{aligned} f: \quad \vec{U} &\to \vec{Y} \\ \mathbf{u}(k) &\mapsto \mathbf{y}(k), \end{aligned} \tag{1.9}$$

where the system maps an input sequence $\mathbf{u}(k)$ to an output sequence $\mathbf{y}(k)$:

$$\mathbf{u}(k) \doteq \langle \mathbf{u}_0, \mathbf{u}_1, \ldots, \mathbf{u}_d \rangle \qquad \mathbf{u}_j \in U, \qquad \mathbf{u}(k) \in \vec{U} \tag{1.10}$$

$$\mathbf{y}(k) \doteq \langle \mathbf{y}_1, \mathbf{y}_2, \mathbf{y}_3, \ldots \rangle \qquad \mathbf{y}_j \in Y, \qquad \mathbf{y}(k) \in \vec{Y}. \tag{1.11}$$

An internal model is then realized by the introduction of a state-space $\vec{X}$, where the *transition map* $g: \vec{U} \to \vec{X}$ (onto) and the *output map* $h: \vec{X} \to \vec{Y}$

(one-to-one) are given as:

$$g: \quad \vec{U} \to \vec{X} \qquad\qquad h: \quad \vec{X} \to \vec{Y} \tag{1.12}$$
$$\mathbf{u}(k) \mapsto \mathbf{x}(k) \qquad\qquad \mathbf{x}(k) \mapsto \mathbf{y}(k)\ .$$

As the mapping refers to sequences, $\vec{U}$, $\vec{X}$, and $\vec{Y}$ denote finite-dimensional vector spaces (linear manifolds) whose elements are sequences of vectors. If the system has n independent inputs and m independent outputs, we assume

$$U \subset \mathbb{R}^n \quad \text{and} \quad Y \subset \mathbb{R}^m\ .$$

The input-output relation provides us with a *global model* of $\mathfrak{S}$:

$$\mathfrak{M}_G = (\vec{X}, \vec{Y}, f)\ . \tag{1.13}$$

Given the input-output relation $f\colon \vec{U} \to \vec{Y}$, in order to construct a more detailed - *internal* model of the system, our task is to construct a state-space $\vec{X}$, an initial state $\mathbf{x}(0)$, *transition map* $g\colon \vec{U} \to \vec{X}$, and *output map* $h\colon \vec{X} \to \vec{Y}$ so that the diagram

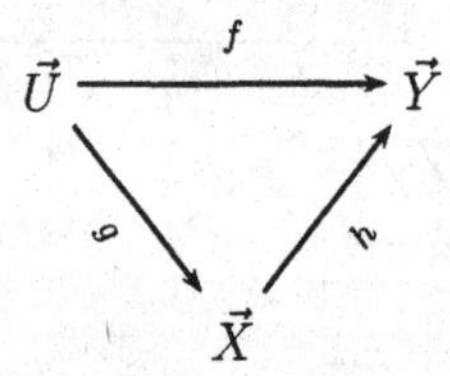

commutes. The *state-vector* $\mathbf{x}(k)$ can be thought of being in an r-dimensional Euclidean space where r is the number of components in vector $\mathbf{x}(k)$. Then, the state-vector $\mathbf{x}(k)$ describes, as a function of time, a curve (*trajectory*) in the r-dimensional state-space. Though this state-space representation is a mathematical operator that maps a sequence of inputs into a sequence of outputs, the *canonical model*

$$\mathfrak{M}_I = (\vec{X}, g, h, \mathbf{x}(0)) \tag{1.14}$$

is, in fact, an *internal model*. Assuming a linear map f in (1.13) we can write the input-output relationship in terms of the input and output sequences $\mathbf{u}(k)$, $\mathbf{y}(k)$ as

$$\mathbf{y}(k) = \sum_{j=0}^{k-1} \mathbf{A}(k-j)\mathbf{u}(j) \qquad k = 1, 2, \ldots \qquad \mathbf{A}(i) \in \mathbb{R}^{m\times n}\ .$$

Consequently, the matrix sequence $\langle A_1, A_2, \ldots \rangle$ describes the behavior of the system and is an equivalent description of the *abstract* map f in (1.13). The behavior sequence $\langle A_1, A_2, \ldots \rangle$ is usually termed the *external* description of the system. It can be shown that we can construct a canonical *internal* description of the system from the sequence. As shown in the next section we need to find matrices $\mathbf{F} \in \mathbb{R}^{r\times r}$, $\mathbf{G} \in \mathbb{R}^{r\times m}$, and $\mathbf{H} \in \mathbb{R}^{r\times m}$ such that

$$A(k) = \mathbf{H}\mathbf{F}^{k-1}\mathbf{G} \qquad k = 1, 2, \ldots$$

An important issue is to determine the system *dimension*, r, for which the sequence $\langle A_1, A_2, \ldots\rangle$ is uniquely determined. In the next section, we will see that processes modelled by differential equations lead directly to the system-theoretic concept of a state-space $\vec{X}$ for an *internal* model. This so-called state-space approach is commonly used in systems theory, control engineering and will form the basis for the Kalman-Bucy filter in Section 2.3. A dynamical system can then be represented in the state-space $\mathbb{R}^r$ by

$$\begin{aligned} \mathbf{x}(k+1) &= \mathbf{F}\mathbf{x}(k) + \mathbf{G}\mathbf{u}(k) \qquad \mathbf{x}(0) = 0,\ \mathbf{x}(k) \in \mathbb{R}^r,\ k = 0, 1, \ldots \\ \mathbf{y}(k) &= \mathbf{H}\mathbf{x}(k) \end{aligned}$$

Apart from the state-space model being suitable to model dynamic systems, we shall see in Section 2.3 that the concept of mapping sequences fits well the formulation of stochastic processes as sequences of random variables.

1.3 DYNAMIC SYSTEMS

All models divide naturally...into two *a priori* parts: one is kinematics, whose aim is to parameterize the forms of the states of the process under consideration, and the other is *dynamics*, describing the evolution in time of these forms.

—Rene Thom

A *dynamical system* is one which changes in time; what changes is the state of the system. The principle of 'mathematical causation' states that there exists some 'law of propagation' which describes how the state of the system evolves. Formally, a dynamical system consists of a pair (X, T) where X is a topological space, and $T = \{T_t\}_{t\in\mathbb{R}_+}$ is a collection of maps indexed by the set $\mathbb{R}_+$ of non-negative reals denoting time[4]. The *state-space* X is usually a manifold or topological vector space. The *dynamic* or *flow* T on X is made up of transformations (maps)

$$T_t : X_t \to X ,$$

[4]The passage of time implies the ideas "before" and "after", stated formally in the *transitive law* $t < t'$ and $t' < t''$ imply $t < t''$ for the binary relation $<$.

where both the set of abstract states X and the time set can be either continuous or discrete and each domain X_t is an open subsets of X. The map

$$T : \{(t,x) \in \mathbb{R}_+ \times X : x \in X_t\} \rightarrow X \qquad (1.15)$$
$$(t,x) \mapsto T_t(x)$$

is usually assumed to be continuous and *determinism* is defined as follows. For each $x \in X$, define the set

$$I(x) = \{t \in \mathbb{R}_+ \ : \ x \in X_t\} \, .$$

If $t \in I(x)$ and $s \in I\big(T_t(x)\big)$, then $s + t \in I(x)$ and

$$T_s\big(T_t(x)\big) = T_{s+t}(x)$$

while

$$\forall x \in X, \ T_0(x) = x \, .$$

When T is understood, we write $T_t(x) \doteq x(t)$. The map

$$I(x) \rightarrow X \qquad (1.16)$$
$$t \mapsto x$$

is the *trajectory* of $x \in X$. Its image is the *orbit* $O(x)$. The previous section introduced the concept of an observable of a system encoded as a mapping $\xi \colon X \rightarrow \mathbb{R}$ from a set X of abstract states into the real numbers $\mathbb{R}$. Given *any* mapping between sets, we can define an equivalence relation E_ξ on its domain, by posing that $E_\xi(x_1, x_2)$ holds if $\xi(x_1) = \xi(x_2)$. Hence if two abstract states x_1 and x_2 generate the same observation they are *indistinguishable* to the measurement. Given an observable ξ on X, we say that ξ is *compatible* with T_t if the dynamic preserves the equivalence classes under ξ, that is, for any two states x and x' such that $\xi(x) = \xi'(x)$, we have $T_t(x) = T_t(x')$ for all $t \in \mathbb{R}$. Let ξ' be another observable on X, then both observables are observing the same system if there exists a mapping

$$\varphi \ : \ X/E_\xi \rightarrow X/E_{\xi'}$$

such that the diagram

$$\begin{array}{ccc} X/E_\xi & \xrightarrow{T_t^{(\xi)}} & X/E_\xi \\ \Big\downarrow \varphi & & \Big\downarrow \varphi \\ X/E_{\xi'} & \xrightarrow{T_t^{(\xi')}} & X/E_{\xi'} \end{array}$$

commutes for all $t \in \mathbb{R}$. Now consider the situation in which the observables ξ and ξ', through their evaluations on abstract states, are themselves the encodings of two formal models. As we have seen, the *linkage*[5] between these two observables describes how much we learn about $\xi(x)$ when $\xi(x)$ is known. In this sense, the linkage between the observables also measures the extent to which the two formal models are equivalent. If, for example, $E_\xi = E_\xi$, that is, every equivalence class in X/E is also an equivalence class for X/E_ξ and there exists an isomorphism between the two formal representations. Or, stated in other words, with our formal models we describe what we observe rather than the actual process. For many physical and engineering systems, we may get the impression to model with differential equations (e.g., Newtons' laws of physics) the actual process "as it is". The subsequent success of such physical models in engineering has led to a focus on simulation and and modelling in the sense of differential equations. Current challenges in the study of complex large-scale systems, in engineering and biology, however demonstrate the limits of this methodology and suggest a focus on "what we observe". For example, in molecular biology, the modelling of dynamic physical processes within a cell are too complex to obtain a sufficiently accurate model of cell interaction. On the other hand, vast amounts of experimental data are available that describe genes and proteins and their 'function'. What seems therefore required is a focus on input-output patterns, on what we can observe, and building models that match data with models – identify models from data. In this section, we first review the conventional engineering perspective to modelling dynamic systems before we start to discuss the identification of models from numerical data and subsequently return to our quest for new modelling paradigms.

A finite number of states leads to concepts such as finite-state machines or Markov chains. A finite-state machine or *automaton* is a triple of finite sets X, U and Y, and a pair of mappings

$$g : X \times U \times K \to X \qquad h : X \times K \to Y , \tag{1.17}$$

where K is the set of integers $K = \{0, 1, 2, \ldots\}$. The elements of X are considered as the *states* of $\mathfrak{S}$; the elements of U are the *inputs* to $\mathfrak{S}$; and the elements of Y are the *outputs* of $\mathfrak{S}$. Time t is assumed to range over K, $k \in K$, with the state, input and output of $\mathfrak{S}$ at time t_k denoted by $x(k)$, $u(k)$, and $y(k)$, respectively. The mappings g and h relate the state at time t_{k+1} and the output at time t_k to the state and the input at time t_k (see

[5]The concept of *linkage* is due to R. Rosen [Ros85] who developed formal models for *anticipatory systems*. An anticipatory system is one in which a present change of state depends upon future circumstances, rather than merely on the present or past. The latter systems, that react on the present or past, are then called *reactive*. The feedback control systems, presented in later sections of this book are reactive. Rosen noted that not only in biology many systems are anticipatory, that is their current behavior (decisions) are based on possible future outcomes.

Figure 1.10):

$$x(k+1) = g\big(x(k), u(k)\big) \quad : \text{state-transition map} \cdot \tag{1.18}$$
$$y(k) = h\big(x(k)\big) \quad : \text{output map} .$$

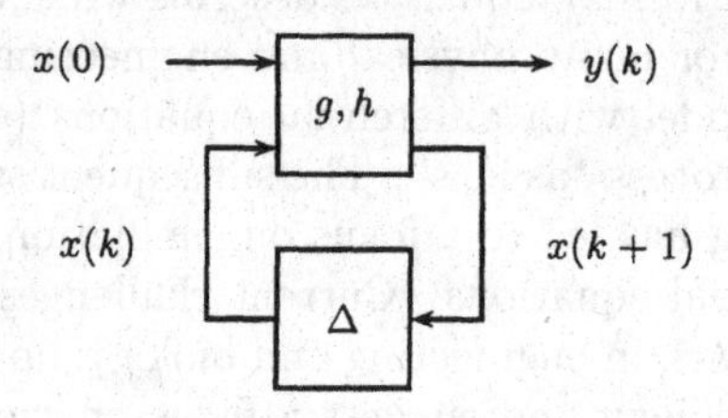

Fig. 1.10 A finite-state machine.

The finite-state machine is therefore a discrete-time dynamical system such that for each state, a number of different transitions may occur. Furthermore, it is assumed that there is a possibility for control action through a supervisor which, at any given point in time, may influence transitions. It is then natural to consider the problem of designing such a supervisor satisfying certain specifications. Intuitively, one would require the supervisor to prohibit the occurrence of certain (undesirable) sequences of events, while at the same time allowing some other (desirable) sequences of events to occur. Such a discrete-event dynamical system is very similar to a discrete-time Markov chain, except that there are no assumptions about the probabilities of the state-transitions. For this reason, the supervisor design problem is a little different from the traditional problems of Markov decision theory, for which dynamic programming provides a solution.

Modelling with Differential Equations

There are numerous phenomena in which the rate of change of some quantity is proportional to the quantity itself and consequently, the differential equation $\frac{dy}{dx} = y$ where $y = e^{ax}$ has served as a model for many biological, chemical and physical systems or processes. Depending on whether a is positive or negative, we obtain models of exponential growth or decay. To name only few examples,

- ▷ If money is compounded continuously at an annual interest rate a, the balance grows exponentially in time.
- ▷ The death rate of bacteria under the action of a disinfectant or heat, follows an exponential function. Note, however, that the bacteria do not

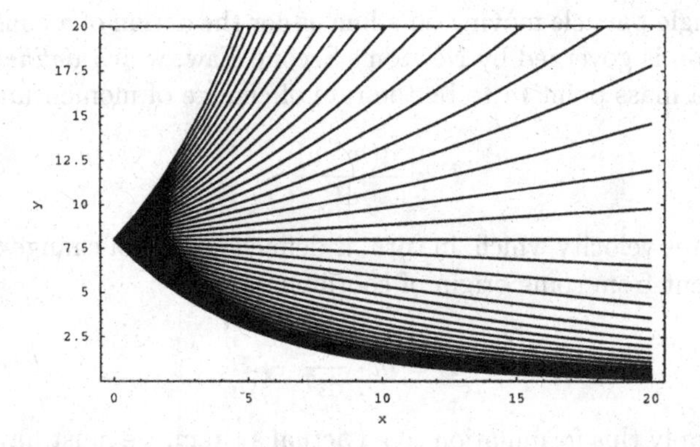

Fig. 1.11 Law of the (average) rate of radioactive decay.

actually die exponentially but the exponential function with parameter a provides a good model, *i.e.* fits the experimental data well.

▷ When sound waves travel through air (or any other medium) their intensity is governed by the simple first order differential equation where y denotes the intensity and x the distance travelled. A similar law, known as Lambert's law, holds for the absorption of light in a transparent medium.

▷ The rate of decay of a radioactive substance is at every moment proportional to its mass. If we denote the mass by x, the model of this process is the differential equation (linkage relation)

$$\frac{\mathrm{d}x}{\mathrm{d}t} = -a \cdot x \ .$$

Integrating the differential equation we obtain the solution $x(t) = x(0) \cdot e^{-at}$, where $x(0)$ is the initial mass of the substance. The solution is of the general form of an equation of state (1.4), $f(x, x(0), a, t) = 0$. (The law of the (average) rate of radioactive decay is derived on the assumption of the mutual independence (or randomness) of the successive individual disintegrations).

The only function which is equal to its own derivative is the *exponential function*: if $y = e^{ax}$, then $\frac{\mathrm{d}y}{\mathrm{d}x} = e^{ax}$. If we solve this simplest of differential equations, $\frac{\mathrm{d}y}{\mathrm{d}x} = ay$, we get the solution $y = e^{ax}$, where a is an arbitrary constant. The solution is a family of exponential curves each corresponding to a different value of a. The solutions for various values of a are plotted in Figure 1.11.

Newton's particle mechanics provides probably the most convincing case for differential equations and state variables. Consider the simplest mechanical system - a single particle moving on a line under the action of a constant force F. The motion is governed by Newton's Second Law, which *defines* the force F acting on a mass point m to be the rate of change of momentum $(m \cdot v)$:

$$F = \frac{\mathrm{d}(m \cdot v)}{\mathrm{d}t} ,$$

where v denotes velocity which, in turn, is defined as rate of change of position or displacement from some origin of coordinates:

$$F = m \cdot \frac{\mathrm{d}^2 x}{\mathrm{d}t^2} .$$

In order to apply this formulation to an actual system, we must have an independent characterization of the force, expressed in terms of varying quantities x and v. For instance, with parameter a,

$$F(x, v) = -a \cdot x .$$

We thus obtain the equation of motion

$$m \cdot \frac{\mathrm{d}^2 x}{\mathrm{d}t^2} = -a \cdot x ,$$

which may be solved for the displacement x as an explicit function of time as shown above. The single second-order equation can be written as a pair of first-order equations:

$$\frac{\mathrm{d}x}{\mathrm{d}t} = v$$

$$\frac{\mathrm{d}(m \cdot v)}{\mathrm{d}t} = -a \cdot x .$$

According to Newton's Laws, a system of particles is then sufficiently characterized by the displacement and momentum of the particles. Knowing these displacements and momenta at an instant of time thus suffices to specify the *state* of the system completely at that instant, and hence the positions and their associated momenta are said to constitute a set of *state variables* (observables) for the system. Furthermore, together with a set of initial values of the state variables, the system behavior is determined for all time. We say that the system is *deterministic*.

At this point we should briefly remind ourselves of the concept of observables introduced earlier. For the simple mechanical system we therefore have two observables providing measurements of the position x and its derivative. Figure 1.12 provides an illustration of how these two observables describe the system. Specifically, only one observable would not be sufficient as it would only lead to one of the horizontal or vertical lines. Only with both observables we can uniquely identify a point in the product space.

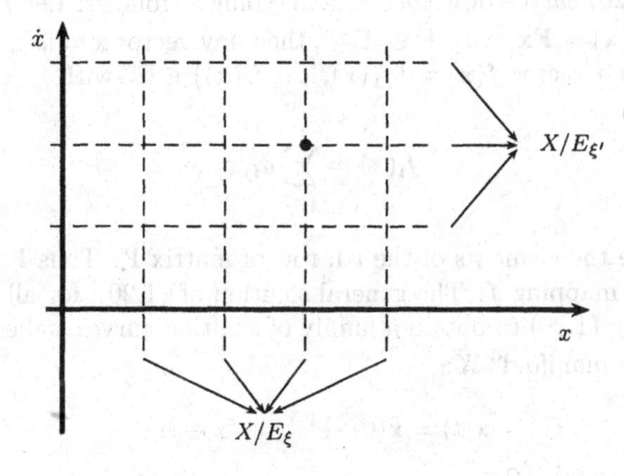

Fig. 1.12 Example of two observables describing a simple mechanical system following Newton's Law.

The system of first-order equations can be generalized to the case where $x_1, \ldots, x_r$ represents a family of physical magnitudes which characterize the states of *any* given system such as chemical processes where x_j may represent the concentrations of reactants. Causality is assumed with the assumption that the rates at which these quantities are changing, at any instant of time, depends only on the present state. We may then write a set of first-order differential equations in the form

$$\frac{\mathrm{d}x_i}{\mathrm{d}t} = f_i\left(x_1, \ldots, x_r\right) \ . \tag{1.19}$$

The mathematical object represented by equation (1.19), has been applied to virtually all areas of scientific investigation. For spatial or distributed systems, equation (1.19) is generalized by an additional variable such that the equations of motion become partial differential equations instead of ordinary differential equations. In many engineering problems it is more appropriate to pass from the continuous-time parameter in (1.19) to a discrete-valued time parameter such that the resulting equations of motion become difference equations. If, in addition, the set of states that such a system can occupy is discrete, the system is described as an *automaton* as already introduced above.

For more than one variable we therefore describe a dynamic system by a finite-dimensional system of first-order differential equations

$$\frac{\mathrm{d}\boldsymbol{x}}{\mathrm{d}t} \doteq \dot{\mathbf{x}} = \mathbf{F}\mathbf{x} \ , \tag{1.20}$$

where $\mathbf{F}$ is a $r \times r$ matrix with constant coefficients and

$$\mathbf{x} = [x_1, \ldots, x_r]^T \in \mathbb{R}^r \ .$$

Equation (1.20) can be described as a mapping as follows: Let $f\colon \mathbb{R}^r \to \mathbb{R}^r$ be defined by $f(\mathbf{x}) = \mathbf{F}\mathbf{x}$, with $\mathbf{F} \in \mathbb{R}^{r\times r}$, then any vector $\mathbf{x} = [x_1, \ldots, x_r]^T \in \mathbb{R}^r$ is mapped to a vector $f(\mathbf{x}) = (f_1(\mathbf{x}), \ldots, f_r(\mathbf{x})) \in \mathbb{R}^r$ with

$$f_i(\mathbf{x}) = \sum_{j=1}^{r} a_{ij} x_j \ ,$$

where a_{ij} are the elements of the ith row of matrix $\mathbf{F}$. Thus $\mathbf{F}$ is a representation of the mapping f. The general solution of (1.20), for all t, is obtained by integrating (1.20) to obtain a family of solution curves, called *trajectories*, drawn on the manifold[6] $\vec{X}$:

$$\mathbf{x}(t) = \mathbf{x}(0) \cdot e^{\mathbf{F}\cdot t} \qquad \mathbf{x} \in \mathbb{R}^r \ .$$

The dynamic system is then described by a linear difference equation

$$\mathbf{x}(k+1) = \mathbf{F}\mathbf{x}(k) \ ,$$

where $\mathbf{F}$ is a $r \times r$ matrix with constant coefficients, $k = 0, 1, \ldots$ and the general solution is denoted by

$$\mathbf{x}(k) = \mathbf{F}^k \mathbf{x}(0) \ .$$

Previous formulation may be generalized to the *time varying* vector differential equation

$$\frac{\mathrm{d}\mathbf{x}(t)}{\mathrm{d}t} = \mathbf{F}(t)\ \mathbf{x}(t) \ ,$$

where $\mathbf{x}(t) \in \mathbb{R}^r$ and $\mathbf{F}(t) \in \mathbb{R}^{r\times r}$. The solution to this equation is then given as

$$\mathbf{x}(t) = \mathbf{\Phi}(t; t_0)\mathbf{x}(0) \ ,$$

where $\mathbf{\Phi}(t; t_0) \doteq e^{\mathbf{F}t}$ denotes the (non-singular) *state-transition matrix*. Now considering the linear time-invariant mapping

$$\begin{aligned} f : \mathbb{R}^r &\to \mathbb{R}^m \\ \mathbf{x} &\mapsto \mathbf{y} = \mathbf{F}\mathbf{x} \ , \end{aligned}$$

where $\mathbf{F}$ is a $r \times r$ non-singular matrix, that is, its determinant is not equal to zero, and

$$f(\mathbf{x}(t)) = [f_1(\mathbf{x}(t)), \ldots, f_r(\mathbf{x}(t))] \in \mathbb{R}^r \ .$$

For a given vector $\mathbf{y}$ a problem arising is to solve the set of r equations given by $\mathbf{F}\mathbf{x} = \mathbf{y}$ for $\mathbf{x}$. The solution is given by $\mathbf{x} = \mathbf{F}^{-1}\mathbf{y}$ where the inverse matrix

[6] The r-dimensional manifold $\vec{X}$ is a space in which it is possible to set up a coordinate system near each point such that locally the space looks like a subset of $\mathbb{R}^r$.

$\mathbf{F}^{-1}$ exists only if $\mathbf{F}$ is non-singular. The equation $\mathbf{x} = \mathbf{F}^{-1}\mathbf{y}$ defines the inverse mapping

$$\begin{aligned} f^{-1} : \mathbb{R}^m &\rightarrow \mathbb{R}^r \\ \mathbf{y} &\mapsto \mathbf{x} = \mathbf{F}^{-1}\mathbf{y} \, . \end{aligned}$$

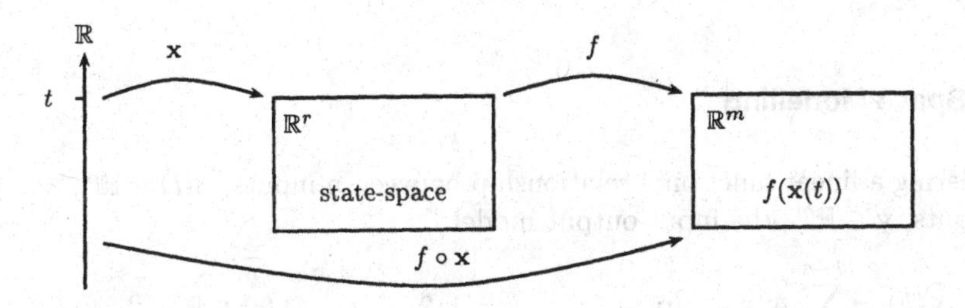

Fig. 1.13 Composite output map.

We should also note that the system

$$\dot{\mathbf{x}}(t) = f\big(\mathbf{x}(t)\big)$$

is, in fact, a composition of two mappings

$$\begin{aligned} \mathbf{x} : \mathbb{R} &\rightarrow \mathbb{R}^r & \qquad f : \mathbb{R}^r &\rightarrow \mathbb{R}^m \\ t &\mapsto \mathbf{x}(t) & \mathbf{x} &\mapsto f(\mathbf{x}) \, . \end{aligned}$$

Such *composite mapping*, applying first $\mathbf{x} \colon \mathbb{R} \rightarrow \mathbb{R}^m$ followed by $f \colon \mathbb{R}^r \rightarrow \mathbb{R}^m$, is denoted as $f \circ \mathbf{x}$, that is,

$$\begin{aligned} (f \circ \mathbf{x}) : \mathbb{R} &\rightarrow \mathbb{R}^r \\ t &\mapsto (f \circ \mathbf{x})(t) \doteq f\big(\mathbf{x}(t)\big) \, . \end{aligned}$$

The same information is given in the following commutative diagram and is also illustrated in Figure 1.13.

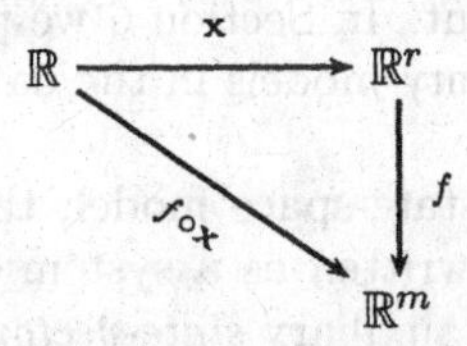

A system is called invertible, if the output uniquely determines the input. That is, if a system is represented by a mapping $\mathbf{y} = f(\mathbf{x})$, there must exist a map g such that $\mathbf{x} = g(f(\mathbf{x}))$. Throughout this text, I emphasize formulations of systems as some kind of mapping and dynamics as compositions. The general motivation is this. Suppose that Ξ is a space on which we have various structures - relations, binary operations, unary operations such as complements in Boolean algebras and so on. If Ξ' is another space with corresponding structures, then the system Ξ with its operations is *isomorphic* to the system Ξ' if there is a one-to-one mapping from Ξ onto Ξ' preserving these structures. A *homomorphism* just preserves the structure, that is, it is not required to be one-to-one.

State-Space Modelling

Considering a linear functional relationship between n inputs, $\mathbf{u}(t) \in \mathbb{R}^n$, and m outputs, $\mathbf{y} \in \mathbb{R}^m$, the input-output model

$$\mathbf{y}(k) = \sum_{j=0}^{k-1} \mathbf{A}(k-j)\mathbf{u}(j) \qquad k = 1, 2, \ldots \qquad \mathbf{A}(i) \in \mathbb{R}^{m \times n}$$

can be shown to be equivalent to the (discrete-time) dynamical system described by

$$\begin{aligned} \mathbf{x}(k+1) &= \mathbf{F}\mathbf{x}(k) + \mathbf{G}\mathbf{u}(k) \\ \mathbf{y}(k) &= \mathbf{H}\mathbf{x}(k)\ , \end{aligned} \tag{1.21}$$

where $\mathbf{x}(k) \in \mathbb{R}^r$, $\mathbf{u}(k) \in \mathbb{R}^n$, and $\mathbf{y}(k) \in \mathbb{R}^m$. In other words, the internal model $\mathfrak{M}_I$ and global model $\mathfrak{M}_G$ are equivalent if and only if

$$\mathbf{A}(k) = \mathbf{H}\mathbf{F}^{k-1}\mathbf{G} \qquad \forall\ k = 1, 2, \ldots .$$

Since there are many models $\mathfrak{M}_I = (\mathbf{F}, \mathbf{G}, \mathbf{H})$ satisfying this relation, the notions of *reachability* and *observability* are introduced (see e.g. [Cas92] for more details). The elements of the state-space may be thought of as being points in $\mathbb{R}^r$ but in general, we can view a state $\mathbf{x}$ to represent an *encoding* of the input $\mathbf{u}$ in the most compact form that is consistent with the generation of output $\mathbf{y}$ via the map f. In other words, a state describes an *equivalence class* $[\mathbf{u}]_f$ of inputs, where we regard two inputs $\mathbf{u}, \mathbf{u}'$ as equivalent if they generate the same output under f, *i.e.* $\mathbf{u} \approx \mathbf{u}'$ if and only if $f(\mathbf{u}) = f(\mathbf{u}')$. Thus, a model may be seen as an encoding $\mathbf{u} \to \mathbf{x} = [\mathbf{u}]_f$, the output of which corresponds to a specific input. In Section 6 we pick this idea up to develop the concept of local uncertainty models in the data space Ξ.

As stated before, in the state-space model, the relationship between the input and output signals is written as a system of first order differential or difference equations using an auxiliary *state-vector*. The state-space representation is especially useful in that physical dependencies can easily be incorporated into the model[7]. For example, consider the following continuous-time

[7]Note that differential equations by themselves do not reflect causation; they do not state that changes are *produced* by anything, but only that they are either *accompanied* or *fol-*

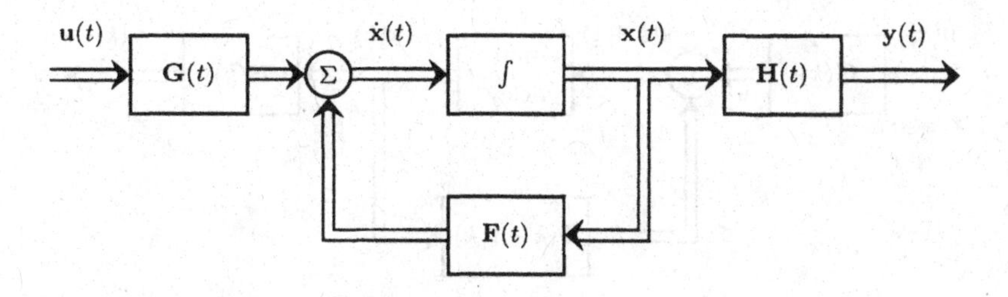

Fig. 1.14 Matrix block diagram of the general linear continuous-time dynamic system.

model, represented by a *state equation* and an *output equation*:

$$\dot{\mathbf{x}}(t) = \mathbf{F}(t)\mathbf{x}(t) + \mathbf{G}(t)\mathbf{u}(t) \tag{1.22}$$
$$\mathbf{y}(t) = \mathbf{H}(t)\mathbf{x}(t)\ , \tag{1.23}$$

where $\mathbf{x}$ denotes the *state* of the system. By a state, it is meant some qualitative information (a set of numbers, a function, etc.) which is the least amount of data one has to know about the past behavior of the system in order to predict its future behavior. The components x_j of r-vector $\mathbf{x}$ are called *state variables*. $\mathbf{u}(t)$ is an n-vector ($n \leq r$) representing the inputs to the system. $\mathbf{F}(t)$ and $\mathbf{G}(t)$ are $r \times r$ and $r \times n$ matrices, respectively. If all coefficients of $\mathbf{F}(t)$, $\mathbf{G}(t)$ and $\mathbf{H}(t)$ are constants, the system is *time-invariant*. $\mathbf{y}(t)$ is a m-vector denoting the outputs of the system; $\mathbf{H}(t)$ is an $r \times m$ matrix where $m \leq r$. The matrix $\mathbf{F}$ represents the dynamics and is therefore frequently called *system matrix*. Matrix $\mathbf{G}$ describes the constraints on affecting the state of the system by the input, and $\mathbf{H}$ the constraints on observing the state of the system from outputs. Figures 1.14 and 1.15 show the general matrix block diagram of a state-space model for continuous- and discrete-time, respectively.

The solution to (1.22) can be written in the form

$$\mathbf{x}(t) = \mathbf{\Phi}(t;t_0)\mathbf{x}(t_0) + \int_{t_0}^{t} \mathbf{\Phi}(t;\tau)\mathbf{G}(\tau)\mathbf{u}(\tau)d\tau\ , \tag{1.24}$$

where $\mathbf{\Phi}(t;t_0) = e^{\mathbf{F}(t-t_0)}$ is called the *transition matrix* of (1.22). Equation (1.24) should be read as describing the system evolving from the initial state $\mathbf{x}(t_0)$ at time t_0 under the action of $\mathbf{u}(t)$.

lowed by certain other changes. Considering $dx/dt = f(t)$ or $dx = f(t)dt$, it merely asserts that the variation dx undergone during the time interval dt, equals $f(t)dt$. The causal problem is not a syntactic but a semantic one; it has to do with the interpretation rather than with the formulations and representations of theories.

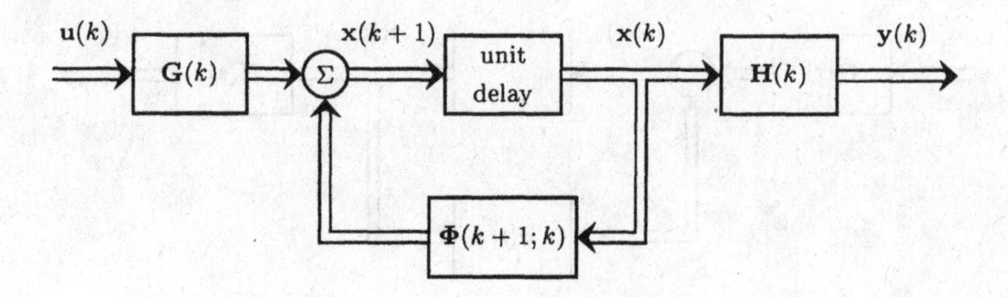

Fig. 1.15 Matrix block diagram of the general linear discrete-time dynamic system.

In *canonical-variable* or *normal-form* representation of a system, the matrix $\mathbf{F}$ turns out to be a diagonal matrix with the r poles λ_i of the system as its diagonal elements. Let us view the equation

$$\mathbf{Fx} = \mathbf{y}$$

as a transformation of vector $\mathbf{x}$ to vector $\mathbf{y}$ by matrix operator $\mathbf{F}$. The question we would like to answer is whether there exists a vector $\mathbf{x}$ such that a matrix operator $\mathbf{F}$ transforms it to a vector $\lambda\mathbf{x}$ (λ is a constant), that is, to a vector having the same direction in state-space as the vector $\mathbf{x}$. Such a vector $\mathbf{x}$, is a solution of the equation

$$\mathbf{Fx} = \lambda\mathbf{x} \; ,$$

which can be rewritten as

$$(\mathbf{F} - \lambda\mathbf{I})\mathbf{x} = 0 \; . \tag{1.25}$$

This set of homogenous equations has a non-trivial solution if and only if

$$|\mathbf{F} - \lambda\mathbf{I}| = 0 \; .$$

This equation may be expressed in expanded form as

$$q(\lambda) = \lambda^r + a_1\lambda^{r-1} + a_2\lambda^{r-2} + \cdots + a_r = 0 \; .$$

The values of λ for which the equation is satisfied are called *eigenvalues* of matrix $\mathbf{F}$ and the last equation is called the *characteristic equation* corresponding to matrix $\mathbf{F}$. The eigenvalues of $\mathbf{F}$ are identical to the poles of the transfer function of the system. The associated vector $\mathbf{x}$ for which a solution exists is then called an *eigenvector*.

The *motion* of the dynamical system (its changing state) describes a trajectory in the state-space. Let us briefly consider the notion of a *space*. Consider the vector $\mathbf{x} = [x_1, x_2]^T \in \mathbb{R}^2$, which we can associate with a point, where x_1 and x_2 are its Cartesian coordinates (see Figure 1.16). Then $\mathbb{R}^2$ may be thought of as the set of all points in the *real plane*. Since each point with coordinates (x_1, x_2) can be associated with a unique vector $\mathbf{x}$, an alternative interpretation of $\mathbb{R}^2$ is as the set of all vectors from the origin to points. The state variables x_1 and x_2 of a dynamical system are sometimes referred to as phase coordinates and $\mathbb{R}^2$ as the *phase-plane*. Any set of elements is called a 'space' rather than a 'set' when some operations on the elements of the set are defined. In other words, a set associated with some *structure* is called a space. The behavior of a second-order system in time is therefore a path or trajectory of a point in $\mathbb{R}^2$ with $\mathbf{x}(t) = (x_1(t), x_2(t))$ varying with time.

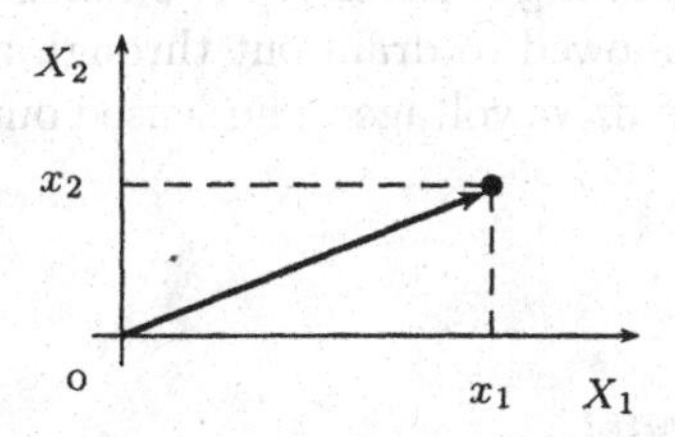

Fig. 1.16 A vector $\mathbf{x} = (x_1, x_2)$ in $\mathbb{R}^2$ represented as a point with cartesian coordinates x_1 and x_2.

With respect to the discrete-time system (1.21) let us first consider the time-invariant case and the homogeneous equation where $\mathbf{x}(k) = 0$:

$$\mathbf{x}(k+1) = \mathbf{F}\mathbf{x}(k)$$

with solution

$$\mathbf{x}(k) = \mathbf{\Phi}(k; k_0)\mathbf{x}(k_0) ,$$

where

$$\mathbf{\Phi}(k; k_0) = \mathbf{F}^{(k-k_0)} .$$

Then, for the time-variant homogeneous equation (1.21) the solution is given as

$$\mathbf{x}(k) = \mathbf{\Phi}(k; k_0)\mathbf{x}(k_0) + \sum_{j=k_0}^{k-1} \mathbf{\Phi}(k; j+1)\mathbf{G}(j)\mathbf{x}(j) .$$

With $k_0 = 0$, instead of (1.21), we write

$$\mathbf{x}(k+1) = \mathbf{\Phi}(k+1; k)\mathbf{x}(k) + \mathbf{G}(k)\mathbf{u}(k) \tag{1.26}$$

$$\mathbf{y}(k) = \mathbf{H}(k)\mathbf{x}(k) . \tag{1.27}$$

If the system is time-invariant we can express $\mathbf{\Phi}$ analytically in terms of the eigenvalues of $\mathbf{F}$.

In Section 2.3, we use the state-space representation (1.26) in modelling stochastic processes. This is followed by a formulation where the encoding of dynamics is transformed into a static regression problem by choosing past or delayed values of y and $\mathbf{x}$ to define the axes of Ξ.

1.4 EXAMPLE: COUPLED TANKS MODEL

In this example we consider a scaled laboratory model of two coupled tanks used to test control algorithms. The coupled tanks apparatus consists of a transparent perspex container divided into two tanks by a center partition. Water is pumped from the reservoir into the first tank by a variable speed pump. Holes at the base of the partition allow fluid flow between the two tanks. Some or all of the holes can be blocked using the rubber bungs provided to change the degree of coupling between the two tanks. The water which flows into the second tank is allowed to drain out through an adjustable tap. The control input is the pump drive voltage. The sensed output is the water depth in tank 2.

Development of the Model

Figure 1.17 shows the diagram of the coupled tanks apparatus used for modelling purposes, where:

Q_i	is the fluid flow rate into tank 1 (from the reservoir).
Q_{12}	is the fluid flow rate from tank 1 into tank 2.
Q_o	is the fluid flow rate out of tank 2 (to the reservoir).
V_1	is the volume of fluid in tank 1.
V_2	is the volume of fluid in tank 2.
H_1	is the level of fluid in tank 1.
H_2	is the level of fluid in tank 2.
A	is the cross-sectional area of each tank.

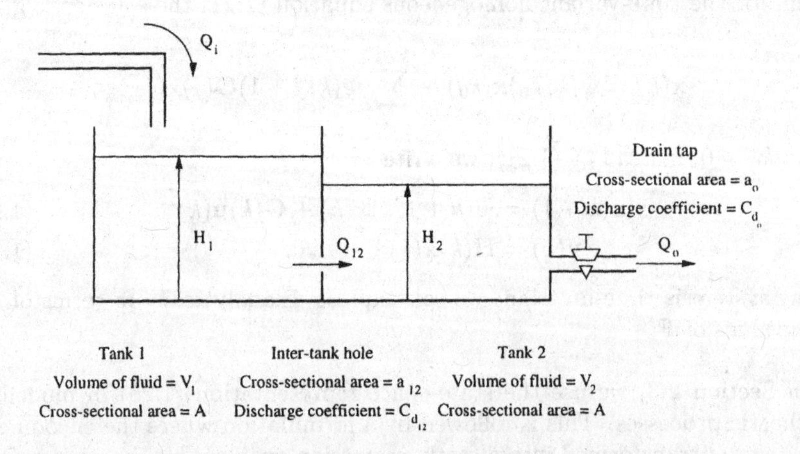

Fig. 1.17 Diagram of the coupled tanks apparatus for modelling purposes.

The dynamic equations of the system may be derived by taking flow balances about each tank. For the first tank we have:

$$Q_i - Q_{12} = \frac{dV_1}{dt} \tag{1.28}$$

$$= A\frac{dH_1}{dt}\,. \tag{1.29}$$

For the second tank:

$$Q_{12} - Q_o = \frac{dV_2}{dt} \tag{1.30}$$

$$= A\frac{dH_2}{dt}\,. \tag{1.31}$$

The inter-tank holes and drain tap are assumed to behave like orifices. The characteristic equation for an orifice gives (N.B. the square root of the difference between the levels of the fluid in tank 1 and tank 2 makes the system nonlinear)

$$Q_{12} = C_{d_{12}} a_{12}\sqrt{2g\,(H_1 - H_2)} \tag{1.32}$$

$$Q_o = C_{d_o} a_o \sqrt{2g\,(H_2 - H_0)}\,, \tag{1.33}$$

where

$C_{d_{12}}$	is the discharge coefficient of the inter-tank holes.
C_{d_o}	is the discharge coefficient of the drain tap.
a_{12}	is the combined cross-sectional area of the inter-tank holes.
a_o	is the cross-sectional area of the drain tap.
H_o	is the height of the drain tap.
g	is the gravitational acceleration constant.

Linear Model

The above equations describe the system in their true nonlinear form. For control system analysis and design it is necessary to linearize these equations by considering small variations q_i in Q_i, q_2 in Q_2, h_1 in H_1, and h_2 in H_2. The linear equations are

$$\frac{dh_1}{dt} = -\frac{k_1}{A}\,h_1 + \frac{k_1}{A}\,h_2 + \frac{1}{A}q_i \tag{1.34}$$

$$\frac{dh_2}{dt} = \frac{k_1}{A}\,h_1 - \frac{k_1 + k_2}{A}\,h_2\,, \tag{1.35}$$

where

$$k_1 = \frac{C_{d_{12}} a_{12}\sqrt{2g}}{2\sqrt{H_1 - H_2}} \tag{1.36}$$

$$k_2 = \frac{C_{d_o} a_o \sqrt{2g}}{2\sqrt{H_2 - H_0}}\,. \tag{1.37}$$

Taking the Laplace transform gives:

$$G(s) = \frac{h_2(s)}{q_i(s)} = \frac{\frac{1}{k_2}}{\left[\frac{A^2}{k_1 k_2}\right] s^2 + \left[\frac{A(2k_1+k_2)}{k_1 k_2}\right] s + 1} . \tag{1.38}$$

The denominator can be factorized giving time constants T_1 and T_2:

$$G(s) = \frac{\frac{1}{k_2}}{(sT_1 + 1)(sT_2 + 1)} . \tag{1.39}$$

Steady-state Operating Conditions

To determine theoretical values for the model we assume a steady-state operating condition with a pump flow rate of $900\text{cm}^3 \cdot \text{min}^{-1}$ ($Q_i = 900\text{cm}^3 \cdot \text{min}^{-1}$). Since this is a steady-state condition the inter-tank flow rate and drain tap flow rates will also be $900\text{cm}^3 \cdot \text{min}^{-1}$. Therefore:

$$Q_i = Q_{12} = Q_o = 900 \text{ cm}^3 \cdot \text{min}^{-1} . \tag{1.40}$$

We will use the coupled tanks apparatus with the largest two inter-tank holes plugged (leaving the smallest two unplugged). The smallest two inter-tank holes are 0.635cm and 0.317cm in diameter. The drain tap orifice is equivalent to an orifice 0.635cm in diameter. Therefore:

$$a_{12} = \pi \left(\frac{0.635}{2}\right)^2 + \pi \left(\frac{0.317}{2}\right)^2 \text{cm}^2 \tag{1.41}$$

$$a_o = \pi \left(\frac{0.635}{2}\right)^2 \text{cm}^2 . \tag{1.42}$$

The inter-tank hole, and drain tap discharge coefficients are:

$$C_{d_{12}} = 0.6 \tag{1.43}$$

$$C_{d_o} = 0.6 . \tag{1.44}$$

Other constants are:

$$g = 9.8 \text{ ms}^{-2} \tag{1.45}$$

$$H_0 = 3 \text{ cm} \tag{1.46}$$

$$A = 100 \text{ cm}^2 . \tag{1.47}$$

The steady-state fluid levels H_1 and H_2 can now be calculated using equations (1.32) and (1.33).

2

Uncertainty Techniques

- ☐ *The expectation operator is a generic concept to summarize information.*
- ☐ *Averaging information leads to probability measures and statistics.*
- ☐ *Aggregating information leads to fuzzy measures and possibility measures.*
- ☐ *The least-squares criterion provides optimal parameter estimates for linear models.*
- ☐ *A geometric representation of the regression problem shows that the optimal solution implies orthogonality.*
- ☐ *The Fourier series is an example using the orthogonality principle.*
- ☐ *The least-squares principle does not require a statistical framework to make sense.*
- ☐ *Maximum likelihood estimation is a statistical framework for parameter estimation.*
- ☐ *Stochastic processes are a probabilistic framework to study time-series.*
- ☐ *The Kalman-Bucy filter is a good example of how a probabilistic framework can be used to develop a new concept to model data.*

It follows that the word probability, in its mathematical acceptance, has reference to the state of our knowledge of the circumstances under which an event may happen or fail. With the degree of information we possess concerning the circumstances of an event, the reason we have to think that it will occur, or, to use a single term, our *expectation* of it will vary. Probability is the expectation founded upon partial knowledge.

—George Boole

In the analysis of complex systems we can expect neither data nor models to be precise and free of uncertainty. This section introduces the *expectation operator* as a generic tool to extract information from variables (*i.e.* signals and data). The expectation of any function h with respect to some function g both defined on a space, say Y, is given by

$$E[h(\cdot)] \doteq \int_Y h(y) \cdot g(y) \, \mathrm{d}y \, . \tag{2.1}$$

The expectation operator may be used in two ways to summarize information: a) *averaging* data to obtain a single measure if the data are uncertain, in particular if they are random; b) *aggregating* information to obtain a consensus between similar pieces of information. The former is primarily dealt with in Probability Theory [Pap91], whereas the latter is considered in Possibility Theory [NW97, Wol98].

If data are considered to be random, the uncertainty of outcomes in Y is characterized by some probability distribution or density $p(y)$ which quantifies the 'likelihood' of whether any particular value in Y will, on average, occur or not. In a probabilistic setting, it is common practice to associate the outcomes in Y with a *random variable*, denoted $\mathbf{y}$. For some *event*, represented by subset $A \subset Y$, the expectation of the *characteristic function* ζ specifying subset A, is defined as the *probability* of event A:

$$\begin{aligned} E[\zeta_A] &= \int_{-\infty}^{+\infty} \zeta_A(y) \, p(y) \, \mathrm{d}y \quad \text{where} \quad \zeta_A(y) = \begin{cases} 1 & \text{if } y \in A \, , \\ 0 & \text{if } y \notin A \, , \end{cases} \\ &= \int_A p(y) \, \mathrm{d}y \\ &\doteq Pr(A) \, . \end{aligned} \tag{2.2}$$

In this context, two kinds of measures are of particular importance: measures of *central tendency* and *dispersion* of values of variable $\mathbf{y}$. A measure of central tendency is the *mean value* defined by

$$\begin{aligned} E[\mathbf{y}] &= \int_Y y \cdot p(y) \, \mathrm{d}y \\ &\doteq \eta \, . \end{aligned} \tag{2.3}$$

From (2.3), the dispersion of data in Y, around η, is quantified by the *variance*

$$E[(y-\eta)^2] = \int_Y (y-\eta)^2 \cdot p(y)\, dy$$
$$\doteq \sigma_{\mathbf{y}}^2 \,. \tag{2.4}$$

The square root of (2.4) is called *standard deviation*. From (2.4), considering two variables $\mathbf{x}$ and $\mathbf{y}$ we define the *covariance* between the two variables as

$$\sigma_{\mathbf{x},\mathbf{y}} \doteq E\big[(\mathbf{x}-\eta_{\mathbf{x}})(\mathbf{y}-\eta_{\mathbf{y}})\big] \,. \tag{2.5}$$

If $\sigma_{\mathbf{x},\mathbf{y}} = 0$, then $\mathbf{x}$ and $\mathbf{y}$ are said to be 'independent'. A bounded measure of this is the *correlation coefficient*:

$$\rho_{\mathbf{x},\mathbf{y}} \doteq \frac{\sigma_{\mathbf{x},\mathbf{y}}}{\sigma_{\mathbf{x}} \cdot \sigma_{\mathbf{y}}} \qquad \text{where} \qquad -1 \leq \rho \leq 1 \,. \tag{2.6}$$

Mean and variance are two important quantities providing a rough summary of the form of the probability distribution. This description is certainly not unique as one could construct many distributions, all of which have the same values of η and σ^2, but with their shapes differing in other respects. In order to characterize distributions with values such as η and σ in a more comprehensive way, the concept of *moments* is introduced.

A 'problem' is that the concepts presented so far are abstract, theoretical models for what happens 'in general'. Considering actual data sets we need to estimate the defined measures by some 'sample statistics'. Given a finite set of data[1], $\mathbf{M} = \{\mathbf{m}_j = x_j\}$, $j = 1, \ldots, d$, we may, for example, use the following estimators of (2.3) and (2.4):

$$\hat{\eta} = \frac{1}{d}\sum_{j=1}^{d} x_j \tag{2.7}$$

$$\hat{\sigma}^2 = \frac{1}{d}\sum_{j=1}^{d} (x_j - \hat{\eta})^2 \tag{2.8}$$

or the unbiased[2] estimator

$$\hat{\sigma}^2 = \frac{1}{d-1}\sum_{j=1}^{d} (x_j - \hat{\eta})^2 \,. \tag{2.9}$$

[1] In subsequent sections, the set of training data $\mathbf{M}$ is a set of objects (vectors) $\mathbf{m}_j$. For example, $\mathbf{m} = (\mathbf{x}, y)$, where $\mathbf{x}$ is a vector of regressors and y denotes the dependent variable. For single-valued outcomes we write $\mathbf{m}_j = x_j$.

[2] Unbiasedness means that the expectation of the estimator, *i.e.* sample mean $\hat{\eta}$ (sample variance $\hat{\sigma}^2$), is the mean η (variance σ^2). A proof can be found in Appendix 13.5.

Note that the expectation operator is not exclusive to probability theory. Viewing the product in the definition of (2.2) as describing a weighting of values y with their likelihood $p(y)$ and the integral as a means to summarize the information across Y, we can generalize the definition of the expectation operator (2.1) in two ways: considering *fuzzy events* $\mu_A\colon Y \to [0,1]$ and generalizing the Riemann integral to the *fuzzy integral*. Let $(Y = \mathbb{R}, \mathcal{F}(Y), Pr)$ be a probability space with the event space $\mathcal{F}(Y)$ and sample space Y and a probability measure $Pr\colon \mathcal{F}(Y) \to [0,1]$. The fuzzy set $A = \{(y, \mu_A(y)) \mid y \in Y\} \in \mathcal{F}(Y)$, where $\mu\colon Y \to [0,1]$, is called a *fuzzy event* in Y and the probability of the fuzzy event A is defined as the expectation of μ_A:

$$
\begin{aligned}
E[\mu_A] &= \int \mu_A(y)\, \mathrm{d}Pr \\
&= \int_{-\infty}^{+\infty} \mu_A(y)\, p(y)\, \mathrm{d}y \qquad (2.10) \\
&\doteq Pr(A)\,.
\end{aligned}
$$

Equation (2.10) evaluates the degree with which space Y has the fuzzy property A. The corresponding experiment is a random selection of elements y more or less belonging to A. At each trial a membership value $\mu_A(y_j)$ is provided and

$$
Pr(A) = \lim_{d\to\infty} \frac{\sum_j^d \mu_A(y_j)}{d}\,,
$$

where d denotes the number of trials.

In (2.1), the Riemann integral accumulates the 'evidence' $g(y)$ of elements in y weighted by $h(y)$. Intuitively we may generalize this idea by replacing the Riemann integral (sum) and the weighting (product) with two more general methods leading to the *fuzzy integral*:

$$
E[h(\cdot)] \doteq \,⨍_Y h(y) \circ g(y)\,,
$$

where the mapping $g\colon \mathcal{F}(Y) \to [0,\infty)$ is referred to as a *fuzzy measure* generalizing the concept of a probability measure. The integral is sometimes referred to as the Sugeno integral since the concept was introduced by Michio Sugeno. If $g(A)$ is considered as a degree of confidence of event A, for consistency, g should be monotone in the sense of set-inclusion:

$$
\text{If } A \text{ implies } B \text{ (}i.e.\ A \subseteq B\text{), then } g(A) \le g(B)\,.
$$

The axiom of additivity for probabilities is relaxed as follows:

$$
\text{If } A \cap B = \emptyset,\ \text{then } g(A \cup B) = g(A) \star g(B)\,. \qquad (2.12)
$$

Probability measures are recovered for $\star = +$. Using $\star = \max$, we obtain possibility measures. Accordingly, the probability distribution function $p(\cdot)$ is replaced by a *possibility distribution* $\pi(\cdot)$ on Y, defined as a mapping from the reference set Y into the unit-interval, that is,

$$\pi : Y \to [0,1] \ ,$$

where $\pi(y)$ is interpreted as the degree of possibility that $y \in Y$ coincides with an existing but inaccessible value. For a given *fuzzy event* represented by fuzzy set A with *membership function*, $\mu_A \colon Y \to [0,1]$, the expectation of μ_A defines the *possibility* of event A:

$$\begin{aligned} E[\mu_A] &= \int_Y \mu_A(y) \circ \pi(y) \\ &= \sup_{y \in Y} \{ \mu_A(y) \wedge \pi(y) \} \qquad (2.13) \\ &\doteq \Pi(A) \ . \end{aligned}$$

For 'crisp' events, where $\zeta_A \colon Y \to \{0,1\}$,

$$\Pi(A) = \sup_{y \in A} \pi(y) \ . \qquad (2.14)$$

Thus, while in (2.2) evidence is accumulated - the probability is determined by measuring the area of $p(y)$ valid w.r.t. A, in (2.14) we pick only the most favorable element in Y. The difference is illustrated in Figure 2.1. In contrast to the 'frequentist interpretation' of probability, describing *whether or not* an event will occur on average, the notion of a degree of possibility is usually used to describe degrees of *feasibility* to which some condition exists. The relationship between possibilities and probabilities will be further discussed in subsequent sections. A more detailed account on how expectation, integrals, probabilities and possibilities are motivated and formalized, can be found in Appendix 13.4. The relationship between probability and possibility measures is further discussed in Section 6. In Section 11, a bijective transformation between both domains is introduced.

There are three important ideas to be remarked in this section. First, fuzzy measures (e.g., possibility measure) are obtained as an expectation using a nonlinear generalization of the Lebesgue integral. We shall use the statistics and measures introduced here throughout as a tool to quantify uncertainty. Other techniques for parameter estimation will be introduced as required in any particular context. Second is the need to accumulate or aggregate (weighted) information into a single object. This operation, formalized by some integral, is applied to a space, general set or over time (sequence of sampled data). Finally, a *criterion* is necessary to quantify the overall properties of such space, set or sequence. In (2.8), the variability of data is

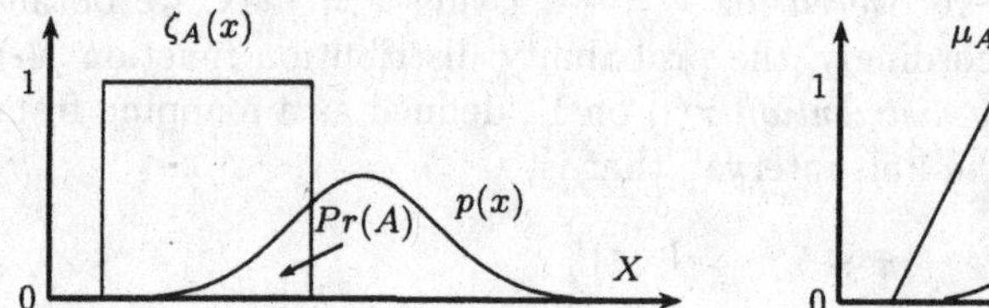

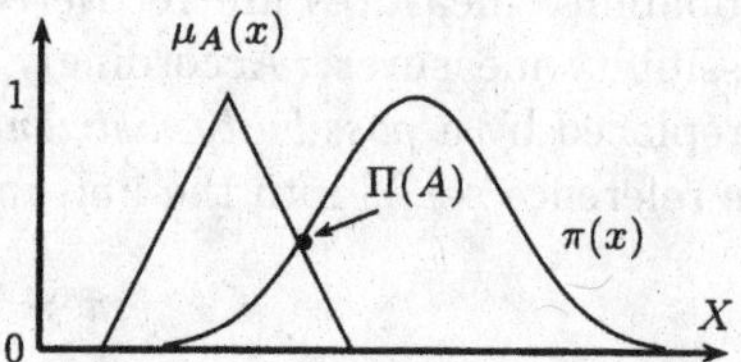

Fig. 2.1 The probability and possibility of event A.

described as the proximity to (similarity with) a reference value. The distance $(x - \eta)$ of a value x to the mean η is squared to eliminate the effect of negative values in accumulating individual qualifications into an overall measure. The 'squares criterion' is fundamental to systems and data analysis. The *least-squares criterion* is commonly used to describe the *objective function* in system identification. It will be introduced in the following section and will play a prominent role throughout this book.

2.1 THE LEAST-SQUARES CRITERION

You've got to draw the line somewhere.

—as they say.

In this section, we introduce a technique which allows us to identify system models, as discussed in Section 1, from sampled data. The most commonly used criterion to quantify the quality of the model fitting the data is called 'least-squares' criterion. It is the basis for regression analysis, hence system identification and plays an important role in fuzzy clustering. The geometrical representation of a least-squares solution provides us with an intuitive approach to fuzzy models as universal approximators of a nonlinear regression surface, that is, the unknown nonlinear function $y = f(\mathbf{x})$ represents a nonlinear (hyper)surface in the product space $X \times Y \subset \mathbb{R}^{r+1}$.

The theory of regression is concerned with prediction of a variable y, on the basis of information provided by variables $\mathbf{x}$. Let $\mathbf{x} \doteq [x_1, x_2 \ldots, x_r]^T$ be the *regression vector* over some domain $X = (X_1 \times \cdots \times X_r) \subset \mathbb{R}^r$, called the *regressor space*. The aim then is to identify the static dependence of a dependent or response variable, $y \in Y \subset \mathbb{R}$, called the *regressand*, on the independent variables x, called *regressors*. The concept of system modelling for regression analysis is summarized in Figure 2.2.

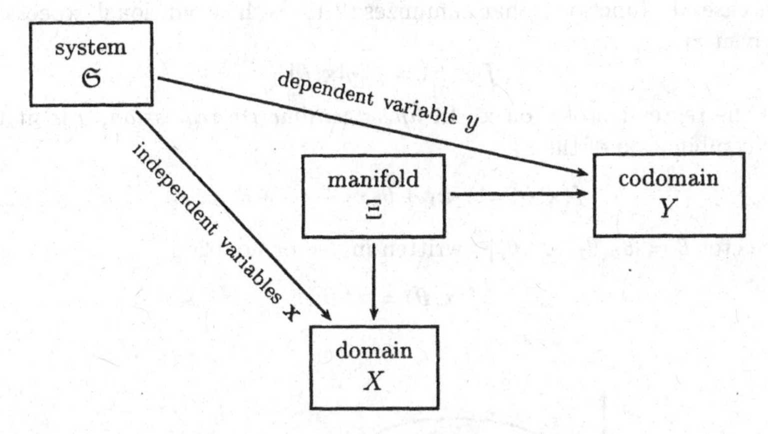

Fig. 2.2 Systems modelling for regression analysis.

The assumption is that $y \approx f(\mathbf{x};\boldsymbol{\theta})$, where the deterministic function $f(\cdot)$ captures the dependence of y on $\mathbf{x}$. The aim of the identification algorithm is to construct a function $f(\mathbf{x};\boldsymbol{\theta})$, from a finite set of data $\mathbf{M} = \{\mathbf{m}_j\}$, in order to find an appropriate representation of $f(\mathbf{x})$. For example, considering input-output models, using an *autoregressive* model structure, the system is described by a finite number of past inputs and outputs:

$$\mathbf{x} \doteq [y(k), \ldots, y(k-n_y+1), u(k), \ldots, u(k-n_u+1)]^T .$$

Linear parametric regression, using the least-squares criterion, provides solutions for linear functions $f(\cdot)$ as discussed in conventional system identification [Lju87]. Note that y and $\mathbf{x}$ may not be related to time at all. On the other hand, y may depend only on time, $y = f(t)$, or y is dependent on some variables which themselves vary in time.

In general, the problem is to find a function of the regressors $f(\mathbf{x};\boldsymbol{\theta})$, called *regression function*, such that the difference, $L\left(y, f(\mathbf{x};\boldsymbol{\theta})\right)$, called *loss* becomes small so that $y = f(\mathbf{x};\boldsymbol{\theta})$ is a good prediction of y. Therefore the *regression model* takes the form

$$y = f(\mathbf{x};\boldsymbol{\theta}) . \tag{2.15}$$

If y and $\mathbf{x}$ are described within a stochastic framework, one could, for example, minimize the expected value of the loss, called the *risk functional* [CM98, Vap98]:

$$E[L] = \int L\left(y, f(\mathbf{x};\boldsymbol{\theta})\right)\, p(\mathbf{x}, y)\, \mathrm{d}\mathbf{x}\mathrm{d}y . \tag{2.16}$$

A common loss function for regression is the squared error (L_2):

$$L\left(y, f(\mathbf{x};\boldsymbol{\theta})\right) = \left(y - f(\mathbf{x};\boldsymbol{\theta})\right)^2 . \tag{2.17}$$

In this case, the function f that minimizes (2.17) is the conditional expectation of y given $x_1, x_2, \ldots, x_r$

$$f(\mathbf{x}; \boldsymbol{\theta}) = E[y|\mathbf{x}; \boldsymbol{\theta}]$$

called the regression of y on $\mathbf{x}$. In *linear parametric regression*, y is fit to a linear combination of the x_i

$$f(\mathbf{x}; \boldsymbol{\theta}) = \theta_1 x_1 + \theta_2 x_2 + \cdots + \theta_r x_r \tag{2.18}$$

with vector $\boldsymbol{\theta} = [\theta_1, \theta_2, \ldots, \theta_r]^T$ written in vector notation

$$f(\mathbf{x}; \boldsymbol{\theta}) = \mathbf{x}^T \boldsymbol{\theta} .$$

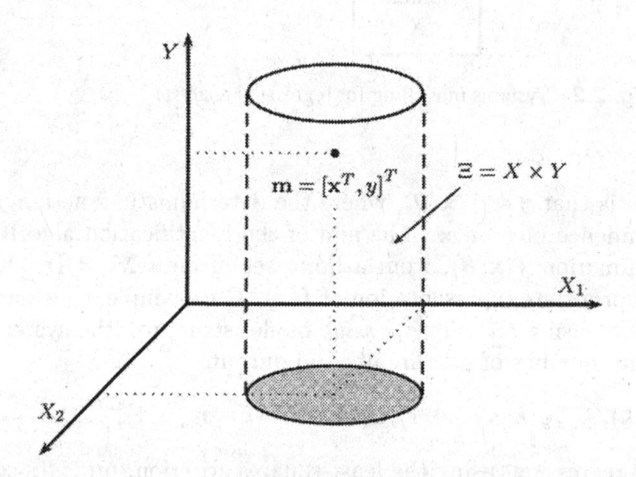

Fig. 2.3 Data in the regression space.

Since only a finite set of sampled data $\mathbf{M} = \{\mathbf{m}_j\}$ is available with

$$\begin{aligned} \mathbf{m}_j &= [\mathbf{x}_j^T, y_j]^T \\ &\doteq [m_{1j}, \ldots, m_{(r+1)j}]^T \in \mathbb{R}^{r+1} , \end{aligned} \tag{2.19}$$

the variance in (2.17) is replaced by the sample variance

$$\frac{1}{d} \sum_{j=1}^{d} \left(y_j - f(\mathbf{x}_j)\right)^2 ,$$

where

$$[m_{1j}, \ldots, m_{rj}]^T \doteq \mathbf{x}_j \in \mathbb{R}^r . \tag{2.20}$$

In the linear case (2.18), we therefore minimize the variance of the residuals

$$\hat{\sigma}_e^2 = \frac{1}{d} \sum_{j=1}^{d} \left(y_j - \mathbf{x}_j^T \boldsymbol{\theta}\right)^2 \tag{2.21}$$

instead of (2.17). In (2.21), using $\mathbf{x}_j^T\boldsymbol{\theta}$ as the prediction of $\hat{y}$,

$$y - \hat{y} \doteq e \tag{2.22}$$

is called the *prediction error* which we aim to minimize. Considering a time-series, that is, y and e depend on time, the variable $e(k)$ thus represents that part of the output $y(k)$ that cannot be predicted from past data. For this reason it is also called the *innovation* at time t_k. A suitable $\boldsymbol{\theta}$ to choose is the minimizing argument of (2.21):

$$\hat{\boldsymbol{\theta}} = \arg\min \frac{1}{d}\sum_{j=1}^{d}\left(y_j - \mathbf{x}_j^T\boldsymbol{\theta}\right)^2 , \tag{2.23}$$

called the *least-squares estimate*. Based on previous observations, we would thus use

$$\hat{y} = \mathbf{x}^T\hat{\boldsymbol{\theta}} \tag{2.24}$$

as a predictor function. Since the loss (2.21) is a quadratic function of $\boldsymbol{\theta}$, it can be minimized analytically. The necessary condition for the minimum of (2.23) is that all derivatives with respect to the parameters θ_i vanish:

$$\frac{\partial\hat{\sigma}_e^2}{\partial\theta_1} = 2\sum_{j=1}^{d} x_1\cdot(\theta_1 x_1 + \cdots + \theta_r x_r - y_j) = 0$$

$$\frac{\partial\hat{\sigma}_e^2}{\partial\theta_2} = 2\sum_{j=1}^{d} x_2\cdot(\theta_1 x_1 + \cdots + \theta_r x_r - y_j) = 0$$

$$\vdots$$

$$\frac{\partial\hat{\sigma}_e^2}{\partial\theta_r} = 2\sum_{j=1}^{d} x_r\cdot(\theta_1 x_1 + \cdots + \theta_r x_r - y_j) = 0 .$$

These conditions can be rewritten in the form of so-called *normal equations*:

$$\theta_1\sum x_1\cdot x_1 + \cdots \theta_d\sum x_1\cdot x_d = \sum y_j\cdot x_1$$

$$\theta_1\sum x_2\cdot x_1 + \cdots \theta_d\sum x_2\cdot x_d = \sum y_j\cdot x_1$$

$$\vdots$$

$$\theta_1\sum x_d\cdot x_1 + \cdots \theta_d\sum x_d\cdot x_d = \sum y_j\cdot x_1 .$$

That is, all $\hat{\boldsymbol{\theta}}$ that satisfy

$$\left[\frac{1}{d}\sum_{j=1}^{d}\mathbf{x}_j\mathbf{x}_j^T\right]\hat{\boldsymbol{\theta}} = \frac{1}{d}\sum_{j=1}^{d}\mathbf{x}_j y_j \tag{2.25}$$

yield a global minimum of (2.23). If the matrix on the left is invertible, we have

$$\hat{\boldsymbol{\theta}} = \left[\frac{1}{d}\sum_{j=1}^{d}\mathbf{x}_j\mathbf{x}_j^T\right]^{-1} \cdot \frac{1}{d}\sum_{j=1}^{d}\mathbf{x}_j y_j \,. \tag{2.26}$$

Rewritten in matrix notation, we define the following $d \times 1$ vector and $d \times r$ matrix

$$\mathbf{Y} = \begin{bmatrix} y_1 \\ y_2 \\ \vdots \\ y_d \end{bmatrix} \qquad \mathbf{X} = \begin{bmatrix} \mathbf{x}_1^T \\ \mathbf{x}_2^T \\ \vdots \\ \mathbf{x}_d^T \end{bmatrix} . \tag{2.27}$$

The normal equations take the form

$$[\mathbf{X}^T\mathbf{X}]\,\hat{\boldsymbol{\theta}} = \mathbf{X}^T\mathbf{Y} \tag{2.28}$$

and the estimate

$$\hat{\boldsymbol{\theta}} = [\mathbf{X}^T\mathbf{X}]^{-1}\mathbf{X}^T\mathbf{Y} \,, \tag{2.29}$$

where $[\mathbf{X}^T\mathbf{X}]^{-1}\mathbf{X}^T$ is known as the Moore-Penrose pseudoinverse and (2.29) thus gives the solution to the overdetermined ($d > r$) system of linear equations

$$\mathbf{Y} = \mathbf{X}\boldsymbol{\theta} \,. \tag{2.30}$$

To ensure that $\mathbf{X}^T\mathbf{X}$ is invertible, one needs to choose inputs to the system so that it is "sufficiently excited". If data are to be weighted, we introduce the weighting matrix

$$\mathbf{W} = \begin{bmatrix} w_1 & & 0 \\ & \ddots & \\ 0 & & w_d \end{bmatrix} \tag{2.31}$$

and write for (2.26) and (2.29),

$$\begin{aligned} \hat{\boldsymbol{\theta}} &= [\mathbf{X}^T\mathbf{W}\mathbf{X}]^{-1}\mathbf{X}^T\mathbf{W}\mathbf{Y} \\ &= \left[\sum_{j=1}^{d} w_j\mathbf{x}_j\mathbf{x}_j^T\right]^{-1} \cdot \sum_{j=1}^{d} w_j\mathbf{x}_j y_j \,. \end{aligned} \tag{2.32}$$

Note that the least-squares fitting makes sense without a probabilistic formulation. However, in order to study properties of least-squares estimates, usually a stochastic framework is used [Lju87]. Then, typical assumptions are that the sequence of regressors $\langle\mathbf{x}(k)\rangle$ is *deterministic*, the output of the system is a random variable that takes on real values and can be interpreted as the sum of a deterministic function and a random error with zero mean, leading to the time-series model

$$y(k+1) = f(\mathbf{x}(k); \boldsymbol{\theta}) + \varepsilon(k) ,$$

where $\varepsilon(k)$ is assumed to be a sequence of independent, identically distributed random variables with zero mean. In a statistical framework, the assumption is that there exists a *population* random variable $\mathbf{y}$ such that $E[\mathbf{y}_j] = f(\mathbf{x}_j)$, and for the *residuals* e_j, $E[e_j] = 0$. Hence, the deterministic function is the mean of the output conditional probability

$$f(\mathbf{x}) = \int y \; p(y|\mathbf{x}) \; \mathrm{d}y \; . \tag{2.33}$$

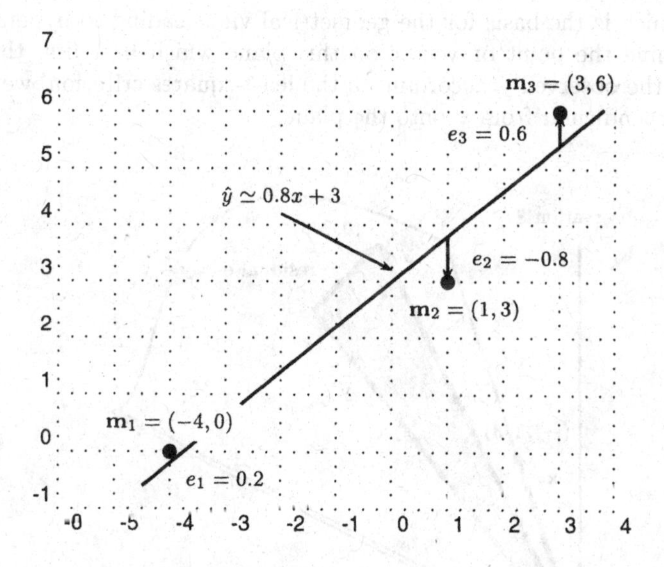

Fig. 2.4 Data scatter plot and least-squares line fit.

From a geometrical perspective, the unknown nonlinear function $y = f(\mathbf{x})$ represents a (non)linear hypersurface in the product space $X \times Y \subset \mathbb{R}^{r+1}$, called *regression surface*. Based on the assumptions that an accurate representation of the dependency $y = f(\mathbf{x})$ is nonlinear, in sections 4 and 5 this hypersurface is decomposed into a set of linear (fuzzy) submodels. The geometrical view on regression is based on the vector representation. Let us consider three data points $\mathbf{m}_1 = (-4, 0)$, $\mathbf{m}_2 = (1,3)$, $\mathbf{m}_3 = (3,6)$ with a regression line fitted through the data as shown in Figure 2.4. The model in vector-matrix notation, $\mathbf{Y} = \mathbf{X}\hat{\boldsymbol{\theta}} + \mathbf{E}$, where $\mathbf{E} = \mathbf{Y} - \hat{\mathbf{Y}}$ denotes the residuals, is defined by

$$\mathbf{Y} = \begin{bmatrix} 1 & -4 \\ 1 & 1 \\ 1 & 3 \end{bmatrix} \cdot \begin{bmatrix} \theta_1 \\ \theta_2 \end{bmatrix} + \begin{bmatrix} e_1 \\ e_2 \\ e_3 \end{bmatrix} .$$

The fitted regression line is a vector denoted by $\mathbf{y}$. The columns of $\mathbf{X}$ are sequences of sampled values from x_1 and x_2 and are therefore also vectors, denoted by $\mathbf{x}_1$ and $\mathbf{x}_2$,

$$\mathbf{y} = \hat{\theta}_1 \cdot \mathbf{x}_1 + \hat{\theta}_2 \cdot \mathbf{x}_2 \qquad \text{or} \qquad \begin{bmatrix} y_1 \\ y_2 \\ y_3 \end{bmatrix} = \hat{\theta}_1 \cdot \begin{bmatrix} 1 \\ 1 \\ 1 \end{bmatrix} + \hat{\theta}_2 \cdot \begin{bmatrix} -4 \\ 1 \\ 3 \end{bmatrix} . \tag{2.34}$$

Considering variables as vectors in a space defined by observations on the axes, we obtain the representation of Figure 2.5. Algebraically, our problem is to find a vector that is a linear combination of the vectors $\mathbf{x}_1$ and $\mathbf{x}_2$, and geometrically this means, we must select an optimum fit, $\mathbf{y}^*$, somewhere on the plane generated by $\mathbf{x}_1$ and $\mathbf{x}_2$. It is the fact that the two variables generate a plane which is the basis for the geometrical view leading to hypersurfaces. To determine the point or vector on this plane which best fits, that is, is closest to the observed $\mathbf{y}$ according to the least-squares criterion, we have to drop a perpendicular from $\mathbf{y}$ onto the plane.

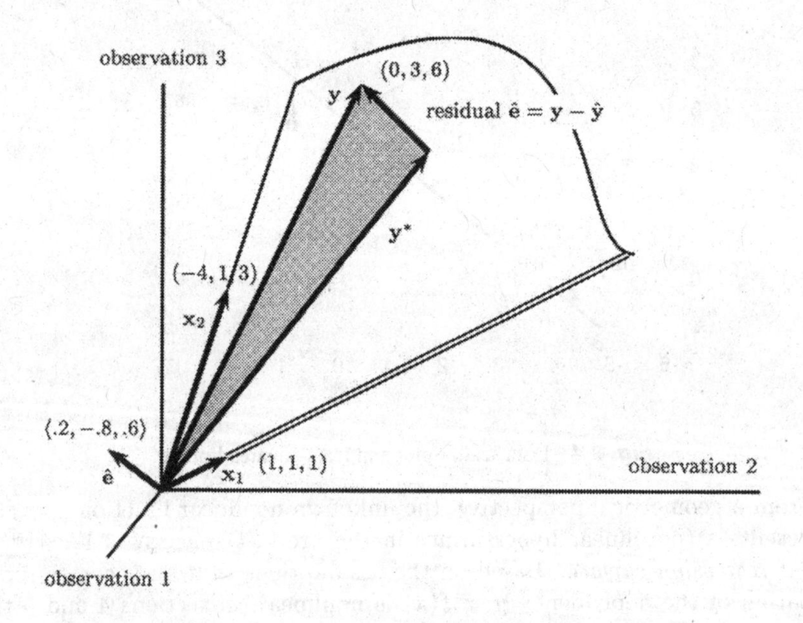

Fig. 2.5 Geometrical representation of a least-squares fit.

If the regressors are orthogonal, as in Figure 2.7, and we denote by $\mathbf{y}_1'$, $\mathbf{y}_2'$ the perpendicular projections of $\mathbf{y}$ onto $\mathbf{x}_1$ and $\mathbf{x}_2$, we have $\mathbf{y}^* = \mathbf{y}_1{}' + \mathbf{y}_2'$, comparing this with (2.34), we can find simple formulas for θ_1 and θ_2. For only two vectors $\mathbf{y}$ and $\mathbf{x}$, the perpendicular (orthogonal) projection of $\mathbf{y}$ onto $\mathbf{x}$ is a scalar multiple of $\mathbf{x}$:

$$\mathbf{y}' = a \cdot \mathbf{x} \tag{2.35}$$

with the problem to determine a such that the inner product $(\mathbf{y} - a \cdot \mathbf{x}) \cdot \mathbf{x} = 0$ is zero, that is, the angle between the two vectors is 90° (see Figure 2.6).

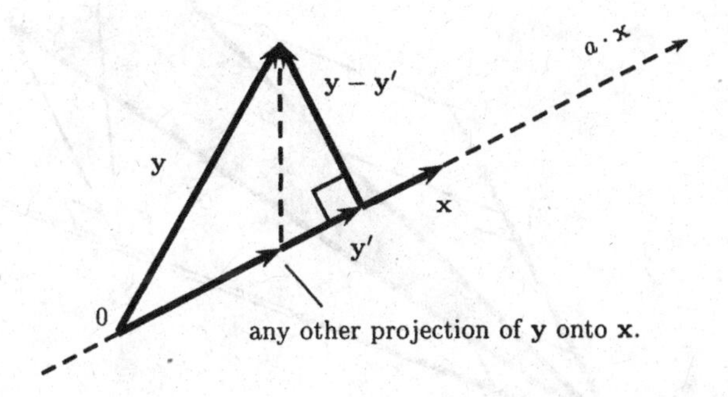

Fig. 2.6 Orthogonal projection $\mathbf{y}'$ of $\mathbf{y}$ onto $\mathbf{x}$.

Hence,

$$a = \frac{\mathbf{y} \cdot \mathbf{x}}{\mathbf{x} \cdot \mathbf{x}} . \tag{2.36}$$

Substituting (2.36) into (2.35) we have as the projection $\mathbf{y}'$ of $\mathbf{y}$ onto $\mathbf{x}$

$$\mathbf{y}' = \left(\frac{\mathbf{y} \cdot \mathbf{x}}{\mathbf{x} \cdot \mathbf{x}} \right) \cdot \mathbf{x} . \tag{2.37}$$

From (2.37), inserted into $\mathbf{y}^* = \mathbf{y}'_1 + \mathbf{y}'_2$, the optimal fit is given as

$$\mathbf{y}' = \left(\frac{\mathbf{y} \cdot \mathbf{x}_1}{\mathbf{x}_1 \cdot \mathbf{x}_1} \right) \cdot \mathbf{x}_1 + \left(\frac{\mathbf{y} \cdot \mathbf{x}_2}{\mathbf{x}_2 \cdot \mathbf{x}_2} \right) \cdot \mathbf{x}_2 . \tag{2.38}$$

Comparing (2.38) with (2.34), we obtain the parameter estimates as

$$\hat{\theta}_i = \frac{\mathbf{y} \cdot \mathbf{x}_i}{\mathbf{x}_i \cdot \mathbf{x}_i} . \tag{2.39}$$

In Section 2.1.2, the least-squares criterion is generalized to a function approximation problem for which the concept of orthogonal functions plays an important role. The Fourier series will serve as an example of a set of orthogonal functions. Further, in Section 2.3, the geometrical view of sequences of random variables as vectors and their orthogonality to ensure optimality is used to derive the Kalman-Bucy filter as a powerful demonstration of how the concept of expectation can be used in system analysis.

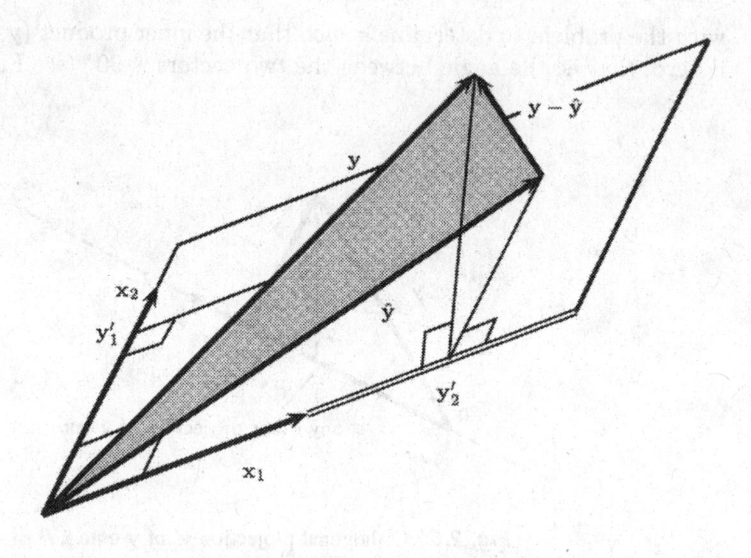

Fig. 2.7 Vector representation of least-squares regression.

2.1.1 Example: Regression Line

The example of a regression line or 'straight-line-fit' provides a solution for the linear parametric model (2.18) simplified to

$$\begin{aligned} y &= \theta_1 x_1 + \theta_2 x_2 \\ &\doteq \theta_1 + \theta_2 x \ . \end{aligned} \tag{2.40}$$

With $x_1 = 1$, we write x for x_2 and use the subscripts for indices of measured values of x. We have the following matrices:

$$\mathbf{X}^T = \begin{bmatrix} 1 & 1 & \cdots & 1 \\ x_1 & x_2 & \cdots & x_d \end{bmatrix} , \qquad \mathbf{X}^T\mathbf{X} = \begin{bmatrix} d & \sum x \\ \sum x & \sum x^2 \end{bmatrix}$$

and for the normal equations (2.28) and the parameter estimate (2.29):

$$\begin{bmatrix} d & \sum x \\ \sum x & \sum x^2 \end{bmatrix} \cdot \begin{bmatrix} \theta_1 \\ \theta_2 \end{bmatrix} = \begin{bmatrix} \sum y \\ \sum xy \end{bmatrix} ,$$

$$\hat{\boldsymbol{\theta}} = \frac{1}{d\sum x^2 - \left(\sum x\right)^2} \cdot \begin{bmatrix} \sum x^2 \sum y - \sum x \sum xy \\ d\sum xy - \sum x \sum y \end{bmatrix} .$$

Considering the two variables y and x, (2.40) defines a straight-line fit through the scatter plot of data for y and x. The slope of the line is determined by parameter θ_2:

$$\hat{\theta}_2 = \frac{d\sum xy - \sum x \sum y}{d\sum x^2 - \left(\sum x\right)^2} . \tag{2.41}$$

The result suggests that the *regression line* (2.40) describes also in some way the correlation between the two variables x and y. Then there should be a relationship to the correlation coefficient (2.6) defined earlier. In a scatter diagram, the 'cloud' of data is characterized by the estimates of mean values and standard deviation of both variables. More specifically, we can draw the σ-line through the point of averages $(\hat{\eta}_x, \hat{\eta}_y)$ with a slope defined by $\hat{\sigma}_y/\hat{\sigma}_x$. In Figure 2.8, the σ-line and least-squares fit of the regression line, together with 95% confidence intervals[3], are shown for the following set of data [FPP97]:

x	300	351	355	421	422	434	448	471	490	528
y	2	2.7	2.72	2.69	2.98	3.09	2.71	3.2	2.94	3.73

The point of averages is found at $(\hat{\eta}_x, \hat{\eta}_y) = (422, 2.876)$, $\hat{\sigma}_x^2 = 65.95$, $\hat{\sigma}_y^2 = 0.42$, and $\theta_1 = 0.56, \theta_2 = 0.0055$. Replacing the covariance and standard

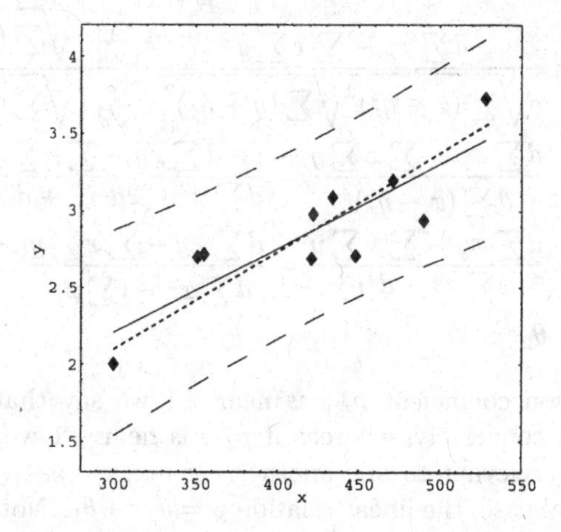

Fig. 2.8 Regression line (solid), σ-line (dotted), 95% confidence interval (dashed).

deviations of x and y by their estimators

$$\hat{\sigma}_x = \sqrt{\frac{1}{d}\sum_{j=1}^{d}(x_j - \hat{\eta}_x)^2}\ ,$$

where

$$\hat{\eta}_x = \frac{1}{d}\sum_{j=1}^{d} x_j\ ,$$

[3]A 95% confidence interval means that we are confident of finding the values in the interval $\pm 2\sigma$ around the regression line. That is, a 95% confidence level means we expect the values to be in the interval $\pm 2\sigma$ 95% of the time.

we obtain the following estimate for the correlation coefficient:

$$\begin{aligned}\hat{\rho}_{x,y} &= \frac{\frac{1}{d}\sum(x-\hat{\eta}_x)(y-\hat{\eta}_y)}{\frac{1}{d}\sqrt{\sum(x-\hat{\eta}_x)^2}\sqrt{\sum(y-\hat{\eta}_y)^2}} = \frac{\sum xy - d\hat{\eta}_x\hat{\eta}_y}{\sqrt{\sum(x-\hat{\eta}_x)^2}\sqrt{\sum(y-\hat{\eta}_y)^2}} \\ &= \frac{\sum xy - d\left[\frac{1}{d}\sum x \cdot \frac{1}{d}\sum y\right]}{\sqrt{\sum(x-\hat{\eta}_x)^2}\sqrt{\sum(y-\hat{\eta}_y)^2}} \\ &= \frac{d\sum xy - \sum x \sum y}{d\sqrt{\sum(x-\hat{\eta}_x)^2}\sqrt{\sum(y-\hat{\eta}_y)^2}} . \end{aligned} \tag{2.42}$$

The numerator already matches the one in (2.41), and we find that if we multiply $\hat{\rho}_{x,y}$ by $\hat{\sigma}_y/\hat{\sigma}_x$, the slope of the σ-line, we have found the slope of the regression line coinciding with θ_2:

$$\begin{aligned}\hat{\rho}_{x,y} \cdot \frac{\hat{\sigma}_y}{\hat{\sigma}_x} &= \frac{d\sum xy - \sum x \sum y}{d\sqrt{\sum(x-\hat{\eta}_x)^2}\sqrt{\sum(y-\hat{\eta}_y)^2}} \cdot \frac{\frac{1}{\sqrt{d}} \cdot \sqrt{\sum(y-\hat{\eta}_y)^2}}{\frac{1}{\sqrt{d}} \cdot \sqrt{\sum(x-\hat{\eta}_x)^2}} \\ &= \frac{d\sum xy - \sum x \sum y}{d\sum(x-\hat{\eta}_x)^2} = \frac{d\sum xy - \sum x \sum y}{d\sum x^2 - 2d^2\hat{\eta}_x^2 + d\sum \hat{\eta}_x^2} \\ &= \frac{d\sum xy - \sum x \sum y}{d\sum x^2 - d^2\hat{\eta}_x^2} = \frac{d\sum xy - \sum x \sum y}{d\sum x^2 - \left(\sum x\right)^2} \\ &= \theta_2 .\end{aligned}$$

If the correlation coefficient $\rho_{x,y}$ is near $+1$ we say that x is favorably relevant to y and conversely, whereas if $\rho_{x,y}$ is near -1 we say that x and y are unfavorably relevant to one another. If $\rho_{x,y}$ is exactly $+1$ or -1, we obtain, as a special case, the linear relation $y = \theta_2 x + \theta_1$. Note, however, that variables that are functionally related among each other are correlated but not conversely: if the correlation coefficient is near $+1$ or -1, we may suspect the existence of a law, but this is all.

2.1.2 Example: Fourier Series

As the previous of a straight-line fit illustrated, the least-squares criterion may in general be used to fit functions to data. Or, in other words, consider a function $y = f(t)$ which we try to approximate by some $\hat{y} = f(t;\theta)$ using parameter θ. The error of approximation, given by

$$\int_{-\infty}^{+\infty} \left(f(t) - f(t;\theta)\right)^2 \, \mathrm{d}t$$

is to be minimized using θ. This is the least-squares cost function (2.17). Now, let $y = f(t)$ only be specified by a sequence $f(t_k)$ sampled at equal distances of time t_k where $k = 1, 2, 3, \ldots$. We define a class of functions $f(t_k; \theta)$, indexed by θ, to minimize the quadratic error

$$\sum_{k=1}^{d} \left(f(t_k) - f(t_k; \theta)\right)^2 .$$

A general linear form of this approximation is

$$f(t; \boldsymbol{\theta}) = \sum_{i=1}^{r} \theta_i \cdot \phi_i(t) , \tag{2.43}$$

a linear combination of a set of functions $\{\phi_1, \phi_2, \ldots, \phi_r\}$ with r parameters $\theta_1, \theta_2, \ldots, \theta_r$ used to minimize the quadratic error. A well-known example is

$$\phi_i = t^i \qquad \text{with} \qquad i = 1, 2, \ldots, r$$

such that

$$f(t; \boldsymbol{\theta}) = \theta_1 + \theta_2 t + \theta_3 t^2 + \cdots + \theta_r t^r$$

is the least-squares polynomial of order r. In the general case the quadratic error takes the form

$$\sum_{k=1}^{d} \left(y_k - \sum_{i=1}^{r} \theta_i \cdot \phi_{ik} \right)^2 \doteq L_2 \quad \text{where} \quad \phi_i(t_k) \doteq \phi_{ik}, \text{ and } f(t_k) \doteq y_k \quad \text{for short.} \tag{2.44}$$

To obtain a minimum, we take the partial derivatives with respect to θ_j and set these equal to zero

$$\frac{\partial\, L_2}{\partial\, \theta_j} = -2 \sum_{k=1}^{d} \phi_{jk} \cdot \left(y_k - \sum_{i=1}^{r} \theta_i \cdot \phi_{ik} \right) = 0 ,$$

leading to

$$\begin{bmatrix} \sum \phi_{1k}\, \phi_{1k} & \cdots & \sum \phi_{1k}\, \phi_{rk} \\ \vdots & \ddots & \vdots \\ \sum \phi_{rk}\, \phi_{1k} & \cdots & \sum \phi_{rk} \cdot \phi_{rk} \end{bmatrix} \cdot \begin{bmatrix} \theta_1 \\ \vdots \\ \theta_r \end{bmatrix} = \begin{bmatrix} \sum y_k\, \phi_{1k} \\ \vdots \\ \sum y_k\, \phi_{rk} \end{bmatrix} . \tag{2.45}$$

The number of solutions for (2.45) depends on the nature of the set of functions $\{\phi_i\}$ as well as the sampled sequence $\langle y_k \rangle$. If the number of samples, d, is larger than the number of functions, r, the quadratic error will be different from zero. For $d = r$, we usually have one solution for which the error is zero. For $d < r$ we have various solutions. For a reasonable approximation, we consider the minimal mean quadratic error which we obtain by extracting

L_2/d from equation (2.44):

$$\frac{L_2}{d} = \frac{1}{d} \sum_{k=1}^{d} \left(y_k - \sum_{i=1}^{r} \theta_i \cdot \phi_{ik} \right)^2$$
$$= \frac{1}{d} \sum_{k=1}^{d} \left(y_k^2 - 2y_k \sum_{i=1}^{r} \theta_i \cdot \phi_{ik} + \sum_{j=1}^{r} \theta_j \sum_{i=1}^{r} \theta_i \cdot \phi_{ik} \cdot \phi_{jk} \right) . \tag{2.46}$$

Using the fact that

$$\sum_{k=1}^{d} \sum_{i=1}^{r} \theta_i \cdot \phi_{ik} \cdot \phi_{jk} = \sum_{k=1}^{d} y_k \cdot \phi_{jk} \qquad j = 1, 2, \ldots, r$$

we obtain

$$\frac{L_{\min}}{d} = \frac{1}{d} \sum_{k=1}^{d} \left(y_k^2 - 2y_k \sum_{i=1}^{r} \theta_i \cdot \phi_{ik} + \sum_{j=1}^{r} \theta_j \cdot y_k \cdot \phi_{jk} \right)$$
$$= \frac{1}{d} \sum_{k=1}^{d} y_k \left(y_k - \sum_{i=1}^{r} \theta_i \cdot \phi_{ik} \right) \tag{2.47}$$

as a general equation of the mean quadratic error.

Looking at the form of the set of functions ϕ in (2.45), it appears particularly useful if (2.45) could be solved explicitly for the coefficients θ without solving the system of $r \times r$ equations beforehand. It turns out that if the set of functions ϕ is orthogonal, we obtain such a simple solution of (2.45). If we view the sampled sequence $\langle y_k = f(t_k) \rangle$ of function $f(t_k)$ as a vector

$$\mathbf{y} = [y_1, y_2, \ldots, y_d]$$

and denote the prediction error (2.22), $f(t_k) - f(t_k; \boldsymbol{\theta})$, by e,

$$\mathbf{e} = [e_1, e_2, \ldots, e_d]$$

then we find the minimum for (2.47) if the two vectors are *orthogonal*, that is, their inner product is zero:

$$\mathbf{y} \cdot \mathbf{e} = \sum_{k=1}^{d} y_k \cdot e_k = 0 . \tag{2.48}$$

Orthogonality of functions is defined analogously. Two functions $g(\cdot)$ and $h(\cdot)$ are orthogonal in the interval $[a, b]$ iff

$$\int_a^b g(t) \cdot h(t) \; dt = 0 . \tag{2.49}$$

Then, the r functions $\phi_i, i = 1, 2, \ldots, r$ are orthogonal over a set of d points t_k, $k = 1, 2, \ldots, d$ if they are pairwise orthogonal, that is,

$$\sum_{k=1}^{d} \phi_{ik} \cdot \phi_{jk} = 0 \qquad i \neq j\ . \tag{2.50}$$

Then, in (2.45), only diagonal elements in the $r \times r$ matrix are different from zero and we have from (2.45) the solutions

$$\theta_j = \frac{\sum\limits_{k=1}^{d} y_k \cdot \phi_{jk}}{\sum\limits_{k=1}^{d} \phi_{jk}^2} \qquad j = 1, 2, \ldots, r\ . \tag{2.51}$$

Note that the denominator is different from zero, as long as at least one ϕ_{jk} is different from zero. In conclusion, we find a simple solution to the least-squares problem if the functions ϕ_i are orthogonal. That is, our approximation problem has a simple solution if the approximation $f(t; \boldsymbol{\theta})$ is a linear combination of orthogonal functions.

Up to now our discussion has been generally about fitting a function $f(t_k; \boldsymbol{\theta})$ through a set of points $f(t_k)$ so as to get a good approximation of $y = f(t)$. Probably the best known example for a specific pair of orthogonal functions leads to the *Fourier series*. The principle idea is to decompose a periodic function $f(t)$ into a linear combination of harmonics of a certain 'fundamental frequency' ω_0:

$$f(t) \doteq \frac{a_0}{2} + \sum_{i=1}^{n_h} \left(a_i \cdot \cos(i\omega_0 t) + b_i \cdot \sin(i\omega_0 t)\right)\ . \tag{2.52}$$

In equation (2.52) the fundamental frequency ω_0 is chosen such that the duration of one period of $f(t)$ equals $2\pi/\omega_0$, and the a_i, b_i are coefficients of the least-squares. Without loss of generality it is usually assumed that the sequence of d points is sampled at $t_k = \frac{-\pi}{\omega_0} + \frac{2\pi k}{\omega_0 d} = \frac{(2k-d)\pi}{\omega_0 d}$, $k = 1, 2, \ldots, d$. The orthogonality of the $\sin(\cdot)$ and $\cos(\cdot)$ terms in (2.52) is given if and only if the number of samples d is at least twice as large as the highest order of harmonics n_h. In this case, the coefficients a_i and b_i in (2.52) are related to the solution (2.51) as follows:

$$\sum_{k=1}^{d} \phi_{jk}^2 = \sum_{k=1}^{d} \cos^2(i\omega_0 t_k) \qquad i = 0, 1, \ldots, n_h$$

$$\text{or} \quad \sum_{k=1}^{d} \phi_{jk}^2 = \sum_{k=1}^{d} \sin^2(i\omega_0 t_k) \qquad i = 1, 2, \ldots, n_h$$

and from (2.51) the least-squares solution for the Fourier coefficients is

$$a_i = \frac{2}{d} \sum_{k=1}^{d} y_k \cdot \cos(i\omega_0 t_k) \qquad i = 0, 1, \ldots, n_h$$

$$b_i = \frac{2}{d} \sum_{k=1}^{d} y_k \cdot \sin(i\omega_0 t_k) \qquad i = 1, 2, \ldots, n_h \; .$$

The extension of the Fourier series leads to the Fourier integral and consequently to the Fourier transform and Laplace transform. The key idea is that with the Fourier series (2.52) we have expressed the *time domain* signal $f(t)$ in terms of its *frequency domain* components. Analyzing signals and systems as a function of frequency rather than time, opens up an alternative framework for systems analysis, which has been very successful in engineering. The practicality of such frequency domain analysis is given by the fact that some operations, nonlinear in the time domain, turn out to be linear operations in the frequency domain. A brief summary of integral transforms is given in Appendix 13.7.

2.2 MAXIMUM LIKELIHOOD ESTIMATION

> An idea which can be used once is a trick. If it can be used more than once it becomes a method.
>
> —G. Polya and S. Szegö (1971)

This section introduces an alternative statistical framework for parameter estimation. The *maximum likelihood* (ML) approach, due to R. A. Fisher, suggests to examine the *likelihood function* of the sample values and to take as the estimates of the unknown parameters those values that maximize the likelihood function.

Let $\mathbf{M} = \{(\mathbf{x}_j, y_j)\}$ also denoted $\{\mathbf{m}_j\}$, (2.19), be a set of d sampled data pairs; the $\mathbf{m}_j$ modelled as outcomes of independent random variables. It is assumed that the data observed is drawn from a distribution with distribution or density $p(\mathbf{M}|\boldsymbol{\theta})$ parameterized by $\boldsymbol{\theta} = [\theta_1, \ldots, \theta_r]^T$. The key idea in ML estimation is to determine the parameter(s) $\boldsymbol{\theta}$ for which the probability of observing the outcome $\mathbf{M}$ is as high as possible. The function

$$\ell\,(\boldsymbol{\theta}; \mathbf{m}_1, \mathbf{m}_2, \ldots, \mathbf{m}_d) = p(\mathbf{M}|\boldsymbol{\theta}) \tag{2.53}$$

is the likelihood function. The ML-estimate of the parameter(s) is that value of parameters which maximizes the likelihood function

$$\boldsymbol{\theta}_{\mathrm{ML}} = \arg\max_{\boldsymbol{\theta}} \; \ell(\boldsymbol{\theta}; \mathbf{M}) \; . \tag{2.54}$$

Since the argument, maximizing ℓ, is of importance - not the actual value of the function at that point, it is common to ignore constants in the likelihood function that do not depend upon the parameter(s). In many applications it is more convenient to consider the logarithm of the likelihood function[4], called the *log-likelihood function*:

$$\mathcal{L}(\boldsymbol{\theta}; \mathbf{M}) \doteq \ln \ \ell(\boldsymbol{\theta}; \mathbf{M}) \ . \tag{2.55}$$

Since the logarithm is monotonically increasing, maximizing the log-likelihood is equivalent to maximizing the likelihood. If the function $\mathcal{L}$ is continuously differentiable, a necessary (but not sufficient) condition to maximize the (log) likelihood is for the gradient to vanish at the value $\boldsymbol{\theta}$ that is the ML value:

$$\nabla_{\boldsymbol{\theta}} \ell(\boldsymbol{\theta} = \boldsymbol{\theta}_{\mathrm{ML}} \mid \mathbf{M}) = \nabla_{\boldsymbol{\theta}} \ln \mathcal{L}(\boldsymbol{\theta} = \boldsymbol{\theta}_{\mathrm{ML}} \mid \mathbf{M}) = 0 \ , \tag{2.56}$$

where

$$\nabla_{\boldsymbol{\theta}} = \left[\frac{\partial}{\partial \theta_1}, \frac{\partial}{\partial \theta_2}, \ldots, \frac{\partial}{\partial \theta_r} \right]^T \ .$$

2.2.1 Example: ML-Estimates for the Normal Distribution

A simple example for the ML approach is estimating the parameters η and σ^2 of the normal distribution from a finite set of training data $\mathbf{M} = \{\mathbf{m}_j = x_j\}$. The basic assumption is that the observed d samples were generated according to the normal distribution

$$p(x; \eta, \sigma) = \frac{1}{\sigma\sqrt{2\pi}} \ e^{-\frac{(x-\eta)^2}{2\sigma^2}} . \tag{2.57}$$

The likelihood function takes the form

$$\begin{aligned} \ell(\boldsymbol{\theta}; \mathbf{M}) &= p(x_1) \cdot p(x_2) \cdot \ldots \cdot p(x_d) \\ &= \frac{1}{\left(\sqrt{2\pi\sigma^2}\right)^d} \exp\left(-\frac{1}{2\sigma^2} \sum_{j=1}^{d} (x_j - \eta)^2 \right) \ . \end{aligned} \tag{2.58}$$

Hence, the log-likelihood function is

$$\begin{aligned} \mathcal{L}(\boldsymbol{\theta}; \mathbf{M}) &= Pr(\mathbf{M}|\eta, \sigma^2) \\ &= -\frac{d}{2} \ \ln(2\pi) - d \ \ln(\sigma) - \frac{1}{2\sigma^2} \sum_{j=1}^{d} (x_j - \eta)^2 \ . \end{aligned} \tag{2.59}$$

[4]For example, if one estimates a parameter of an exponential probability law, taking the natural logarithm simplifies the maximization to taking the derivative of a sum of values.

We maximize the log-likelihood function by taking the partial derivatives, and equating these to zero

$$\frac{\partial \mathcal{L}}{\partial \eta} = \frac{1}{\sigma^2} \sum_{j=1}^{d} (x_j - \eta) = 0 \tag{2.60}$$

$$\frac{\partial \mathcal{L}}{\partial \sigma^2} = -\frac{d}{2\sigma^2} + \frac{1}{2\sigma^4} \sum_{j=1}^{d} (x_j - \eta)^2 = 0 \,. \tag{2.61}$$

From (2.60) and (2.61) we obtain the ML-estimates as

$$\hat{\eta} = \frac{1}{d} \sum_{j=1}^{d} x_j \tag{2.7}$$

$$\hat{\sigma}^2 = \frac{1}{d} \sum_{j=1}^{d} (x_j - \hat{\eta})^2 \,. \tag{2.8}$$

For more complicated likelihood functions numerical methods are required for an iterative optimization. A well-established example is the Expectation Maximization (EM) algorithm introduced by A. Dempster to problems with a many-to-one mapping from an underlying distribution to the distribution governing the observations.

2.2.2 The EM Algorithm

A common task in data analysis or signal processing is the estimation of the parameters of a probability distribution function. In many practical situations this is a non-trivial problem because direct access to the data necessary to estimate the parameters is impossible; some of the data are missing. Such difficulties arise when an observed outcome is a result of an accumulation of simpler outcomes. The Expectation Maximization (EM) algorithm, introduced by A.P. Dempster, is commonly used to estimate parameters of a mixture or missing data model via the maximum likelihood principle. The algorithm is ideally suited to problems where there is a many-to-one mapping from an underlying distribution to the distribution governing the observation.

The EM-algorithm 2.1 consists of two major steps: an expectation step, followed by a maximization step. The expectation is with respect to the unknown underlying variables, using the current estimate of the parameters and conditioned upon the observations. The maximization step then provides a new estimate of the parameters. These two steps are iterated until convergence. Section 3.4 provides an example for the EM-algorithm used to identify mixture density models.

Set loop counter $l = 0$; choose the termination tolerance $\delta > 0$ and initialize parameter(s) $\boldsymbol{\theta}^{(0)}$.

Repeat for $l = 1, 2, \ldots$:

Step 1: E-Step: Estimate *unobserved information* using $\boldsymbol{\theta}^{(l-1)}$. The unobserved probability density function is

$$p(\mathbf{x}; \boldsymbol{\theta}) \ ,$$

where $\boldsymbol{\theta} \in \Theta$ is the set of parameters of the density. Because we do not have the information of $\mathbf{x}$ to maximize $\ln p(\mathbf{m}; \boldsymbol{\theta})$, we instead maximize the expectation of $\ln p(\mathbf{x}; \boldsymbol{\theta})$ given the data $\mathbf{M}$ and our current estimate of $\boldsymbol{\theta}$:

$$E[\ln p(\mathbf{x}; \boldsymbol{\theta}) | \mathbf{m}, \boldsymbol{\theta}^{(l)}] \doteq Q(\boldsymbol{\theta} | \boldsymbol{\theta}^{(l)}) \ .$$

Step 2: M-Step: Compute the ML-estimate of parameter(s) $\boldsymbol{\theta}^{(l+1)}$ using information estimated from the E-step:

$$\boldsymbol{\theta}^{(l+1)} = \arg\max_{\boldsymbol{\theta}} Q(\boldsymbol{\theta} | \boldsymbol{\theta}^{(l)}) \ .$$

Analytically, the ML-estimate is obtained by taking the derivative of $\ln p(\mathbf{x}; \boldsymbol{\theta})$ with respect to $\boldsymbol{\theta}$, equating it to zero, and solving for $\boldsymbol{\theta}$.

Until $\left\| \boldsymbol{\theta}^{(l)} - \boldsymbol{\theta}^{(l-1)} \right\| < \delta$.

Algorithm 2.1 The EM-algorithm.

2.3 STOCHASTIC PROCESSES

A random variable is neither random nor variable

— it is simply a function.

A *time-series* is a sequence of observations taken sequentially in time. Time-series analysis is concerned with techniques for the analysis of the dependence among observations and thus to build dynamic models for time-series data. A model that describes the probability structure of a sequence of observations is called *stochastic process*. It is assumed that the generating mechanism is probabilistic and that the observed series $\langle y_1, y_2, \ldots, y_d \rangle$ is a realization of a stochastic process $\langle \mathbf{y}_1, \mathbf{y}_2, \ldots, \mathbf{y}_d \rangle \doteq \vec{Y}(t)$[5].

Considering a signal that varies over time, we write $\mathbf{y}(t)$ to denote observations changing over time. Then at discrete instances of time, t_k, $\mathbf{y}(t_k) \doteq \mathbf{y}(k)$

[5] We use the notation $\vec{Y}$ to denote a vector space in general and $\vec{Y}(t)$ to denote a vector space induced by the sequence of random variables $\mathbf{y}(0), \ldots, \mathbf{y}(t)$ up to time t. This notation will prove useful in deriving the Kalman-Bucy filter.

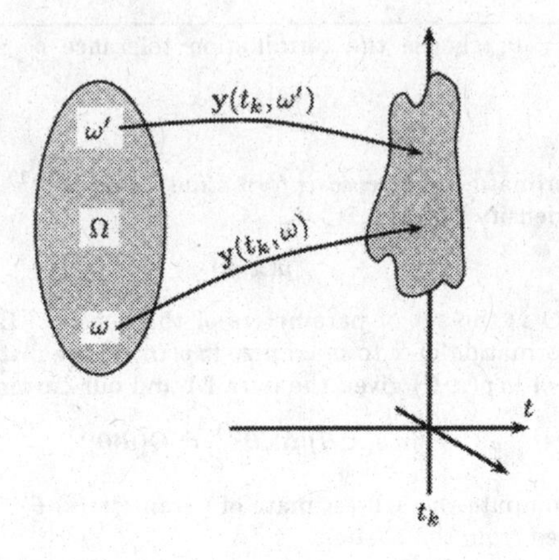

Fig. 2.9 Stochastic process $\mathbf{y}(t, \omega)$ as t-dependent random variable.

is a *random variable*

$$\begin{aligned} \mathbf{y} : \Omega &\to Y \\ \omega &\mapsto \mathbf{y}(\omega) . \end{aligned}$$

Consequently, for each t, the random variable $\mathbf{y}(t)$ has a probability density function and hence mean and variance as functions of time, given by

$$\eta_{\mathbf{y}}(t) = E[\mathbf{y}(t)] = \int_{-\infty}^{+\infty} y \, p(y;t) \, \mathrm{d}y$$

$$\sigma^2_{\mathbf{y}}(t) = E\left[\left(\mathbf{y}(t) - \eta_{\mathbf{y}}(t)\right)^2\right] = \int_{-\infty}^{+\infty} (y - \eta)^2 p(y;t) \, \mathrm{d}y .$$

Considering a discrete-time process, that is, values are sampled at equally spaced instances of time t_k, a *stochastic process* is then a sequence of random variables

$$\begin{aligned} \mathbf{y}(k) : K \times \Omega &\to Y \\ (k, \omega) &\mapsto \mathbf{y}(k, \omega) , \end{aligned}$$

where $k \in K$ is an index set. A stochastic process can be viewed as a t-dependent random variable (Figure 2.9), as a joint function of t and ω (Figure 2.10) or for every $\omega \in \Omega$ the mapping from K into Y is called a *realization* (or *sample function*) of the process (Figure 2.11). The collection of all possible realizations is called *ensemble*. As we are dealing with a collection of random variables, one at each time point, the stochastic process is characterized by its joint distributions. Individual distributions for any instant of time are then referred to as *marginal distributions*. The problem is that, in general, we cannot construct the joint distribution from its marginals since these functions tell us nothing about the joint variation. So, in principle, a stochastic process is a non-countable sequence of random variables, one for each t_k. In general, it seems necessary to consider an infinite-dimensional probability distribution. Fortunately, it can be shown, that if we are given the joint probability distribution for d values we should have sufficient information to enable us to calculate the probability of any event associated with the complete overall behavior of the process.

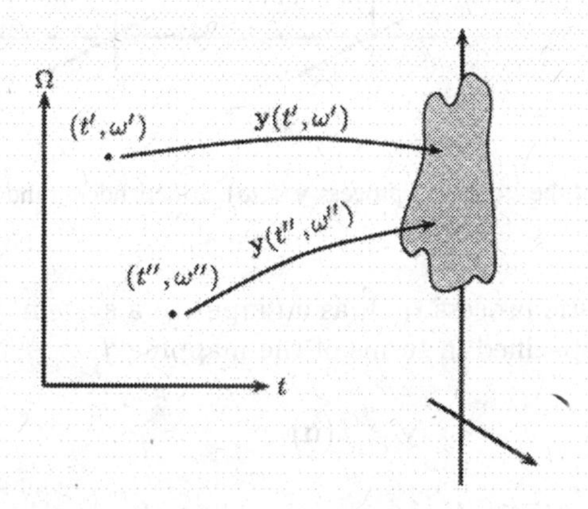

Fig. 2.10 Stochastic process $\mathbf{y}(t,\omega)$ as a joint function of t and ω.

A fundamental assumption of time-series analysis is that the value of the series at time t_k, $y(k)$, depends only on its previous values (deterministic part) and on a random disturbance (stochastic part). Furthermore, this dependence is assumed to be linear, leading to a class of linear models called autoregressive moving average (ARMA) models. An ARMA model of order n_y and n_d, ARMA(n_y,n_d), is defined by

$$\mathbf{y}(k) = \sum_{i=1}^{n_y} a_i \cdot \mathbf{y}(k-i) + \sum_{i=1}^{n_d} b_i \cdot \varepsilon(k-i+1) \ , \tag{2.62}$$

where $\{a_i\}$ and $\{b_i\}$ are the coefficients of the autoregressive (AR) and moving average (MA) parts, respectively. The disturbance $\{\varepsilon(k)\}$ is white noise with zero mean and variance σ^2 usually assumed normally distributed. In general, a *white noise process* $\{\varepsilon(t)\}$ is defined by

$$\begin{aligned} E[\varepsilon(t)] &= 0 \\ E[\varepsilon^2(t)] &= \sigma^2 \qquad \forall\, t \\ E[\varepsilon(t)\varepsilon(t+\tau)] &= 0 \qquad \text{if} \quad \tau > 0 \,. \end{aligned}$$

In other words, a white noise process is a sequence of uncorrelated random variables.

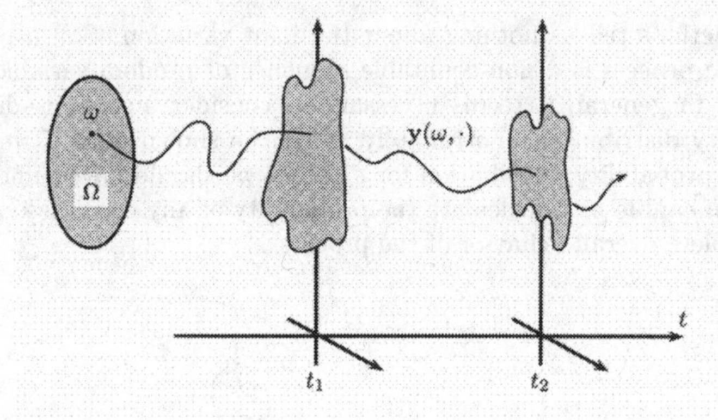

Fig. 2.11 Realization of the stochastic process $\mathbf{y}(t, \omega)$, ω-dependent random variable.

Consider a stochastic process $\mathbf{u}(t)$, as an input to a system, with output $\mathbf{y}$ where the system is specified in terms of the mapping f

$$\mathbf{y} = f(\mathbf{u}) \ . \tag{2.63}$$

The system is *deterministic* if it operates only on the variable t, treating ω ('selecting realizations') as a parameter. This implies that if two realizations of the input are identical in t, then the corresponding realizations of the output are also identical in t. The system is called *stochastic* if f operates on both variables t and ω. Hereafter we shall only consider deterministic systems. For any linear system $E[f(\mathbf{u}(t))] = f(E[\mathbf{u}(t)])$.

In order to make any kind of statistical inference from a single realization of a stochastic process, we are forced to make simplifying assumptions. The most important restriction leads to the class of *stationary* processes. It is assumed that the process is in some steady-state mode or equilibrium, that is, for these processes statistical properties do not change in time, the probability distributions are time-invariant, $p(y;t) = p(y)$. In practice, a much weaker definition of stationarity, called second-order or weak stationarity, is employed. Stationarity up to order 2 implies that $E[\mathbf{y}(t)] = \eta$, a constant independent of time and the variance of $\mathbf{y}(t)$ is equal to σ^2, also a constant, independent of time and $E[\varepsilon(t)\varepsilon(t+\tau)]$ is a function of τ only.

By definition, stationarity implies that the process has a constant mean η. The fact that the covariance of a stationary process is only a function of the time difference τ (called *lag*) allows the introduction of two important functions used to characterize stochastic processes: the *autocovariance function* of $\mathbf{y}(t)$

$$\sigma_{\mathbf{y},\mathbf{y}}(\tau) = E[(\mathbf{y}(t) - \eta)\ (\mathbf{y}(t+\tau) - \eta)] \tag{2.64}$$

and the autocorrelation of $\mathbf{y}(t)$ is the expected value of the product $\mathbf{y}(t) \cdot \mathbf{y}(t+\tau)$, leading to the *autocorrelation function*

$$\rho_{\mathbf{y},\mathbf{y}}(\tau) = \frac{\sigma_{\mathbf{y},\mathbf{y}}(\tau)}{\sigma_{\mathbf{y},\mathbf{y}}(0)} = \frac{E[(\mathbf{y}(t)-\eta)(\mathbf{y}(t+\tau)-\eta)]}{\sqrt{E[(\mathbf{y}(t)-\eta)^2]\,E[(\mathbf{y}(t+\tau)-\eta)^2]}} \,. \tag{2.65}$$

For each τ, $\rho_{\mathbf{y},\mathbf{y}}(\tau)$ defines the correlation coefficient between pairs of values of $\mathbf{y}(t)$ separated by an interval of length τ. Since the expectation operator is a theoretical construct, estimators are required to implement these functions for finite sequences of measured data. Given d observations $\langle y_1, y_2, \ldots, y_d \rangle$ of a process $\mathbf{y}(t)$, and assuming stationarity, the sample mean

$$\hat{\eta}(\tau) = \frac{1}{d} \sum_{j=1}^{d} y_j$$

is used to estimate the *sample autocovariance sequence*

$$\hat{\sigma}_{\mathbf{y},\mathbf{y}}(\tau) = \frac{1}{d} \sum_{j=1}^{d-\tau} (y_{j+\tau} - \hat{\eta})(y_j - \hat{\eta}) \,. \tag{2.66}$$

That is, the expectation is replaced by an average over the series at different times. It should be noted that $\hat{\eta}$ and $\hat{\gamma}$ are random variables. The *sample autocorrelation sequence* $\hat{\rho}_{\mathbf{y},\mathbf{y}}(\tau)$ is defined to be the normalized sample covariance sequence

$$\hat{\rho}_{\mathbf{y},\mathbf{y}}(\tau) = \frac{\hat{\sigma}_{\mathbf{y},\mathbf{y}}(\tau)}{\hat{\sigma}_{\mathbf{y},\mathbf{y}}(0)} \,. \tag{2.67}$$

As in Section 1, a sequence is a vector and so is the sequence of random variables up to time t, $\mathbf{y}(t) = \langle \mathbf{y}(0), \ldots, \mathbf{y}(t) \rangle$, (1.11). The set of all linear combinations of these random variables forms a vector space (linear manifold), $\vec{Y}(t)$, which is a finite-dimensional subspace of the space of all observations. The scalar product of any two vectors (sequences), say $\mathbf{x}(t)$ and $\mathbf{y}(t)$, then defines the covariance $\sigma_{\mathbf{x},\mathbf{y}}$. Then the autocovariance, given by the Euclidean norm, corresponds to the length of the vector:

$$\sigma_{\mathbf{y},\mathbf{y}} = \|\mathbf{y}(t)\|^2 \,.$$

The angle between two vectors $\mathbf{x}(t)$ and $\mathbf{y}(t)$ is given by

$$\begin{aligned}\cos(\alpha) &= \frac{\sigma_{\mathbf{x},\mathbf{y}}}{\|\mathbf{x}(t)\| \cdot \|\mathbf{y}(t)\|} \\ &= \frac{\sigma_{\mathbf{x},\mathbf{y}}}{\sqrt{\sigma_{\mathbf{x},\mathbf{x}} \cdot \sigma_{\mathbf{y},\mathbf{y}}}} \,.\end{aligned} \tag{2.68}$$

From this, two sequences are uncorrelated if their vectors are orthogonal. In terms of expectations, a stochastic process is *uncorrelated* if

$$E[\mathbf{y}(t)\mathbf{y}(t+\tau)] = E[\mathbf{y}(t)] \cdot E[\mathbf{y}(t+\tau)] \qquad \tau \geq 1 \,. \tag{2.69}$$

If in addition

$$E[\mathbf{y}(t)\mathbf{y}(t+\tau)] = 0 , \tag{2.70}$$

then the stochastic process is said to be *orthogonal*. In general, a *random vector* $\mathbf{x}$ is a vector of random variables

$$\mathbf{x} = [\mathbf{x}_1, \mathbf{x}_2, \ldots, \mathbf{x}_n]^t , \tag{2.71}$$

characterized by a probability distribution function

$$Pr(x_1, \ldots, x_n) = Pr(\mathbf{x}_1 \leq x_1, \ldots, \mathbf{x}_n \leq x_n) .$$

The mean of the random vector $\mathbf{x}$ is then defined by

$$\eta_{\mathbf{x}} = E[\mathbf{x}] \tag{2.72}$$

$$\eta_i = \int x_i p_i(x_i) \mathrm{d}x_i ,$$

where $p_i(x_i)$ is the *marginal density* of the ith component of x. An important set of parameters is that which indicates the dispersion of the distribution. The *covariance matrix* of $\mathbf{x}$ is defined by

$$\begin{aligned}
\mathbf{\Sigma} &\doteq E\left[(\mathbf{x}-\eta)(\mathbf{x}-\eta)^T\right] \\
&= E\left[\begin{bmatrix} \mathbf{x}_1-\eta_1 \\ \vdots \\ \mathbf{x}_n-\eta_n \end{bmatrix} [\mathbf{x}_1-\eta_1 \ldots \mathbf{x}_n-\eta_n]\right] \\
&= E\left[\begin{bmatrix} (\mathbf{x}_1-\eta_1)(\mathbf{x}_1-\eta_1) & \cdots & (\mathbf{x}_1-\eta_1)(\mathbf{x}_n-\eta_n) \\ \vdots & & \vdots \\ (\mathbf{x}_n-\eta_n)(\mathbf{x}_1-\eta_1) & \cdots & (\mathbf{x}_n-\eta_n)(\mathbf{x}_n-\eta_n) \end{bmatrix}\right] \\
&= \begin{bmatrix} E[(\mathbf{x}_1-\eta_1)(\mathbf{x}_1-\eta_1)] & \cdots & E[(\mathbf{x}_1-\eta_1)(\mathbf{x}_n-\eta_n)] \\ \vdots & & \vdots \\ E[(\mathbf{x}_n-\eta_n)(\mathbf{x}_1-\eta_1)] & \cdots & E[(\mathbf{x}_n-\eta_n)(\mathbf{x}_n-\eta_n)] \end{bmatrix} \\
&\doteq \begin{bmatrix} c_{11} & \cdots & c_{1n} \\ \vdots & \ddots & \vdots \\ c_{n1} & \cdots & c_{nn} \end{bmatrix} .
\end{aligned} \tag{2.73}$$

The components σ_{ij} of this matrix are

$$c_{ij} = E[(\mathbf{x}_i - \eta_i)(\mathbf{x}_j - \eta_j)] \qquad i, j = 1, \ldots, n .$$

The diagonal components of the covariance matrix are the variances of individual random variables, and the off-diagonal components are the covariances of two random variables $\mathbf{x}_i$ and $\mathbf{x}_j$. The covariance matrix is symmetric and can be rewritten in the following form:

$$\begin{aligned}\mathbf{\Sigma} &= E[\mathbf{x}\mathbf{x}^T] - E[\mathbf{x}]\eta^T - \eta E[\mathbf{x}^T] + \eta\eta^T \\ &= \mathbf{S} - \eta\eta^T ,\end{aligned} \tag{2.74}$$

where

$$\mathbf{S} = E[\mathbf{x}\mathbf{x}^T] = \begin{bmatrix} E[\mathbf{x}_1\mathbf{x}_1] & \cdots & E[\mathbf{x}_1\mathbf{x}_n] \\ \vdots & \ddots & \vdots \\ E[\mathbf{x}_n\mathbf{x}_1] & \cdots & E[\mathbf{x}_n\mathbf{x}_n] \end{bmatrix} . \tag{2.75}$$

Equation (2.74) describes the relationship between the covariance and autocorrelation matrices, demonstrating that both essentially contain the same amount of information. The matrix $\mathbf{S}$ of (2.75) is called the *autocorrelation matrix* of $\mathbf{x}$. We can then replace the elements in (2.73) by the variances σ_i^2, standard deviations and correlation coefficients ρ_{ij}:

$$c_{ii} = \sigma_i^2 \quad \text{and} \quad c_{ij} = \rho_{ij}\sigma_i\sigma_j . \tag{2.76}$$

We can then express $\mathbf{\Sigma}$ as a combination of two matrices:

$$\mathbf{\Sigma} = \mathbf{\Upsilon}\mathbf{R}\mathbf{\Upsilon} , \tag{2.77}$$

where

$$\mathbf{\Upsilon} = \begin{bmatrix} \sigma_1 & 0 & \cdots & 0 \\ 0 & \sigma_2 & & \\ \vdots & & \ddots & \vdots \\ 0 & & \cdot & \sigma_n \end{bmatrix} \quad \text{and} \quad \mathbf{R} = \begin{bmatrix} 1 & \rho_{12} & \cdots & \rho_{1n} \\ \rho_{12} & 1 & & \\ \vdots & & \ddots & \vdots \\ \rho_{1n} & & \cdots & 1 \end{bmatrix} . \tag{2.78}$$

Matrix $\mathbf{R}$ is called *correlation matrix*. In contrast to variance or standard deviation which depend on the scales of coordinate systems, the correlation matrix captures the relationship between random variables independent of scale.

The basic ideas of random variables, random vectors and the parameters characterizing them, have been developed into a comprehensive framework to analyze signals and systems, in particular, with respect to filtering and control. Consider a signal $x(t)$ and noise $\varepsilon(t)$ where only the sum $y(t) = x(t) + \varepsilon(t)$ can be observed. In a stochastic framework the (real-valued) variables y, x, ε are considered as random variables[6] $\mathbf{y}$, $\mathbf{x}$ and $\boldsymbol{\varepsilon}$. Sampling the signal gives a realization of the sequence of random variables $\langle \mathbf{y}(0), \mathbf{y}(1), \ldots, \mathbf{y}(t) \rangle$. The

[6] Note that we use bold letters to describe random variables as well as vector(s). See also Figure 2.12.

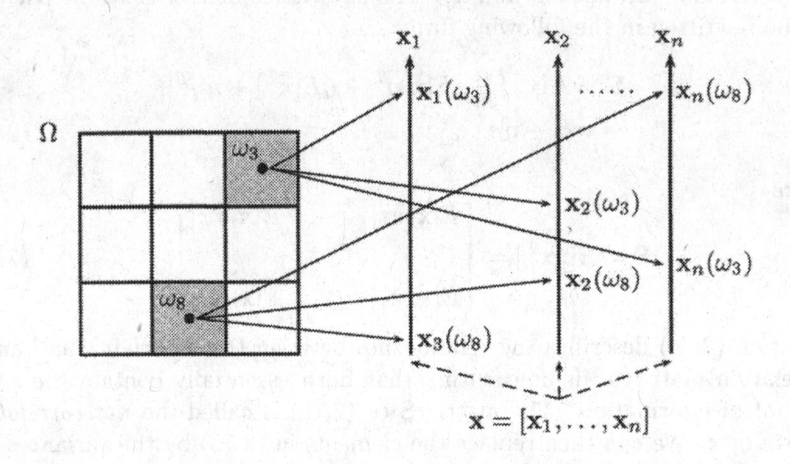

Fig. 2.12 Vector-valued random variable $\mathbf{x}$. For simplicity some results in this section are derived for single-valued random variables but can be extended to vectors.

objective is, on the basis of observations, to make some inference about the value of the signal at $t = t'$, where the following cases can occur[7]:

$t' \leq t$:	*data smoothing (interpolation)*
$t' = t$:	*filtering*
$t' > t$:	*prediction*

For any given sequence of measured values $\langle y_1, y_2, \ldots, y_d \rangle$ of the (real-valued) random variable $\mathbf{y}(t)$, the conditional probability distribution function

$$Pr\big(\mathbf{x}(t) \leq \delta \mid \mathbf{y}(0) = y_1, \ldots, \mathbf{y}(t) = y_d\big)$$

represents all information about random variable $\mathbf{x}(t)$ obtained from the measurements of random variables $\mathbf{y}(0), \ldots, \mathbf{y}(t)$. Any statistical estimate of random variable $\mathbf{x}$ will be some function of this conditional distribution and hence be a function of random variables $\mathbf{y}(0), \ldots, \mathbf{y}(t)$. The estimate itself will therefore be a random variable too and the quality of the estimate $\hat{\mathbf{x}}(t'|t)$ is quantified by a loss function $L(\cdot)$ where $L(\cdot)$ is positive, $L(0) = 0$, and is a non-decreasing function of the *estimation error* $\mathbf{x}(t') - \hat{\mathbf{x}}(t')$. Commonly, the estimate is required to minimize the average loss

$$E\big[L\big(\mathbf{x}(t') - \hat{\mathbf{x}}(t')\big)\big] = E\big[E[L(\mathbf{x}(t') - \hat{\mathbf{x}}(t')) \mid \mathbf{y}(0), \ldots, \mathbf{y}(t)]\big] .$$

[7] All three cases of data smoothing, filtering and prediction were summarized by R. E. Kalman as the *estimation problem*.

Since the first expectation on the right-hand side does not depend on $\hat{\mathbf{x}}$ we may only minimize

$$E\big[L(\mathbf{x}(t') - \hat{\mathbf{x}}(t')) \mid \mathbf{y}(0), \ldots, \mathbf{y}(t)\big] .$$

Making some assumptions about the properties of the stochastic processes $\mathbf{x}(t)$, $\mathbf{y}(t)$ and $\varepsilon(t)$, one can show that the random variable $\hat{\mathbf{x}}(t'|t)$ which minimizes the average loss is the conditional expectation

$$\hat{\mathbf{x}}(t'|t) = E\big[\mathbf{x}(t') \mid \mathbf{y}(0), \ldots, \mathbf{y}(t)\big] .$$

One of the most significant application of these ideas to engineering was presented by R. E. Kalman and R. S. Bucy. On the basis of expectations alone, without actually calculating any integrals, using the geometric view and the orthogonality condition, they were able to derive an optimal (least-squares) estimate of $\mathbf{x}(t')$ as an *orthogonal projection* of $\mathbf{x}(t')$ onto the vector space $\vec{Y}(t)$ formed by the random variables $\mathbf{y}(0), \ldots, \mathbf{y}(t)$ (up to time t) and their linear combinations

$$\sum_{i=t_0}^{t} a_i \mathbf{y}(i) .$$

Let $\mathbf{e}_{t_0}, \ldots, \mathbf{e}_t$ be an *orthogonal basis*, that is, any vector (sequence) in $\vec{Y}(t)$ is given by $\sum_{i=t_0}^{t} a_i$. We denote these vectors in $\vec{Y}(t)$ by

$$\bar{\mathbf{x}}(t) \doteq \sum_{i=t_0}^{t} a_i \cdot \mathbf{e}_i .$$

Any vector $\mathbf{x}(t)$, not necessarily in $\vec{Y}(t)$, can be decomposed into two parts: a part $\bar{\mathbf{x}}(t) \in \vec{Y}(t)$ and a part $\tilde{\mathbf{x}}(t)$ orthogonal to $\vec{Y}(t)$, *i.e.* orthogonal to every vector in $\vec{Y}(t)$:

$$\begin{aligned} \mathbf{x}(t) &= \bar{\mathbf{x}}(t) + \tilde{\mathbf{x}}(t) \\ &= \sum_{i=t_0}^{t} E[\mathbf{x}(t) \cdot \mathbf{e}_i] \cdot \mathbf{e}_i + \tilde{\mathbf{x}}(t) , \end{aligned} \tag{2.79}$$

where $\bar{\mathbf{x}}(t)$ is called the *orthogonal projection* of $\mathbf{x}(t)$ on $\vec{Y}(t)$. In the next section we will derive the equations leading to what is known as the *Kalman filter* as an example of how to use expectations and the geometrical view of sequences.

2.3.1 Example: Kalman-Bucy Filtering

The *assumption* of randomness is another mode of abstraction, a constraint which will be satisfied in certain kinds of situations and which will fail to be satisfied in others.

—Robert Rosen

In this section, we derive in detail the basic equations of the Kalman-Bucy filter as presented in the first paper [Kal60] that initiated decades of research in extending its basic ideas. The Kalman filtering approach may be briefly summarized as follows:

- Prior to Kalman's paper, Wiener's work in time-series analysis led to complex integral equations. Kalman and Bucy converted the problem into a nonlinear differential equation whose solution yields the covariance matrix of the minimum filtering error. This matrix contains all information required in designing an optimal filter.

- A ("random") signal is modelled as a dynamic system excited by white noise.

- The optimal filter generates a best linear estimate of the actual (noise-free) signal.

- The variance equations are of Ricatti type which occur in the calculus of variations and are closely related to the canonical differential equations of Hamilton. This establishes some duality between estimation and control theory.

Kalman's ideas established a novel, mathematically more tractable and elegant approach to prediction and filtering by using conditional distributions and expectations. We derive his equations as an example of how the geometrical perspective - viewing sequences of random variables as vectors, and orthogonality - ensuring optimality can be used effectively. The overall structure of the Kalman-Bucy filter - the signal model and the filter itself, are illustrated in Figure 2.13.

For the signal, we consider the dynamic model (1.26) illustrated in Figure 1.15, where the input $\mathbf{u}(k)$ to the system is assumed to be an independent Gaussian vector-valued stochastic process with zero mean and covariance matrix $\mathbf{Q}(k)$:

$$\begin{aligned} E[\mathbf{u}(k)] &= 0 \quad \forall\ k \\ E[\mathbf{u}(k)\mathbf{u}'(k+\tau)] &= 0 \quad \text{if} \quad \tau > 0 \\ E[\mathbf{u}(k)\mathbf{u}(k)] &= \mathbf{Q}(k)\ . \end{aligned} \tag{2.80}$$

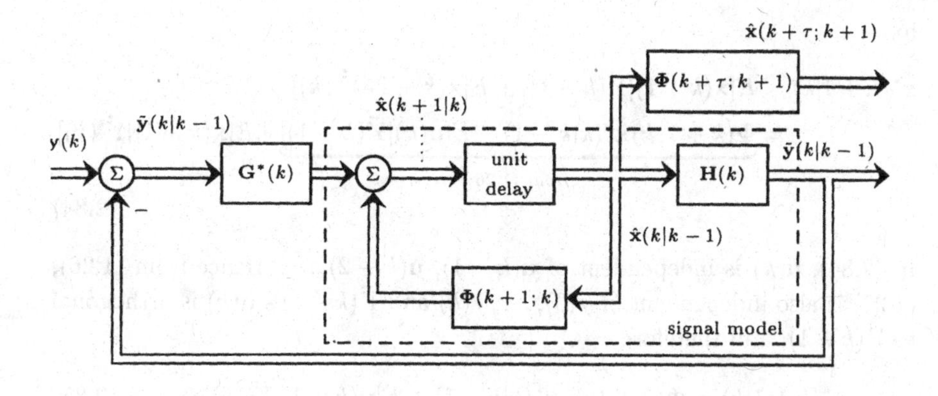

Fig. 2.13 Matrix block diagram of the Kalman-Bucy filter.

Here, $\mathbf{x}(k)$ is an r-vector, and $\mathbf{y}(k)$ is an m-vector. Here the elements of matrix $\mathbf{G}(k)$ are all equal to one. Given the observed values of $\mathbf{y}(0), \ldots, \mathbf{y}(k)$, we aim to find an estimate $\mathbf{x}^*(k+1|k)$ of $\mathbf{x}(k)$ which minimizes the expected loss. From the previous section and the geometrical view of least-squares estimation in Section 2.1, the solution to the filtering problem is the orthogonal projection of $\mathbf{x}(k)$ onto the linear manifold $\vec{Y}(k)$ generated by the observed sequence of random variables:

$$\begin{aligned}\mathbf{x}^*(k+1|k) &= E[\mathbf{x}(k+1)|\mathbf{y}(0), \ldots, \mathbf{y}(k)] \\ &= E[\mathbf{x}(k+1)|\vec{Y}(k)] \\ &= \bar{\mathbf{x}}(k+1|k) \in \vec{Y}(k)\ !\end{aligned} \tag{2.81}$$

In other words, $\bar{\mathbf{x}}$ is that linear function of random variables $\mathbf{y}(0), \ldots, \mathbf{y}(k)$ which minimizes a quadratic loss function. Vector $\mathbf{y}(k)$ is composed of two parts, $\bar{\mathbf{y}}(k|k-1) \in \vec{Y}(k-1)$ and *residual* $\tilde{\mathbf{y}}(k|k-1)$ orthogonal to $\vec{Y}(k-1)$:

$$\underbrace{\mathbf{y}(k)}_{\in \vec{Y}(k)} = \underbrace{\bar{\mathbf{y}}(k|k-1)}_{\in \vec{Y}'(k)} + \underbrace{\tilde{\mathbf{y}}(k|k-1)}_{\in \vec{Y}(k-1)}\ . \tag{2.82}$$

The orthogonal spaces $\vec{Y}(k-1)$ (known from, or induced by, $\langle \mathbf{y}(0), \ldots, \mathbf{y}(k-1)\rangle$) and $\vec{Y}'(k)$ make up $\vec{Y}(k)$. The component of $\mathbf{y}(k)$ lying in $\vec{Y}(k-1)$, denoted by $\bar{\mathbf{y}}(k|k-1)$, using (1.27), is specified by

$$\bar{\mathbf{y}}(k|k-1) = \mathbf{H}(k)\mathbf{x}^*(k|k-1) \qquad \text{inserted into (2.82) leads to}$$

$$\begin{aligned}\tilde{\mathbf{y}}(k|k-1) &= \mathbf{y}(k) - \bar{\mathbf{y}}(k|k-1) \\ &= \mathbf{y}(k) - \mathbf{H}(k)\mathbf{x}^*(k|k-1)\ .\end{aligned} \tag{2.83}$$

Equation (2.83) describes the "not expected" part of the measurement. Assuming by induction that $\mathbf{x}^*(k|k-1)$ is known, from (2.81) and (2.82) we have

$$\mathbf{x}^*(k+1|k) = E[\mathbf{x}(k+1)|\vec{Y}(k-1)] + E[\mathbf{x}(k+1)|\vec{Y}'(k)]$$
$$= \underbrace{\mathbf{\Phi}(k+1;k)\mathbf{x}^*(k|k-1) + E[\mathbf{u}(k)|\vec{Y}(k-1)]}_{\text{from (1.26)}} + E[\mathbf{x}(k+1)|\vec{Y}'(k)] \;. \tag{2.84}$$

In (2.84), $\mathbf{u}(k)$ is independent of $\mathbf{u}(k-1), \mathbf{u}(k-2), \ldots$. Hence from (1.26), (1.27) is also independent of $\mathbf{y}(0), \ldots, \mathbf{y}(k)$ and $\vec{Y}(k-1)$ - $\mathbf{u}(k)$ is orthogonal to $\vec{Y}(k-1)$, and therefore

$$\mathbf{x}^*(k+1|k) = \mathbf{\Phi}(k+1;k)\mathbf{x}^*(k|k-1) + E[\mathbf{x}(k+1)|\vec{Y}'(k)] \;, \tag{2.85}$$

where the term $E[\mathbf{x}(k+1)|\vec{Y}'(k)]$ is assumed to be a linear operation on $\tilde{\mathbf{y}}(k|k-1)$. Introducing the gain (matrix) $\mathbf{G}^*(k)$ of the optimal filter (cf. Figure 2.13),

$$E[\mathbf{x}(k+1)|\vec{Y}'(k)] = \mathbf{G}^*(k)\tilde{\mathbf{y}}(k|k-1) \;. \tag{2.86}$$

Combining (2.83), (2.84) and (2.86),

$$\mathbf{x}^*(k+1|k) = \mathbf{\Phi}(k+1;k)\mathbf{x}^*(k|k-1) + \mathbf{G}^*(k)\big[\mathbf{y}(k) - \mathbf{H}(k)\mathbf{x}^*(k|k-1)\big]$$
$$= \mathbf{\Phi}(k+1;k)\mathbf{x}^*(k|k-1) + \mathbf{G}^*(k)\mathbf{y}(k) - \mathbf{G}^*(k)\mathbf{H}(k)\mathbf{x}^*(k|k-1)$$
$$= \underbrace{\big[\mathbf{\Phi}(k+1;k) - \mathbf{G}^*(k)\mathbf{H}(k)\big]}_{\doteq \mathbf{\Phi}^*(k+1;k)}\mathbf{x}^*(k|k-1) + \mathbf{G}^*(k)\mathbf{y}(k) \;. \tag{2.87}$$

With the new notation,

$$\mathbf{\Phi}^*(k+1;k) \doteq \mathbf{\Phi}(k+1;k) - \mathbf{G}^*(k)\mathbf{H}(k) \;, \tag{2.88}$$

the optimal state estimate, given $\vec{Y}(k)$, is

$$\mathbf{\Phi}^*(k+1;k) = \mathbf{\Phi}^*(k+1;k)\mathbf{x}^*(k|k-1) + \mathbf{G}^*(k)\mathbf{y}(k) \;. \tag{2.89}$$

From (2.89), it becomes apparent that signal and filter are governed by the same model structure (1.26) and $\mathbf{x}^*(k|k-1)$ is the state of the optimal filter while the observations $\mathbf{y}(k)$ are the inputs to the optimal filter (see Figure 2.13). Next, we consider the estimation error

$$\tilde{\mathbf{x}}(k+1|k) = \mathbf{x}(k+1) - \mathbf{x}^*(k+1|k)$$
$$= \underbrace{\mathbf{\Phi}(k+1;k)\mathbf{x}(k) + \mathbf{u}(k)}_{\text{from (1.26)}} - \underbrace{\big[\mathbf{\Phi}^*(k+1;k)\mathbf{x}^*(k|k-1) + \mathbf{G}^*(k)\mathbf{y}(k)\big]}_{\text{from (2.89)}}$$
$$= \mathbf{\Phi}(k+1;k)\mathbf{x}(k) + \mathbf{u}(k) - \mathbf{\Phi}^*(k+1;k)\mathbf{x}^*(k|k-1)$$
$$- \mathbf{G}^*(k)\underbrace{\mathbf{H}(k)\mathbf{x}(k)}_{\text{from (1.27)}} \;.$$

Since $\mathbf{x}^*(k|k-1) = \mathbf{x}(k) - \tilde{\mathbf{x}}(k|k-1)$,

$$\begin{aligned}
\tilde{\mathbf{x}}(k+1|k) &= \mathbf{\Phi}(k+1;k)\mathbf{x}(k) + \mathbf{u}(k) - \mathbf{\Phi}^*(k+1;k)\mathbf{x}(k) \\
&\quad + \mathbf{\Phi}^*(k+1;k)\tilde{\mathbf{x}}(k|k-1) - \mathbf{G}^*(k)\mathbf{H}(k)\mathbf{x}(k) \\
&= \mathbf{\Phi}^*(k+1;k)\tilde{\mathbf{x}}(k|k-1) + \mathbf{u}(k) + \mathbf{\Phi}(k+1;k)\mathbf{x}(k) \\
&\quad - \big[\mathbf{\Phi}(k+1;k) - \mathbf{G}^*(k)\mathbf{H}(k)\big]\mathbf{x}(k) - \mathbf{G}^*(k)\mathbf{H}(k)\mathbf{x}(k) \\
&\quad + \mathbf{\Phi}(k+1;k)\mathbf{x}(k) - \mathbf{\Phi}(k+1;k)\mathbf{x}(k) + \mathbf{G}^*(k)\mathbf{H}(k)\mathbf{x}(k) \\
&\quad - \mathbf{G}^*(k)\mathbf{H}(k)\mathbf{x}(k)\ .
\end{aligned}$$

These re-arrangements and substitutions lead eventually to

$$\tilde{\mathbf{x}}(k+1|k) = \mathbf{\Phi}^*(k+1;k)\tilde{\mathbf{x}}(k|k-1) + \mathbf{u}(k)\ . \tag{2.90}$$

Equation (2.90) shows that the estimation error is governed by a linear dynamic system with transition matrix $\mathbf{\Phi}^*$. Next, we derive an expression for the covariance matrix of the error $\tilde{\mathbf{x}}(k+1|k)$. Inserting (2.88) into (2.90), we obtain

$$\begin{aligned}
\tilde{\mathbf{x}}(k+1|k) &= \big[\mathbf{\Phi}(k+1;k) - \mathbf{G}^*(k)\mathbf{H}(k)\big]\tilde{\mathbf{x}}(k|k-1) + \mathbf{u}(k) &(2.91)\\
&= \mathbf{\Phi}(k+1;k)\tilde{\mathbf{x}}(k|k-1) - \mathbf{G}^*(k)\mathbf{H}(k)\tilde{\mathbf{x}}(k|k-1) + \mathbf{u}(k) \\
&= \mathbf{\Phi}(k+1;k)\tilde{\mathbf{x}}(k|k-1) - \mathbf{G}^*(k)\tilde{\mathbf{y}}(k|k-1) + \mathbf{u}(k)\ . &(2.92)
\end{aligned}$$

Using (2.92), the covariance of the error is specified by

$$\begin{aligned}
\mathbf{P}(k+1) &\doteq E\Big[\tilde{\mathbf{x}}(k+1|k)\tilde{\mathbf{x}}^T(k+1|k)\Big] \\
&= E\Big[\tilde{\mathbf{x}}(k+1|k)\big(\mathbf{\Phi}(k+1;k)\tilde{\mathbf{x}}(k|k-1) - \mathbf{G}^*(k)\tilde{\mathbf{y}}(k|k-1) + \mathbf{u}(k)\big)^T\Big].
\end{aligned} \tag{2.93}$$

From the definition of optimality, that is, $\mathbf{x}^*(k+1|k) \doteq \bar{\mathbf{x}}(k+1|k) \in \vec{Y}(k)$, the optimal estimate is the orthogonal projection of $\mathbf{x}(k+1)$ onto $\vec{Y}(k)$. Therefore,

$$E\left[\tilde{\mathbf{x}}(k+1|k)\tilde{\mathbf{y}}^T(i)\right] = 0 \qquad \forall\, i = 0, \ldots, k$$

and hence

$$E\left[\tilde{\mathbf{x}}(k+1|k)\tilde{\mathbf{y}}^T(k|k-1)\right] = 0 \tag{2.94}$$

simplifies (2.93) to

$$\mathbf{P}(k+1) = E\Big[\tilde{\mathbf{x}}(k+1|k)\big(\mathbf{\Phi}(k+1;k)\tilde{\mathbf{x}}(k|k-1) + \mathbf{u}(k)\big)^T\Big]\ .$$

Replacing $\tilde{\mathbf{x}}(k+1|k)$ by (2.91) in the last equation,

$$\begin{aligned}
\mathbf{P}(k+1) &= E\Big[\big\{(\boldsymbol{\Phi}(k+1;k) - \mathbf{G}^*(k)\mathbf{H}(k))\tilde{\mathbf{x}}(k|k-1) + \mathbf{u}(k)\big\} \\
&\qquad \cdot \big\{\boldsymbol{\Phi}(k+1;k)\tilde{\mathbf{x}}(k|k-1) + \mathbf{u}(k)\big\}^T\Big] \\
&= E\Big[\big\{\boldsymbol{\Phi}^*(k+1;k)\tilde{\mathbf{x}}(k|k-1) + \mathbf{u}(k)\big\} \\
&\qquad \cdot \big\{\tilde{\mathbf{x}}^T(k|k-1)\boldsymbol{\Phi}^T(k+1;k) + \mathbf{u}^T(k)\big\}\Big] \\
&= E\Big[\boldsymbol{\Phi}^*(k+1;k)\tilde{\mathbf{x}}(k|k-1)\tilde{\mathbf{x}}^T(k|k-1)\boldsymbol{\Phi}^T(k+1;k) \\
&\qquad + \underbrace{\boldsymbol{\Phi}^*(k+1;k)\tilde{\mathbf{x}}(k|k-1)\mathbf{u}^T(k)}_{\to 0} \\
&\qquad + \underbrace{\mathbf{u}(k)\tilde{\mathbf{x}}^T(k|k-1)\boldsymbol{\Phi}^T(k+1;k)}_{\to 0} + \mathbf{u}(k)\mathbf{u}^T(k)\Big] .
\end{aligned}$$

Leading to

$$\mathbf{P}(k+1) = \boldsymbol{\Phi}^*(k+1;k)E\big[\tilde{\mathbf{x}}(k|k-1)\tilde{\mathbf{x}}^T(k|k-1)\big]\boldsymbol{\Phi}^T(k+1;k) + \mathbf{Q}(k)$$

or

$$\mathbf{P}(k+1) = \boldsymbol{\Phi}^*(k+1;k)\mathbf{P}(k)\boldsymbol{\Phi}^T(k+1;k) + \mathbf{Q}(k) . \tag{2.95}$$

Terms involving the product of $\mathbf{u}(k)$ and $\tilde{\mathbf{x}}(k|k-1)$ vanish since $\mathbf{u}(k)$ is independent of $\mathbf{x}(k)$ and therefore of $\tilde{\mathbf{x}}(k|k-1)$. Independence implies uncorrelatedness, that is,

$$\begin{aligned}
E[(\cdot)\cdot\mathbf{u}(k)] &= E[(\cdot)]\cdot E[\mathbf{u}(k)] \\
&= 0 ,
\end{aligned}$$

since $E[\mathbf{u}(k)] = 0$, by definition. If, in the derivation of $\mathbf{P}(k+1)$, we use (2.90), replacing both terms $\tilde{\mathbf{x}}(k+1|k)$ instead of using (2.91) for one and (2.90) for the other, we have

$$\begin{aligned}
\mathbf{P}(k+1) &= E\Big[\tilde{\mathbf{x}}(k+1|k)\tilde{\mathbf{x}}^T(k+1|k)\Big] \\
&= E\Big[\big\{\boldsymbol{\Phi}^*(k+1;k)\tilde{\mathbf{x}}(k|k-1) + \mathbf{u}(k)\big\} \\
&\qquad \cdot \big\{\tilde{\mathbf{x}}^T(k|k-1)\boldsymbol{\Phi}^{*T}(k+1;k) + \mathbf{u}^T(k)\big\}\Big] .
\end{aligned}$$

Multiplying the terms within the curly brackets gives us

$$\begin{aligned}
\mathbf{P}(k+1) &= E\Big[\boldsymbol{\Phi}^*(k+1;k)\tilde{\mathbf{x}}(k|k-1)\tilde{\mathbf{x}}^T(k|k-1)\boldsymbol{\Phi}^{*T}(k+1;k) \\
&\quad + \boldsymbol{\Phi}^*(k+1;k)\tilde{\mathbf{x}}(k|k-1)\mathbf{u}^T(k)\mathbf{u}(k)\tilde{\mathbf{x}}^T(k|k-1)\boldsymbol{\Phi}^{*T}(k+1;k) \\
&\quad + \mathbf{u}(k)\mathbf{u}^T(k)\Big] \\
&= \boldsymbol{\Phi}^*(k+1;k)E\Big[\tilde{\mathbf{x}}(k|k-1)\tilde{\mathbf{x}}^T(k|k-1)\Big]\boldsymbol{\Phi}^{*T}(k+1;k) \\
&\quad + \underbrace{E\Big[\boldsymbol{\Phi}^*(k+1;k)\tilde{\mathbf{x}}(k|k-1)\mathbf{u}^T(k)\Big]}_{\to 0} \\
&\quad + \underbrace{\mathbf{u}(k)\tilde{\mathbf{x}}^T(k|k-1)\boldsymbol{\Phi}^{*T}(k+1;k)}_{\to 0} + E\Big[\mathbf{u}(k)\mathbf{u}^T(k)\Big] \\
&= \boldsymbol{\Phi}^*(k+1;k)\mathbf{P}(k)\boldsymbol{\Phi}^{*T}(k+1|k) + \mathbf{Q}(k)\ . \qquad (2.96)
\end{aligned}$$

What remains to be found is an expression for the gain $\mathbf{G}^*(k)$ of the optimal filter. First note that we found two equivalent recursive expressions (2.95) and (2.96) for $\mathbf{P}(k+1)$:

$$\begin{aligned}
\mathbf{P}(k+1) &= \big[\boldsymbol{\Phi}(k+1;k) - \mathbf{G}^*(k)\mathbf{H}(k)\big]\mathbf{P}(k)\boldsymbol{\Phi}^T(k+1;k) + \mathbf{Q}(k) \qquad (2.95) \\
&= \big[\boldsymbol{\Phi}(k+1;k) - \mathbf{G}^*(k)\mathbf{H}(k)\big] \\
&\quad \cdot \mathbf{P}(k)\big(\boldsymbol{\Phi}(k+1;k) - \mathbf{G}^*(k)\mathbf{H}(k)\big)^T + \mathbf{Q}(k)\ . \qquad (2.96)
\end{aligned}$$

For both expressions to hold true,

$$\begin{aligned}
&\big[\boldsymbol{\Phi}(k+1;k) - \mathbf{G}^*(k)\mathbf{H}(k)\big]\mathbf{P}(k)\mathbf{H}^T(k)\mathbf{H}(k)^T = 0 \\
\textit{i.e.}\quad &\big[\boldsymbol{\Phi}(k+1;k)\mathbf{P}(k)\mathbf{H}^T(k) - \mathbf{G}^*(k)\mathbf{H}(k)\mathbf{P}^*(k)\mathbf{H}^T(k)\big]\,\mathbf{G}^{*T}(k) = 0\ .
\end{aligned}$$

Therefore, we obtain for the optimal gain the expression

$$\mathbf{G}^*(k) = \boldsymbol{\Phi}(k+1;k)\mathbf{P}(k)\mathbf{H}^T(k)\big[\mathbf{H}(k)\mathbf{P}(k)\mathbf{H}^T(k)\big]^{-1}\ . \qquad (2.97)$$

For more details and extensions of the Kalman-Bucy filter, the reader is referred to the vast literature available on the subject. Part of the impact Kalman's original paper had, comes from the fact that though conceptually very different to the filtering problem described here, Kalman identified a dual optimal control problem which merely requires a change of interpretation with most equations remaining unchanged. Consequently almost every introductory book on control engineering will mention Kalman filtering.

3

Learning from Data: System Identification

> ☐ *The identification of a model is an approximation of the function which relates independent (e.g., input-) and dependent (e.g., output-) variables.*
>
> ☐ *Linear parametric regression, employing the least-squares principle, is an efficient tool to identify parameters from data - to learn linear functional relationships.*
>
> ☐ *In a probabilistic framework data are assumed to be distributed according to some unknown probability density function.*
>
> ☐ *Statistical learning can be seen as a generalization of density estimation.*
>
> ☐ *Like the Fourier series, Kernel density estimation provides another example of the approximation of an unknown function by means of so-called basis functions.*

That all our knowledge begins with experience, there is indeed no doubt ... but although our knowledge originates *with* experience, it does not all arise *out of* experience.

—Immanuel Kant

In order to use regression techniques in identifying models from data, we now introduce a vector of *independent variables*. Let the observables $\xi_1, \ldots, \xi_n$ be maps from the set of abstract states Ω into an n-dimensional manifold Ξ, that is,

$$\begin{aligned} \xi : \Omega &\to \Xi \\ \omega &\mapsto \mathbf{o} = (\mathbf{x}, y) \, , \end{aligned}$$

where $m = 1$ (single-output system) and the vector of independent variables be denoted by

$$\mathbf{x} \doteq [x_1, x_2, \ldots, x_r] \ .$$

We can now describe the manifold Ξ or *data space* as the Cartesian product of the input and output spaces

$$\Xi = X \times Y \qquad \text{where} \qquad X \doteq X_1 \times X_2, \times \cdots \times X_r$$

for short. In Section 1 we have seen that a formal model $\mathfrak{M}$ of a system $\mathfrak{S}$ is described by some mapping[1] $f(\cdot)$ which relates variables $\mathbf{x}$ with y:

$$\begin{aligned} f : X &\rightarrow Y \\ \mathbf{x} &\mapsto y \ . \end{aligned} \tag{3.1}$$

A specific model thus describes a *graph* $F \subset X \times Y$, (1.8), of the mapping which represents system $\mathfrak{S}$. In time-series analysis, y refers to the output whereas $\mathbf{x}$ describes some inputs and the vector of variables $\boldsymbol{\varepsilon}$ is used to denote all other factors that affect the output but whose values are not observed or controlled. Uncertainty in the output reflects the lack of knowledge of the unobserved factors $\boldsymbol{\varepsilon}$. In this context, we express the transformation T_t, (1.15), by the dynamical law

$$y(k+1) = f(\mathbf{x}, k) \ , \qquad \mathbf{x} \in X \ ,$$

describing y at time t_{k+1} depending upon previous states of observables which are summarized in vector $\mathbf{x}$. It will be this class of systems we focus upon in subsequent sections.

The *identification* of a model $\mathfrak{M}$ is an approximation of $f\colon X \rightarrow Y$, based on a sampled set of training data, that is, measurements $\mathbf{m}_j = (\mathbf{x}_j, y_j)$, $j = 1, 2, \ldots, d$ of the observations or objects $\mathbf{o}$.[2] The identified dependency between $\mathbf{x}$ and y is described by means of a parameter vector $\boldsymbol{\theta}$, that is, we assume that $f(\cdot)$ is an appropriate representation of $\mathfrak{S}$ and using a finite set of sampled data we implement a function $f(\mathbf{x}; \boldsymbol{\theta})$ which we hope is as close to $f(\mathbf{x})$ as desirable. *Learning*, in general, means the selection of a set of functions $f(\mathbf{x}; \boldsymbol{\theta})$, where $\boldsymbol{\theta}$ is a set of abstract parameters used only to index the set of functions. It follows that the most general form of *prediction* concerning

[1] A mapping or function f on X to Y is a set $F \subset X \times Y$ of ordered pairs which for each $\mathbf{x} \in X$ contains exactly one ordered pair $\langle \mathbf{x}, y \rangle$. X is called the *domain* and Y the *codomain* of f. Visualizing X and Y as "spaces" of some sort, $X \times Y$ is called *Cartesian product space*. Since F is a subset of the product space which meets each "vertical" subspace $\{x\} \times Y$ in exactly one point, it describes a curve and is therefore often called the *graph* of the function.
[2] With respect to dynamic systems, $\mathbf{o}$ may be regarded as the *state* of the system and Ξ as the state-space. On the other hand, in clustering $\mathbf{o}$ is an abstract *object*. Since regression and clustering are generic tools for pattern recognition, we refer to $\mathbf{o}$ as an *'observation'*.

$\mathfrak{S}$ is that the 'point' $\mathbf{o}$, determined by measurements, will lie in some subset A of Ξ.

In *linear parametric regression*, the set of functions is specified as a polynomial of fixed degree, and the set of functions implemented is

$$f(\mathbf{x};\boldsymbol{\theta}) = \sum_{i=1}^{r} \theta_i \cdot x_i \tag{3.2}$$

such that an appropriate set of θ_i can be found using least-squares (2.29). For example, considering input-output "black-box" models, using an *autoregressive* model structure, the system is described by a finite number of past inputs and outputs:

$$\mathbf{x} \doteq [y(k),\ldots,y(k-n_y+1),u(k),\ldots,u(k-n_u+1)]^T \ . \tag{3.3}$$

Considering system identification [Lju87], this leads to the ARX (AutoRegressive with eXogenous input) model structure:

$$y(k+1) = \sum_{i=1}^{n_y} \theta_i \cdot y(k-i+1) + \sum_{i=1}^{n_u} \theta_{n_y+i} \cdot u(k-i+1) \ . \tag{3.4}$$

Equation (3.4) is commonly called the *predictor* for model

$$y(k) = \sum_{i=1}^{n_y} a_i \cdot y(k-i) + \sum_{i=1}^{n_u} b_i \cdot u(k-i) \ ,$$

where the adjustable parameters are denoted $\boldsymbol{\theta} = [a_1,\cdots,a_{n_y},b_1,\cdots,b_{n_u}]^T$ and $r = n_y + n_u$. In Sections 4 and 5 we describe fuzzy models which approximate a nonlinear regression surface by linear submodels. These models can be described as NARX (Nonlinear AutoRegressive with eXogenous[3] input) models which establish a relation between the past input-output data and the predicted output

$$\begin{aligned} y(k+1) &= f(\mathbf{x},k) + \varepsilon(k) \\ &= f\big(y(k),\ldots,y(k-n_y+1),u(k),\ldots,u(k-n_u+1)\big) + \varepsilon(k) \ , \end{aligned} \tag{3.5}$$

where k denotes discrete time samples, n_u and n_y are integers related to the system's order and f is some nonlinear mapping.

[3]Variables in models for forecasting are usually classified as *endogenous* or *exogenous*. Endogenous variables are those values the model is built to explain, while the exogenous variables are those that are not determined by the model but nonetheless impact on it.

3.1 THE PROBABILISTIC PERSPECTIVE

Due to non-observed and uncontrolled variables, the knowledge of observed values $\mathbf{x}$ does not uniquely specify the output y. The consequence is some uncertainty in y. This leads to the formulation of a 'statistical dependency' between $\mathbf{x}$ and y.

A stochastic framework describes the $\mathbf{x} \in \mathbb{R}^r$ as random vectors drawn independently from a fixed probability density $p(\mathbf{x})$ which is unknown. The system under study produces a value y for every input vector $\mathbf{x}$ according to the fixed conditional density $p(y|\mathbf{x})$ which is unknown. The problem of learning is to select a function that best approximates the system's response. The training data $\mathbf{M} = \{\mathbf{m}_j\}$, $\mathbf{m}_j = (\mathbf{x}_j, y_j)$ are assumed to be *independent*[4] and *identically distributed*, following the joint probability density function

$$p(\mathbf{x}, y) = p(\mathbf{x}) \cdot p(y|\mathbf{x}) \ , \tag{3.6}$$

where $(\mathbf{x}, y) \in X \times Y$. A finite sample from this distribution is denoted by $\{\mathbf{m}_j\}$, $j = 1, 2, \ldots, d$. Consider the regression model

$$y = f(\mathbf{x}) + \varepsilon \ , \tag{3.7}$$

where ε is some zero mean random noise[5]. Then, we find

$$f(\mathbf{x}) = \int y \cdot p(y|\mathbf{x}) \, \mathrm{d}y \tag{3.8}$$

such that the graph (1.8) of the mapping which describes the system is specified by

$$F = \{(f(\mathbf{x}), \mathbf{x}) \colon f(\mathbf{x}) = (3.8)\} \ . \tag{3.9}$$

Within a stochastic framework, in order to estimate $f(\mathbf{x}; \boldsymbol{\theta})$, one minimizes the expected value of the loss

$$\begin{aligned} E[L] &= \int L\big(y, f(\mathbf{x}; \boldsymbol{\theta})\big) \, p(\mathbf{x}, y) \, \mathrm{d}\mathbf{x}\mathrm{d}y \\ &\doteq R(\boldsymbol{\theta}) \ , \end{aligned} \tag{2.16}$$

where $p(\mathbf{x}, y)$ is unknown. Hence, density estimation is the most important learning problem and is therefore further discussed in the next section.

[4]The assumption of independence is based on the view that microscopic phenomena have much smaller time constants and therefore can be seen as independent of the primary signal.
[5]Assuming the process is linear with noise being added to the signal, probability distributions are usually assumed to be Gaussian with the following argument. Macroscopic random effects are thought of as a superposition of microscopic random effects. Under the central limit theorem the accumulated effect tends to be Gaussian regardless the distribution of the microscopic effects.

At this point, it might be useful to clarify some terminology. *Learning* is probably the most general term describing methods that estimate the unknown mapping (dependency) $\mathbf{y} = f(\mathbf{x})$ between a system's inputs and outputs from a finite set of *training data*, that is, input-output samples $\mathbf{m}_j$. Common learning tasks are classification, regression, density estimation, and clustering. We use the terms 'learning' and 'system identification' interchangeably, though system identification is usually associated with linear parametric regression [Lju87]. In a probabilistic setting, all these tasks learn, that is, estimate $f(\mathbf{x};\boldsymbol{\theta})$, by minimizing the *risk functional* $E[L]$, (2.16). In each case, however, the loss function and the output differ. In *statistical learning theory* [Vap98, CM98] the general *predictive learning* problem is formalized as follows: Let $\mathbf{M} = \{\mathbf{m}_j\}$, and $\mathbf{m} = (\mathbf{x}, y)$ denote and input-output pair of which we are given d samples as training data. The data are assumed to be distributed according to some unknown probability density function $p(\mathbf{m})$. The objective of predictive learning is to find a loss function $Q(\mathbf{m}, \boldsymbol{\theta})$, $\boldsymbol{\theta} \in \Theta$ that minimizes the risk functional

$$R(\boldsymbol{\theta}) \doteq \int Q(\mathbf{m}, \boldsymbol{\theta}) \, \mathrm{d}F(\mathbf{m})$$

$$= \int Q(\mathbf{m}, \boldsymbol{\theta}) \, p(\mathbf{m}) \, \mathrm{d}\mathbf{m} \,, \tag{3.10}$$

where

$$Q(\mathbf{m}, \boldsymbol{\theta}) \doteq L\big(y, f(\mathbf{x};\boldsymbol{\theta})\big) \tag{3.11}$$

denotes the loss function as applicable for any of the learning problems (classification, regression, density estimation, and clustering). In regression (2.16), we used the least-squares principle, that is, the *quadratic loss*

$$Q(\mathbf{m}, \boldsymbol{\theta}) = \big(y - f(\mathbf{x};\boldsymbol{\theta})\big)^2 \,. \tag{2.17}$$

At the root of statistical learning theory is the idea to estimate the risk functional (3.10), by taking an average of the risk over the training data:

$$R_{\mathrm{emp}}(\boldsymbol{\theta}) = \frac{1}{d} \sum_{j=1}^{d} Q(\mathbf{m}_j, \boldsymbol{\theta}) \,. \tag{3.12}$$

As the optimal estimate, obtained by minimizing the risk functional (3.10), depends on the cumulative distribution function $F(\mathbf{m})$, $p(\mathbf{m})$, respectively. One usually therefore first has to estimate the unknown probability density function from the available data $\mathbf{M}$ and then find an optimal estimate for $f(\mathbf{x};\boldsymbol{\theta})$. Alternatively, we use the estimated risk (3.12) as a substitute for the unknown true risk. This approach is called the *Empirical Risk Minimization* (ERM) inductive principle. The main claim of statistical learning theory is that the ERM inductive principle is not only a general framework for learning but also should be preferred to density estimation in case of small, finite samples. With ERM the goal is to find values for $\boldsymbol{\theta}$ that minimize the empirical risk. Then, the solution to the learning problem is the approximating function of $f(\mathbf{x};\boldsymbol{\theta})$, minimizing (3.12) with respect to the parameters. A nonlinear parameterization of a set of approximating functions $f(\mathbf{x};\boldsymbol{\theta})$ leads to a nonlinear optimization problem. One commonly used optimization approach is the EM-algorithm. As parameters are estimated iteratively, the value of the empirical risk is minimized. As illustrated in Figure 3.1, in [CM98], predictive learning is described as a two-step inference:

1. **Induction:** Learning (estimation) of unknown dependency from data.

2. **Deduction:** Using the identified model for prediction.

Such conventional formulation of identification algorithm implies that we are estimating the unknown function $f(\cdot)$ everywhere in Ξ, that is, for all possible input values. Such global function approximation has obvious disadvantages and has led to various concepts that support 'localized predictive learning'. For instance, in [CM98], the *transductive* approach of *support vector machines*, rooted in statistical learning theory, is outlined. The ideas put forward in Section 4 and Section 6, will also support *localized* modelling and prediction - though motivated by a very different philosophy.

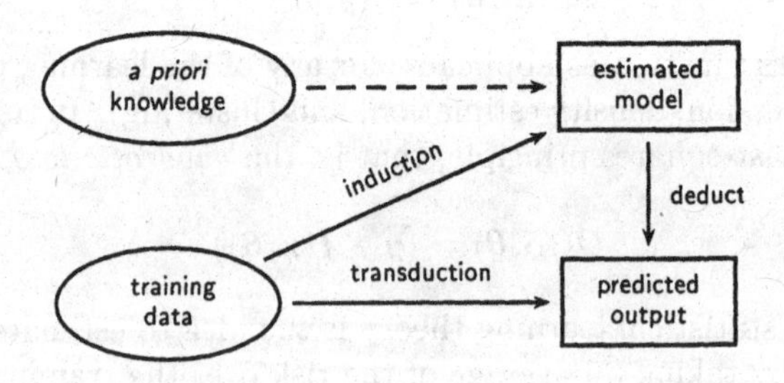

Fig. 3.1 Predictive learning [CM98].

I shall argue that the nature of the uncertainty involved in describing $\mathfrak{S}$ can have different forms (fuzziness, randomness, ambiguity, vagueness, imprecision) depending on the nature of the system, the nature of the data, necessary *assumptions* and personal *preference* for a particular mathematical framework. It is therefore important to note that a statistical/probabilistic formulation is not the only valid framework. It is surprising that despite the number of assumptions required, the number of parameters not estimated but chosen, numerical compromises and the consequences of our foregoing discussion we still find statements in the literature such as *"Probability measures are adequate for describing all types of uncertainty."* or

"In contrast to the classical statistics [..] using various types of a priori information, the new theory [..] does not rely on a priori knowledge about a problem to be solved."

One can only have the impression that there exists an unreasonable respect for statistics and probabilistic modelling which may inhibit creativity in the progress of systems theory. In subsequent sections, I try to demonstrate that fuzzy mathematical objects *occur naturally*; that is, not by applying some *extension principle* to mathematical objects but by necessary generalization when considering uncertainty in systems engineering. In combination, fuzzy mathematics, statistics and possibility theory should provide us with a flexible set of tools, necessary to analyze complex systems.

3.2 KERNEL DENSITY ESTIMATION

The previous sections highlighted the importance of density estimation to statistical learning theory. In this section, we first introduce the problem with Parzen's kernel estimators, and then extend the results to function approximation in general.

The most basic approach to density estimation is the histogram. Its main disadvantage is that it is discontinuous and has two parameters (the number of bins, and bin width) which heavily depend on the nature and number of data available. E. Parzen [Par62] suggested a class of smooth estimates of probability density functions which we now discuss in more detail. Let $\mathbf{x}_1, \mathbf{x}_2, \ldots, \mathbf{x}_d$ be independent random variables identically distributed with cumulative distribution function

$$\begin{aligned} F(x') &= Pr(\mathbf{x} \leq x') \\ &= \int_{-\infty}^{x'} p(x)\, dx \,. \end{aligned} \tag{3.13}$$

Given a set of training data $x_1, \ldots, x_d$, an empirical estimate of (3.13) is

$$\hat{F}(x') = \frac{1}{d}\sum_{j=1}^{d} \zeta(x_j \leq x') \,, \tag{3.14}$$

where $\zeta(\cdot)$ is the indicator function taking the values 0 and 1 depending on whether the argument is false or true. The estimator (3.14) is itself a binomially distributed random variable with mean $E[\hat{F}(x)] = F(x)$ and variance equal to $\frac{1}{d}F(x)\cdot(1-F(x))$. From (3.14), a simple candidate to estimate $p(x)$ is

$$\hat{p}(x) = \frac{\hat{F}(x+h) - \hat{F}(x-h)}{2\cdot h} \,, \tag{3.15}$$

where h is a suitably chosen positive number. Introducing the *kernel function* $K(\cdot)$ defined by

$$K(x') = \begin{cases} 0.5 & \text{if } |x'| \leq 1 \\ 0 & \text{if } |x'| > 1\,, \end{cases} \tag{3.16}$$

we can rewrite (3.15) as a weighted average over the sample distribution function:

$$\begin{aligned} \hat{p}(x) &= \int_{-\infty}^{+\infty} \frac{1}{h}\, K\left(\frac{x - x'}{h}\right) \mathrm{d}\hat{F}(x') \\ &= \frac{1}{d \cdot h} \sum_{j=1}^{d} K\left(\frac{x - x_j}{h}\right). \end{aligned} \tag{3.17}$$

Equation (3.17) is usually referred to as *kernel estimator*. Apart from the naive kernel estimator (3.16), a Gaussian kernel estimator is frequently used:

$$K(x') = \frac{1}{\sqrt{2\pi}} e^{-0.5(x')^2}\,. \tag{3.18}$$

To illustrate density estimation, we use the well-known data set of observations of eruptions of the Old Faithful geyser [Sil86]. Figure 3.2 shows the results for the 'naive' estimator (3.16) and the Gaussian estimator (3.18) with $h = 0.5$.

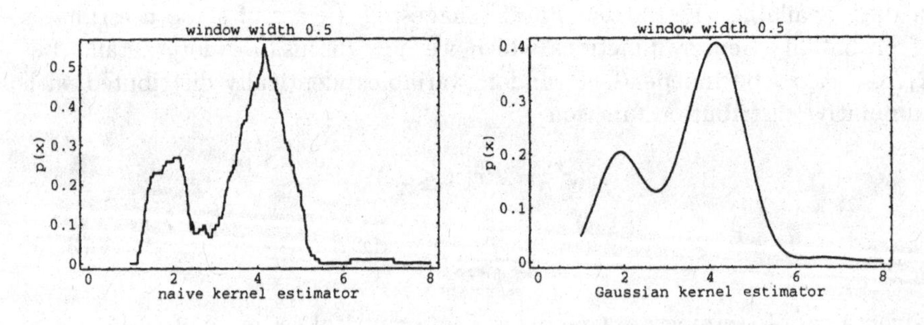

Fig. 3.2 Kernel density estimates for the 'Old Faithful' data set.

Multi-variate kernel estimators are straightforward generalizations of the univariate case. Using the same kernel in each dimension but with a different smoothing parameter for each dimension the estimate is defined pointwise as

$$\hat{p}(\mathbf{x}) = \frac{1}{dh_1 \cdots h_r} \sum_{j=1}^{d} \left\{ \prod_{i=1}^{r} K\left(\frac{x_i - x_{ij}}{h_i}\right) \right\}\,, \tag{3.19}$$

where $\mathbf{x} = [x_1, \ldots, x_r]^T \in \mathbb{R}^r$ and the data x_{ij} come from a $r \times d$ matrix. Geometrically, the estimate places a probability mass of size $1/d$ centered on each sample point, exactly as in the univariate case. In Section 9.1 we will demonstrate the equivalence of fuzzy and statistical classifiers where the optimal Bayesian classifier uses a multi-variate kernel density estimate.

3.3 BASIS FUNCTION APPROXIMATION

Returning from basic density estimation to the more general estimation of the mapping $y = f(\mathbf{x})$, we can make the following useful observation. In a stochastic framework, the estimation of the density $p(\mathbf{x}, y)$ is the most important learning task since it is essential to minimizing the risk functional $R(\boldsymbol{\theta})$, (3.10). A loss function for density estimation is

$$L\big(f(\mathbf{x};\boldsymbol{\theta})\big) = -\ln f(\mathbf{m};\boldsymbol{\theta}) \; . \tag{3.20}$$

Inserting (3.20) into (3.10) gives the risk functional

$$R(\boldsymbol{\theta}) = \int -\ln f(\mathbf{x};\boldsymbol{\theta}) p(\mathbf{x}) \, \mathrm{d}\mathbf{x} \; . \tag{3.21}$$

$\boldsymbol{\theta}$ is an l-dimensional vector in $\Theta \subset \mathbb{R}^l$ and it is assumed that the unknown density is captured by this parameterized class of functions. Given a set of independent identically distributed training data $\mathbf{M} = \{\mathbf{m}_j\}$, $j = 1, \ldots, d$, the likelihood function (2.53) describes the probability of the data set $\mathbf{M}$ conditioned on $\boldsymbol{\theta}$:

$$\begin{aligned} Pr(\mathbf{M}|\boldsymbol{\theta}) &= \prod_{j=1}^{d} f(\mathbf{m}_j;\boldsymbol{\theta}) \\ &= \ell(\boldsymbol{\theta};\mathbf{M}) \; . \end{aligned} \tag{2.53}$$

As shown in Section 2.2, a maximum likelihood estimate is obtained by maximizing the log-likelihood function, (2.55), $\mathcal{L}(\boldsymbol{\theta};\mathbf{M}) = \ln \; \ell(\boldsymbol{\theta};\mathbf{M})$ which is equivalent to minimizing the risk functional (3.21).

Estimating the density of $\mathbf{x}$, the output represents the density and hence $f(\mathbf{x};\boldsymbol{\theta})$, $\boldsymbol{\theta} \in \Theta$ becomes a set of densities. With respect to the more general learning problem, the representation of the kernel estimator, (3.17) suggests a generalization called *basis function approximation*:

$$f(\mathbf{x};\boldsymbol{\theta}) = \sum_{i=1}^{r} \theta_i \cdot \phi_i(\mathbf{x}) \; , \tag{3.22}$$

which is a linear combination of so-called *basis functions* ϕ_i. The θ_i are weights which sum up to one for some fixed r. Equation (3.22) describes a large class of approximating functions we can choose from in order to model the mapping $y = f(\mathbf{x})$. Algebraic polynomials lead to the linear regression models introduced in Section 2.1:

$$f(\mathbf{x};\boldsymbol{\theta}) = \sum_{i=1}^{r} \theta_i \cdot x_i \;. \tag{2.18}$$

In Section 2.1.2, the Fourier series was derived with the same objective - to find a good approximation of $f(t)$ using a linear combination of trigonometric functions

$$f(t;\boldsymbol{\theta}) = \sum_{i=1}^{r} \theta_i \cdot \phi_i(t) \;, \tag{2.43}$$

$$\doteq \frac{a_0}{2} + \sum_{i=1}^{n_h} \left(a_i \cdot \cos(i\omega_0 t) + b_i \cdot \sin(i\omega_0 t)\right) \;. \tag{2.52}$$

From density estimation, we have seen that basis function approximation (3.22) not only provides linear models but also generalizes to nonlinear universal approximator. This class of universal approximators also includes some neural network approaches (such as Radial Basis Function (RBF) networks) and in Section 5 we will show that fuzzy models are another example from this class.

3.4 EXAMPLE: EM ALGORITHM FOR MIXTURE-DENSITY ESTIMATION

In this section, we introduce the Expectation Maximization (EM) algorithm as a tool to estimate parameters of a mixture of probability density functions via the maximum likelihood method (Section 2.2).

Let $\mathbf{M} = \{\mathbf{m}_1, \mathbf{m}_2, \ldots, \mathbf{m}_d\}$ denote data which are assumed to be generated independently from some unknown mixture of density functions. This unknown mixture is estimated using the following general class of approximating functions:

$$f(\mathbf{x};\boldsymbol{\theta},\mathbf{w}) = \sum_{i=1}^{c} w_i \cdot \phi_i(\mathbf{x},\boldsymbol{\theta}_i) \;, \tag{3.23}$$

where the θ_i are the parameters of the individual densities in the mixture and the w_i are weights summing up to one. From Section 2.2 on ML-estimation we know that the best estimator is the mixture density (chosen from the class (3.23)) maximizing the log-likelihood function. This density is denoted as

$$p(\mathbf{m}) \doteq \sum_{i=1}^{c} p(\mathbf{m}|i,\boldsymbol{\theta}_i) \cdot Pr(i) \qquad \text{where} \qquad \sum_{i=1}^{c} Pr(i) = 1 \;. \tag{3.24}$$

Individual densities in the mixture are indexed by i and parameterized by $\boldsymbol{\theta}_i$. In (3.24), $Pr(i)$ denotes the probability that a given data sample came from density i. Hence, the log-likelihood function for (3.24) is

$$(\boldsymbol{\theta};\mathbf{M}) = \sum_{j=1}^{d} \ln \sum_{i=1}^{c} Pr(i) \cdot p(\mathbf{m}_j|i,\boldsymbol{\theta}_i) \; . \tag{3.25}$$

According to the maximum likelihood principle, we should use the parameters $\boldsymbol{\theta}$ that maximize (3.25). This is numerically difficult to achieve and it would be easier if the data were labelled, that is, if it would be known which component of the mixture generated any given data point. In this case, $u_{ij} \in \mathbf{U}$ would denote whether sample j originated from density component i and the log-likelihood function for *complete* information would be

$$(\boldsymbol{\theta};\mathbf{M},\mathbf{U}) = \sum_{j=1}^{d} \sum_{i=1}^{c} u_{ij} \cdot \ln p(\mathbf{m}_j|\mathbf{u}_j,\boldsymbol{\theta}_i) \cdot Pr(\mathbf{u}_j) \; . \tag{3.26}$$

Given the complete information, the maximization problem could be split into a set of simpler problems, estimating densities independently using the associated data samples selected by $u_{ij} \in \mathbf{U}$. However, we assumed that the data are unlabelled, meaning the the u_{ij} are unknown and we have to operate with *incomplete data*. Since it is impossible to work with (3.26) directly, A.P. Dempster showed that we can use the expected value of (3.26) instead. That is, it can be shown that if a certain value of parameter $\boldsymbol{\theta}$ increases the expected value of (3.26), then the log-likelihood function (3.26) will also increase. This leads to the formulation of the EM-algorithm (see Section 2.2.2) which we here state for density estimation and basis function approximation in general. Iterating with loop counter l, the two steps characterizing the EM-algorithm are:

- **E-Step:** Compute the expectation of the *complete* data log-likelihood function $f(\mathbf{x};\mathbf{w},\boldsymbol{\theta})$

$$Q(\boldsymbol{\theta},\boldsymbol{\theta}^{(l)}) = \sum_{j=1}^{d} \sum_{i=1}^{c} p_{ij} \cdot \left(\ln \phi_i \left(\mathbf{m}_j, \boldsymbol{\theta}_i^{(l)} \right) + \ln w_i^{(l)} \right) \; , \tag{3.27}$$

  where p_{ij} is the probability that density i generated data point j, calculated by

$$\begin{aligned} p_{ij} &\doteq E[u_{ij}|\mathbf{m}_j] \\ &= \frac{w_i^{(l)} \cdot \phi_i \left(\mathbf{m}_j, \boldsymbol{\theta}_i^{(l)} \right)}{\sum_{k=1}^{c} w_k^{(l)} \cdot \phi_k \left(\mathbf{m}_j, \boldsymbol{\theta}_k^{(l)} \right)} \; . \end{aligned} \tag{3.28}$$

- **M-Step:** Find the parameters $\mathbf{w}^{(l+1)}$ and $\boldsymbol{\theta}^{(l+1)}$ that maximize the expected complete data log-likelihood:

$$w_i^{(l+1)} = \frac{1}{d} \sum_{j=1}^{d} p_{ij} \tag{3.29}$$

$$\boldsymbol{\theta}_i^{(l+1)} = \arg\max_{\boldsymbol{\theta}_i} \sum_{j=1}^{d} p_{ij} \cdot \ln \phi_i \left(\mathbf{m}_j, \boldsymbol{\theta}_i^{(l)} \right) . \tag{3.30}$$

A common assumption on the density functions of the mixtures is that they are Gaussian; each density described by a mean value η_i and covariance matrix $\boldsymbol{\Sigma}_i^{-1}$ with variance σ_i^2 on the diagonal. In this case, the approximating density function $f(\mathbf{x})$ is

$$f(\mathbf{x}) = \sum_{i=1}^{c} w_i \cdot \frac{1}{(2\pi\sigma_i^2)^{\frac{r}{2}}} \cdot e^{-\frac{\|\mathbf{x}-\eta_i\|^2}{2\sigma_i^2}} , \tag{3.31}$$

where the w_i are unknown weights. Hence, for the Gaussian mixture model the expectation and minimization steps are:

- **E-Step:** Compute $p_{ij} = E[u_{ij} | \mathbf{m}_j, \boldsymbol{\theta}^{(l)}]$ as

$$p_{ij} = \frac{\left(\sigma_i^{-n}\right)^{(l)} \cdot e^{-\frac{\left\|\mathbf{x}_i - \eta_i^{(l)}\right\|^2}{2(\sigma_i^2)^{(l)}}}}{\sum\limits_{k=1}^{c} \left(\sigma_i^{-n}\right)^{(l)} \cdot e^{-\frac{\left\|\mathbf{x}_i - \eta_k^{(l)}\right\|^2}{2(\sigma_k^2)^{(l)}}}} . \tag{3.32}$$

- **M-Step:** Estimate new weights and parameters of the Gaussians:

$$w_i^{(l+1)} = \frac{1}{d} \sum_{j=1}^{d} p_{ij} , \tag{3.33}$$

$$\eta_i^{(l+1)} = \frac{\sum\limits_{j=1}^{d} p_{ij} \cdot \mathbf{m}_j}{\sum\limits_{j=1}^{d} p_{ij}} , \tag{3.34}$$

$$\left(\sigma_i^2\right)^{(l+1)} = \frac{\sum\limits_{j=1}^{d} p_{ij} \cdot \left\| \mathbf{m}_j - \eta_i^{(l+1)} \right\|^2}{\sum\limits_{j=1}^{d} p_{ij}} . \tag{3.35}$$

3.5 DISCUSSION: MODELLING AND IDENTIFICATION

Truth can never be told so as to be understood, and not believ'd.

—William Blake

With finite data we cannot expect to find $f(\mathbf{x};\boldsymbol{\theta})$ exactly. To obtain a useful solution, one needs to incorporate *prior* knowledge in addition to data; the joint density $p(\mathbf{x}, y)$ is not known. However, density estimation requires, in general, a large number of samples, but even more with respect to the order of Ξ (a problem called the *curse of dimensionality*[6]). Similarly, the notion of minimizing *expected* risk implies large numbers of samples to have the averaging as a good approximation of the expectation.

A stochastic framework usually assumes stationarity, that is, the process is assumed to be in a particular state of *statistical equilibrium*. Strict stationarity is the assumption that statistical properties of process do not change over time, that is, the joint probability distribution $p(\mathbf{x}, y)$ of any set of observations must be unaffected by shifting all the times of observation forward or backward by any integer amount. The concept of stationarity is a mathematical convenience as it allows us to estimate properties (mean, variance) of the distribution, associated with a particular observation/instant of time, by averaging over time, that is, the sequence of observations. Similarly, the assumption of *ergodicity* allows us to replace *ensemble averages* by their corresponding *time averages*. A process is then called 'ergodic'. For some practical problems, the assumption of each observation (sequentially in time) being independent, identically distributed random variables may appear somewhat unrealistic or non-intuitive.

We can conclude, that a stochastic framework may appear more rigorous, thus more desirable than heuristic approaches but the many assumptions made (linearity, large samples, stationarity, ergodicity, independent identically distributed Gaussian variables) and the necessity to include prior knowledge, make any nonlinear, possibly heuristic, approach appear equally useful. The main advantage of a stochastic framework is that it provides a rich set of mathematical tools for an analysis on paper.

In subsequent sections, I outline a fuzzy mathematical perspective which clusters training data in $\mathbf{M}$, in the product space Ξ. As a consequence the regression concept in system identification finds its analogy in clustering. Using *fuzzy clustering* algorithms, local linear submodels are represented by fuzzy

[6]The phrase 'curse of dimensionality' was first coined by R. Bellman to describe the exponential growth in combinatorical optimization as the dimension increases.

sets (fuzzy relations, possibility distributions), allowing an interpretation of cluster as (fuzzy) propositions. These ideas are described in greater detail in Section 4.

Fuzzy clustering divides the available data into groups in which local linear relations exist between regressors and the regressand: $\mathbf{y} = f_i(\mathbf{x}; \boldsymbol{\theta}_i) + \varepsilon_i$, where each $\boldsymbol{\theta}_i$ is a vector of parameters to be determined, and each ε_i is a random vector with zero mean vector and some covariance matrix. The principle of fuzzy identification by product space clustering is therefore to approximate a nonlinear regression problem by decomposing it into several local linear submodels. The obtained *fuzzy partition matrix* is subsequently used to define an *if-then* rule-based fuzzy model. The decomposition of a global nonlinear mapping into a set of locally linear models is based on a geometrical interpretation of the regression problem. The unknown nonlinear function $y = f(\mathbf{x})$, (3.1), represents a nonlinear regression (hyper)surface in the product space $X \times Y \subset \mathbb{R}^{r+1}$. Fuzzy clustering as a method to identify fuzzy if-then rule-based systems from data is introduced in Section 4. The conceptual relationship between multiple or switching regression models and fuzzy clustering, attributed to Hathaway and Bezdek, is reviewed in Section 5.7.

A rule-based fuzzy model has a number of advantages in comparison with global nonlinear models, such as neural networks. The model structure is easy to understand and is in some cases interpretable. Various types of knowledge can be integrated in the model, including statistical objects and empirical knowledge. Furthermore, fuzzy models allow, in principle, for input data being imprecise (intervals) or vague (fuzzy sets).

Typical structures for fuzzy models are the so-called 'Takagi-Sugeno Model', 'Linguistic Model', and the 'Relational Model'. In Sections 5, 6 and Section 7.2, we describe fuzzy models which generalize the graph F, (1.8), of the mapping $f: X \to Y$ into a fuzzy graph $\widetilde{F}$.

4

Propositions as Subsets of the Data Space

- ☐ *A more general concept to represent data sampled from a system is that of a data space.*
- ☐ *System properties and behavior are reflected by clusters of data.*
- ☐ *Clusters may be interpreted as linear submodels of an overall nonlinear system.*
- ☐ *Clusters may also be interpreted as if-then rules relating properties of the variables that form the data space.*
- ☐ *Fuzzy clustering provides least-squares solutions to identify clusters, to partition the data space into clusters or classes.*
- ☐ *Fuzzy boundaries between clusters are differentiable functions and hence are computationally attractive.*
- ☐ *For many real-world problems a fuzzy partitioning of the underlying space is more realistic than 'hard clustering'.*

For the things we have to learn before we can do them, we learn by doing them.
—Aristotle

An alternative view of a 'functional' representation of the system $\mathfrak{S}$, is to identify subsets of the data space $\Xi = X \times Y$, with *propositions* in order to formulate a calculus of 'logical' propositions that leads to a rule-based representation of the data. Modelling dynamic systems, time-series data may then be transformed to points in the data space. Choosing variables, representing the axis of Ξ carefully, the data may exhibit 'pattern' (clusters) in Ξ (see Figure 4.1).

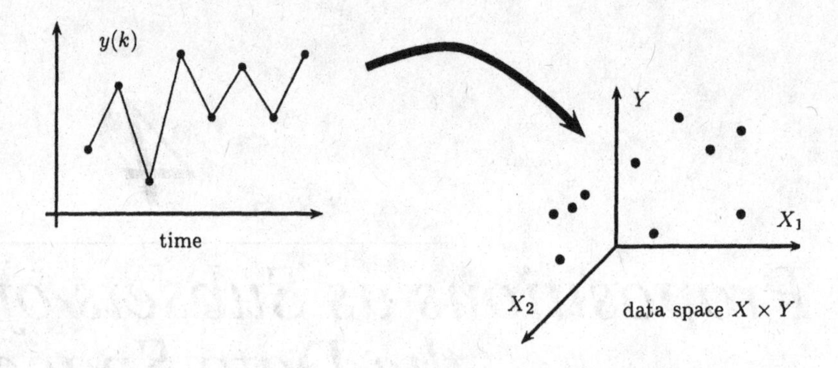

Fig. 4.1 From time-series data to the data space representation.

If, as in previous sections, $\mathbf{x} = [x_1, \ldots, x_r]^T$ and y are characterizing the system $\mathfrak{S}$, then the most general form of prediction concerning $\mathfrak{S}$ is that the point $(x_1, x_2, \ldots, x_r, y) \in \Xi$, determined by actually measuring $\mathbf{x}$ and y, will lie in a subset A of the $(x_1, \ldots, x_r, y)$-spaces. Hence, we call subsets of space Ξ 'experimental propositions' concerning $\mathfrak{S}$. The state-space is however a mathematical concept which needs to be related to some 'experimental reality'. In general, (*i.e.*, not only in quantum theory) a one-to-one correspondence of measurements and knowledge of the state of $\mathfrak{S}$ is unattainable. Such *certain uncertainty* motivates some calculus of experimental propositions that generalizes relations of points in $\mathbb{R}^n$ to (fuzzy) subsets. We are going to use notions of equivalence relations and equivalence classes to explain clustering. The reason is that, although no proofs are provided, the connection to fuzzy systems and uncertainty modelling is intuitive.

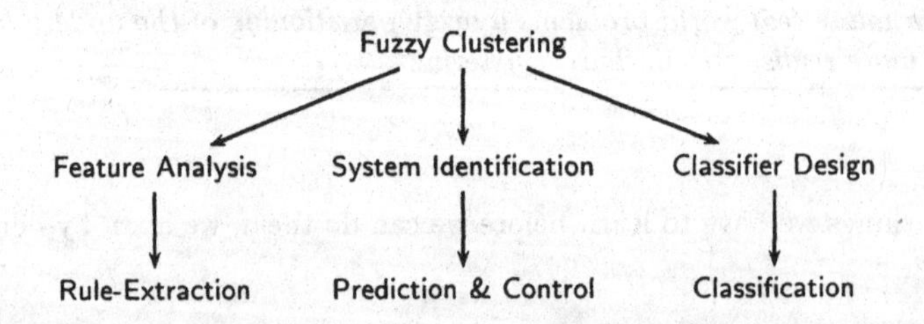

Fig. 4.2 Numerical pattern recognition.

The following section describes clustering techniques to identify subsets of the data space from sampled data. Clustering is a generic tool for *pattern recognition* - the search for structure in data, with a wide range of applications including fuzzy-system identification, classification and image analysis (see Figure 4.2). As cluster analysis is concerned with methods that identify *pattern* in data, it can form the basis for a refined analysis of dependency relationships. Using a proximity measure between *objects* (points) in Ξ, pattern in data are found by partitioning the set of objects into separate clusters (*classes*). Clusters form connected (dependent) regions of a multi-dimensional space Ξ in which objects (sampled data) are clustered with varying density across the space Ξ. Let the data space Ξ be non-empty and define by $\mathcal{C} = \{C_i\}$ the "cluster space" describing the outcome of clustering, that is, a set of clusters (classes) C_i with which elements $\mathbf{o} \in \Xi$ are to be associated. By identifying these clusters, partitioning Ξ, the algorithms implicitly identify a function

$$f : \Xi \to \mathcal{C} ,$$

which for any observation (object) $\mathbf{o} \in \Xi$ assigns a set $C_i \in \mathcal{C}$ of possible results. The 'correct' function f is identified (learned from a set of training data) by specifying an *objective function* or mapping

$$J : \{f\} \to \mathbb{R} ,$$

which measures the quality or error we aim to optimize.

Let an observation consist of n measured variables describing an *object* $\mathbf{o} \in \mathbb{R}^n$. The set of d measurements $\mathbf{m}_j = [m_{1j}, \ldots, m_{nj}]^T$ of $\mathbf{o}$ forms a $n \times d$ *data matrix*[1] $\mathbf{M}$:

$$\mathbf{M} = \begin{bmatrix} m_{11} & m_{12} & \cdots & m_{1d} \\ m_{21} & m_{22} & \cdots & m_{2d} \\ \vdots & \vdots & \ddots & \vdots \\ m_{n1} & m_{n2} & \cdots & m_{nd} \end{bmatrix} . \tag{4.1}$$

For instance, with regard to the model structure, defined by (3.3) and (3.5), we consider the following regressor variables:

$$y(k+1) = f\big(y(k), y(k-1), y(k-2), u(k)\big) .$$

We therefore have for the regression vector

$$\mathbf{x}^T = [y(k), y(k-1), y(k-2), u(k)] .$$

[1] As clustering has wide-ranging applications various sets of notation are in use. Often, $\mathbf{m}_j \in \mathbf{M}$ are referred to as objects or *feature vectors*. Since clustering algorithms attempt to organize unlabelled feature vectors into 'natural groups', clustering is also called 'unsupervised learning' and columns of the obtained partition matrix $\mathbf{U}$ are referred to as 'label vectors'.

Since the data are provided in pairs $\{(u(j), y(j))\}$, $j = 1, \ldots, d$, we have

$$\mathbf{y} = \begin{bmatrix} y(4) \\ y(5) \\ \vdots \\ y(d) \end{bmatrix}, \quad \mathbf{X} = \begin{bmatrix} y(3) & y(2) & y(1) & u(3) \\ y(4) & y(3) & y(2) & u(4) \\ \vdots & \vdots & \vdots & \vdots \\ y(d-1) & y(d-2) & y(d-3) & u(d-1) \end{bmatrix}.$$

From $\mathbf{X}$ and $\mathbf{y}$ we form the data matrix $\mathbf{M}^T = [\mathbf{X}, \mathbf{y}]$, with $n = r+1$, suitable for fuzzy clustering:

$$\mathbf{M} = \begin{bmatrix} y(3) & y(4) & \cdots & y(d-1) \\ y(2) & y(3) & \cdots & y(d-2) \\ y(1) & y(2) & \cdots & y(d-3) \\ u(3) & u(4) & \cdots & u(d-1) \\ y(4) & y(5) & \cdots & y(d) \end{bmatrix}.$$

Note that y and each variable in $\mathbf{x}$ represent an axis of the product space Ξ. Here we consider *intrinsic partitional classification* as opposed to hierarchical clustering[2]. Whereas hierarchical clustering leads to a nested sequence of partitions by imposing the ultra-metric on the set of objects, frequently visualized as a dendrogram, partitional clustering aims at a single partition. We obtain such a partition for a fixed number of clusters by minimizing a *clustering criterion*, for example, 'squared errors' (4.8).

Given a set of d objects $\mathbf{m}_j$, that is, points in the n-dimensional space $\Xi \subset \mathbb{R}^n$, we wish to partition them into c clusters by means of a (reflexive, symmetric, transitive) *equivalence relation*, that is, the map

$$E : \Xi \times \Xi \quad \to \quad \{0, 1\}$$

or, informally, the subset of $\Xi \times \Xi$ defined by $\{(\mathbf{o}, \mathbf{o}') : E(\mathbf{o}, \mathbf{o}') = 1\}$. Given $\mathbf{M}$, a clustering algorithm generates a *partition matrix*

$$\mathbf{U} = \begin{bmatrix} E(\mathbf{c}^{(1)}, \mathbf{m}_1) & E(\mathbf{c}^{(1)}, \mathbf{m}_2) & \cdots & E(\mathbf{c}^{(1)}, \mathbf{m}_d) \\ \vdots & \vdots & \ddots & \vdots \\ E(\mathbf{c}^{(c)}, \mathbf{m}_1) & E(\mathbf{c}^{(c)}, \mathbf{m}_2) & \cdots & E(\mathbf{c}^{(c)}, \mathbf{m}_d) \end{bmatrix}. \tag{4.2}$$

An element $u_{ij} \in \mathbf{U}$ is the equivalence of any $\mathbf{m}_j$ with the cluster prototype $\mathbf{c} \in \mathbf{C}$ or, in other words, it is the membership in the ith cluster, that is, the membership in $[\mathbf{c}^{(i)}]_E$. Let the *prototypes* of the clusters be denoted by

[2]For an overview of clustering and a description of the single-linkage algorithm see Appendix 13.3.

$\mathbf{C} = [\mathbf{c}^{(1)}, \ldots, \mathbf{c}^{(c)}]$, where $\mathbf{c}^{(i)}$ is commonly defined as the *centroid* - the mean vector

$$\mathbf{c}^{(i)} = \frac{\sum_{j=1}^{d} u_{ij} \cdot \mathbf{m}_j}{\sum_{j=1}^{d} u_{ij}} \qquad i = 1, 2, \ldots, c . \tag{4.3}$$

In general, for any $\mathbf{c}^{(i)} \in \Xi$ the equivalence relation E induces the *equivalence class*

$$[\mathbf{c}^{(i)}]_E \doteq \left\{ \mathbf{o} : \ \mathbf{o} \in \Xi, \ E(\mathbf{c}^{(i)}, \mathbf{o}) = 1 \right\} . \tag{4.4}$$

With $\mathbf{c}^{(i)}$ on both sides, equation (4.4) suggests the need for an iterative solution to the clustering problem. Hence, a cluster, $[\mathbf{c}^{(i)}]_E$, describes the set of elements $\mathbf{m}_j \in \Xi$ satisfying equivalence relation E. The collection of equivalence classes $\{[\mathbf{c}^{(i)}]_E\}$ forms a *hard-c-partition*, a field or more specifically a *quotient* or *factor set*, Ξ/E, of Ξ by E:

$$\Xi/E \doteq \left\{ [\mathbf{c}^{(i)}]_E \right\} . \tag{4.5}$$

The *natural map* of Ξ onto Ξ/E is defined as

$$\begin{aligned} \psi : \ \Xi \ &\rightarrow \ \Xi/E \\ \mathbf{o} \ &\mapsto \ [\mathbf{o}']_E . \end{aligned} \tag{4.6}$$

The natural map associates an element with the equivalence class it is in. It is thus a *classifier* which in condition monitoring could be used to identify and classify faults within a system.

4.1 HARD-C-MEANS CLUSTERING

Let $\mathbf{M} = \{\mathbf{m}_1, \mathbf{m}_2, \ldots, \mathbf{m}_d\}$ be a finite set and $2 \leq c < d$ be an integer. Then the *hard partitioning space* for $\mathbf{M}$ is the set

$$M_{hc} = \left\{ \mathbf{U} \in V_{cd} : \ u_{ij} \in \{0,1\}, \forall (i,j); \ \sum_{i=1}^{c} u_{ij} = 1; \ 0 < \sum_{j=1}^{d} u_{ij} < d, \forall i \right\} . \tag{4.7}$$

where $u_{ij} \in \mathbf{U}$ and V_{cd} is the vector space of real $c \times d$ matrices - the set of admissible solutions for the (hard) clustering problem. The partition (4.2) requires the definition of the cluster center (4.3) and vice versa. For this reason clustering algorithms iteratively optimize a *clustering criterion* (*objective function, cost function*) such as

$$J_{hc}(\mathbf{M}; \mathbf{U}, \mathbf{C}) = \sum_{i=1}^{c} \sum_{j=1}^{d} u_{ij} \, d_{\mathbf{A}}^2 \left(\mathbf{m}_j, \mathbf{c}^{(i)} \right) , \tag{4.8}$$

where $\mathbf{U} \in M_{hc}$ and $d^2_{\mathbf{A}}\left(\mathbf{m}_j, \mathbf{c}^{(i)}\right)$ is the distance, proximity, of object $\mathbf{m}_j$ with respect to prototype $\mathbf{c}^{(i)}$, calculated by the squared inner-product norm

$$d^2_{\mathbf{A}}\left(\mathbf{m}_j, \mathbf{c}^{(i)}\right) \doteq \left\|\mathbf{m}_j - \mathbf{c}^{(i)}\right\|^2_{\mathbf{A}} = \left(\mathbf{m}_j - \mathbf{c}^{(i)}\right)^T \mathbf{A}\left(\mathbf{m}_j - \mathbf{c}^{(i)}\right) . \tag{4.9}$$

If the positive definite matrix $\mathbf{A}$ is the identity matrix $\mathbf{I}$ (ones on the diagonal and all other elements zero), the distance measure induced on $\mathbb{R}^n$, is the Euclidean metric (L_2 norm). The 'Mahalonobis norm' is obtained by choosing $\mathbf{A}$ to be the inverse of the $n \times n$ sample covariance matrix $1/d\sum_{j=1}^{d}(\mathbf{m}_j - \hat{\eta}_{\mathbf{m}})(\mathbf{m}_j - \hat{\eta}_{\mathbf{m}})^T$ of $\mathbf{M}$, where $\hat{\eta}_{\mathbf{m}}$ is the sample mean (2.7) of the data. Whereas the Euclidean norm imposes hyperspherical clusters on $\mathbb{R}^n$, the Mahalonobis norm generates hyperellipsoidal clusters.

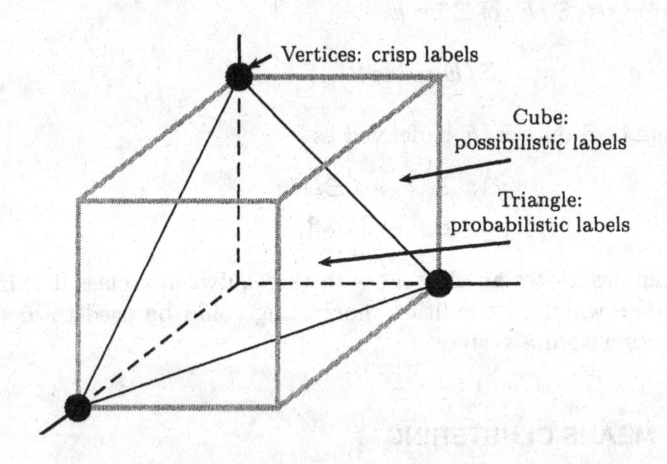

Fig. 4.3 Options for cluster labels.

From (4.8), the objective is to minimize the 'within-cluster-overall-variance', where the 'within-cluster variance' is the squared error with respect to the prototypical $\mathbf{c}^{(i)}$. The iterative algorithm starts with an initial partition and assigns objects to clusters so as to optimize the objective function (4.8). If the value of the objective function obtained is sufficiently small, the clustering is assumed to be valid:

$$(\mathbf{U}, \mathbf{C}) = \arg \min_{M_{hc} \times \mathbb{R}^{d \times c} \times V_{dd}} J_{hc}(\mathbf{M}; \mathbf{U}, \mathbf{C}, \mathbf{A}) ,$$

where V_{dd} is the set of $d \times d$ positive definite matrices. Clustering the data assigns a *label* to each object $\mathbf{m}_j$. The label is the jth column from $\mathbf{U}$ describing its membership of $\mathbf{m}_j$ in the c clusters (classes). In hard-c-means clustering only one element of the label vector can be equal to one and all other elements must be zero. Obvious generalizations are shown in Figure 4.3 and followed up further below.

Each hard-c-partition $\mathbf{U}$ of $\mathbf{M}$ induces on $\Xi \times \Xi$ a unique hard equivalence relation E, the elements of which are defined by $e_{ij} = 1$ iff $\mathbf{m}_i$ and $\mathbf{m}_j$ belong to the same hard subset of $\mathbf{U}$, and $e_{ij} = 0$ otherwise:

$$e_{ij} = \begin{cases} 1 & \text{iff} \quad \mathbf{m}_{j=1,\dots,d} \in [\mathbf{c}^{(i=1,\dots,c)}]_E \\ 0 & \text{otherwise}. \end{cases}$$

The (hard) equivalence relation E satisfies the following conditions:

$$\begin{array}{lrl} e_{ii} = 1, & 1 \le i \le d & \text{(reflexivity)} \\ e_{ij} = e_{ji}, & 1 \le i \ne j \le d & \text{(symmetry)} \\ \text{if } e_{ij} = 1 \text{ and } e_{jk} = 1, \text{ then } e_{ik} = 1 & \forall(i,j,k) & \text{(transitivity)}. \end{array}$$

Hard clustering has been criticized with various examples in which objects should have partial membership in clusters in order to achieve a realistic (fuzzy) partitioning in the given context (see example 4.3 below). But there are not only conceptual reasons to introduce fuzzy partitions. A problem with the hard-c-partition is that the space M_{hc} is too large to search for an optimal partition. The number of distinct ways to partition the data space into c non-empty subsets is

$$|M_{hc}| = \frac{1}{c!}\left[\sum_{i=1}^{c} \binom{c}{i} (-1)^{c-i} \cdot i^d\right].$$

The problem is due to the discrete nature of the characteristic function taking the u_{ij} as its values. If the u_{ij} are continuous, taking any value in the unit-interval $[0,1]$, then we can determine the derivatives of some objective function $J(\cdot)$ with respect to u_{ij} in order to analytically derive an optimal solution. Many of those problems were solved with the introduction of fuzzy partition spaces (which are differentiable). The study of least-squares solutions to fuzzy objective functions has been largely progressed by J. Bezdek [Bez81].

4.2 LEAST-SQUARES FUNCTIONALS: FUZZY CLUSTERING

The generalization of the hard-c-partition to a fuzzy partition follows directly by allowing degrees of membership in the unit-interval. From (4.7) the *fuzzy partition space* of $\mathbf{M}$ is the set

$$M_{fc} = \left\{ \mathbf{U} \in V_{cd} : \; u_{ij} \in [0,1], \forall(i,j); \; \sum_{i=1}^{c} u_{ij} = 1; \; 0 < \sum_{j=1}^{d} u_{ij} < d, \forall i \right\}. \quad (4.10)$$

Other partition spaces may be obtained by relaxing the condition that the sum of membership values across cluster must add up to one [Bab98]. The objective function (4.8) is generalized into the fuzzy-c-means functional

$$J_{fc}(\mathbf{M};\mathbf{U},\mathbf{C}) = \sum_{i=1}^{c}\sum_{j=1}^{d} (u_{ij})^{w}\, d^2_{\mathbf{A}}\left(\mathbf{m}_j, \mathbf{c}^{(i)}\right) , \tag{4.11}$$

where $\mathbf{U} \in M_{fc}$ is a fuzzy partition of $\mathbf{M}$ and $w \in [1, \infty)$ is a weighting exponent which determines the fuzziness of the resulting clusters. For $w \to 1$, the clusters tend to be crisp, hard, *i.e.* either $u_{ij} \to 1$ or $u_{ij} \to 0$, for $w \to \infty$ we have $u_{ij} \to 1/c$. A typical value for w is 2. The objective function (4.11) minimizes the sum over the quadratic distances of the data to the prototypes weighted by their membership degrees and is therefore a least-squares criterion. Hence, the value of (4.11) can be seen as a measure of the total variation of $\mathbf{m}_j$ from $\mathbf{c}^{(i)}$. The minimization of the c-means functional represents a nonlinear optimization problem that can be solved by a variety of methods. The most popular method is an iteration through the first-order stationary conditions of the functional (4.11), known as the fuzzy-c-means algorithm [Bez81, Bab98, HKKR99]. To obtain necessary conditions for (local) minima we somehow have to take partial derivative of (4.11). In [Bez81], the constraint $\sum_{i=1}^{c} u_{ij} = 1$ is added to the objective function by means of Lagrange multipliers[3] to obtain an expression like

$$\sum_{i=1}^{c}\sum_{j=1}^{d} (u_{ij})^{w} d^2_{\mathbf{A}}\left(\mathbf{m}_j, \mathbf{c}^{(i)}\right) + \sum_{j=1}^{d} \lambda_j \left(\sum_{i=1}^{c} u_{ij} - 1\right) .$$

We can then relax the search space for $\mathbf{U}$ to allow $0 \leq \sum_{j=1}^{d} u_{ij} \leq d$ so that minimization can be done on uncoupled columns of $\mathbf{U}$. This leads to the Lagrangian

$$L_j(\lambda, \mathbf{u}_j) = \sum_{i=1}^{c} (u_{ij})^{w} d^2_{ij} - \lambda \left(\sum_{i=1}^{c} u_{ij} - 1\right) ,$$

where we write $d^2_{ij} \doteq d^2_{\mathbf{A}}(\mathbf{m}_j, \mathbf{c}^{(i)})$ for short and $\mathbf{u}_j$ denotes the jth column of $\mathbf{U}$. Stationary points are found for this expression by setting the gradient equal to zero. Therefore, the following partial derivatives must be zero:

$$\frac{\partial}{\partial \lambda} L_j(\lambda, \mathbf{u}_j) = \sum_{i=1}^{c} u_{ij} - 1 \tag{4.12}$$

$$\frac{\partial}{\partial u_{st}} L_j(\lambda, \mathbf{u}_j) = w \cdot u_{st}^{w-1} \cdot d^2_{st} - \lambda . \tag{4.13}$$

[3] *Lagrange multipliers* are a sequence of real numbers λ such that a point x_0 that minimizes $J(x)$ subject to $g_1 = 0, \ldots, g_m = 0$ will be a stationary point of the *Lagrangian* $L(\lambda, x) = J(x) + \sum_{i=1}^{m} \lambda_i g_i(x)$.

From (4.13),

$$u_{st} = \left(\frac{\lambda}{w \cdot d_{st}^2}\right)^{\frac{1}{w-1}}. \tag{4.14}$$

From (4.12) set equal to zero, and (4.14), we have

$$\begin{aligned}1 &= \sum_{i=1}^{c} u_{ij} \\ &= \sum_{i=1}^{c} \left(\frac{\lambda}{w}\right)^{\frac{1}{w-1}} \left(\frac{1}{d_{st}^2}\right)^{\frac{1}{w-1}} \\ &= \left(\frac{\lambda}{w}\right)^{\frac{1}{w-1}} \cdot \left(\sum_{i=1}^{c} \left(\frac{1}{d_{st}^2}\right)^{\frac{1}{w-1}}\right)\end{aligned}$$

and thus

$$\left(\frac{\lambda}{w}\right)^{\frac{1}{w-1}} = \frac{1}{\sum_{i=1}^{c} \left(\frac{1}{d_{st}^2}\right)^{\frac{1}{w-1}}}. \tag{4.15}$$

Inserting (4.15) into (4.14),

$$\begin{aligned}u_{st} &= \frac{1}{\sum_{i=1}^{c} \left(\frac{1}{d_{it}^2}\right)^{\frac{1}{w-1}}} \cdot \left(\frac{1}{d_{it}^2}\right)^{\frac{1}{w-1}} \\ &= \frac{1}{\sum_{i=1}^{c} \left(\frac{d_{st}^2}{d_{it}^2}\right)^{\frac{1}{w-1}}}.\end{aligned}$$

Or switching back to our usual notation, within an optimization loop, we would determine for any given object $\mathbf{m}_j$ the degree of membership in the ith cluster by

$$u_{ij} = \frac{1}{\sum_{k=1}^{c} \left(\frac{d_{\mathbf{A}}^2(\mathbf{m}_j, \mathbf{c}^{(i)})}{d_{\mathbf{A}}^2(\mathbf{m}_j, \mathbf{c}^{(k)})}\right)^{\frac{1}{w-1}}} \tag{4.16}$$

and

$$\mathbf{c}^{(i)} = \frac{\sum_{j=1}^{d} (u_{ij})^w \, \mathbf{m}_j}{\sum_{j=1}^{d} (u_{ij})^w}, \quad 1 \le i \le c. \tag{4.17}$$

In equation (4.16), it is possible that $\exists i \colon d_{\mathbf{A}}(\mathbf{m}_j, \mathbf{c}^{(i)}) = 0$ for some $\mathbf{m}_j$ and one or more cluster prototypes $\mathbf{c}^{(s)}$, $s \in S \subset \{1, 2, \ldots, c\}$. This problem, for which the new u_{ij} is undefined, is called a *singularity*. To fix it, we assign to each u_{ij}, $i \in S^c$ the value 0 subject to the constraint $\sum_{s \in S} u_{sk} = 1, \forall j$. The solution, (4.16) and (4.17) also satisfies the constraints given in (4.10). The fact that the $\mathbf{c}^{(i)}$ are determined as the (membership) weighted mean of the data gives the name *fuzzy-c-means*. Algorithm 4.1 is terminated if changes in the partition matrix are negligible, that is, if for some convenient matrix norm $\|\mathbf{U}^{(l)} - \mathbf{U}^{(l-1)}\| < \delta$. Examples for such a criterion are

$$\left\|\mathbf{U}^{(l)} - \mathbf{U}^{(l-1)}\right\| \doteq \sum_{j=1}^{d} \sum_{i=1}^{c} \left|u_{ij}^{(l)} - u_{ij}^{(l-1)}\right|$$

$$\doteq \max_{ij} \left\{\left|u_{ij}^{(l)} - u_{ij}^{(l-1)}\right|\right\} .$$

Fix c, $2 \leq c < d$, and choose any inner-product norm metric for $\mathbb{R}^n$, the termination tolerance $\delta > 0$, e.g., between 0.01 and 0.001, and fix w, $1 \leq w < \infty$, e.g., 2. Initialize $\mathbf{U}^{(0)} \in M_{fc}$, (e.g., randomly).

Repeat for $l = 1, 2, \ldots$:

Step 1: Compute cluster prototypes:

$$\mathbf{c}_l^{(i)} = \frac{\sum_{j=1}^{d} \left(u_{ij}^{(l-1)}\right)^w \mathbf{m}_j}{\sum_{j=1}^{d} \left(u_{ij}^{(l-1)}\right)^w}, \quad 1 \leq i \leq c .$$

Step 2: Compute distances:

$$d_{\mathbf{A}}^2\left(\mathbf{m}_j, \mathbf{c}_l^{(i)}\right) = \left(\mathbf{c}_l^{(i)} - \mathbf{m}_j\right)^T \mathbf{A}\left(\mathbf{c}_l^{(i)} - \mathbf{m}_j\right), \quad 1 \leq i \leq c, \quad 1 \leq j \leq d .$$

Step 3: Update the partition matrix:

If $d_{\mathbf{A}}\left(\mathbf{m}_j, \mathbf{c}_l^{(i)}\right) > 0$ for $1 \leq i \leq c$, $1 \leq j \leq d$,

$$u_{ij}^{(l)} = \frac{1}{\sum_{k=1}^{c} \left(d_{\mathbf{A}}^2(\mathbf{m}_j, \mathbf{c}^{(i)}) / d_{\mathbf{A}}^2(\mathbf{m}_j, \mathbf{c}^{(k)})\right)^{1/(w-1)}}$$

otherwise

$$u_{ij}^{(l)} = 0 \text{ if } d_{\mathbf{A}}\left(\mathbf{m}_j, \mathbf{c}_l^{(i)}\right) > 0, \text{ and } u_{ij}^{(l)} \in [0,1] \text{ with } \sum_{i=1}^{c} u_{ij}^{(l)} = 1 .$$

Until $\|\mathbf{U}^{(l)} - \mathbf{U}^{(l-1)}\| < \delta$.

Algorithm 4.1 Fuzzy-c-means algorithm (FCM).

The generalization of the binary set-membership $E(\mathbf{o}, \mathbf{o}') \in \{0,1\}$ to a fuzzy set-membership function $\widetilde{E}(\mathbf{o}, \mathbf{o}') \in [0,1]$ reformulates the clustering problem as follows: The problem consists of characterizing a *fuzzy equivalence relation*, also called *similarity relation*, $\widetilde{E}$, by the set of its clusters in the same way as equivalence relations E are uniquely determined by their sets of equivalence classes (4.4). In analogy to the partition matrix (4.2), we find

$$\mathbf{U} = \begin{bmatrix} \widetilde{E}(\mathbf{c}^{(1)},\mathbf{m}_1) & \widetilde{E}(\mathbf{c}^{(1)},\mathbf{m}_2) & \cdots & \widetilde{E}(\mathbf{c}^{(1)},\mathbf{m}_d) \\ \vdots & \vdots & \ddots & \vdots \\ \widetilde{E}(\mathbf{c}^{(c)},\mathbf{m}_1) & \widetilde{E}(\mathbf{c}^{(c)},\mathbf{m}_2) & \cdots & \widetilde{E}(\mathbf{c}^{(c)},\mathbf{m}_d) \end{bmatrix}, \tag{4.18}$$

where the fuzzy equivalence relation is the map

$$\widetilde{E} : \Xi \times \Xi \quad \to \quad [0,1] \tag{4.19}$$

and $u_{ij} \in \mathbf{U}$ describes the degree of similarity of object $\mathbf{m}_j$ with prototype $\mathbf{c}^{(i)}$, $\widetilde{E}(\mathbf{c}^{(i)},\mathbf{m}_j)$, or the membership of the data samples in the clusters. That is, fuzzy sets are induced by 'crisp' sets when taking a similarity relation into account. In other words, the point $\mathbf{c}^{(i)} \in \Xi$ induces a fuzzy set $\mu_{\mathbf{c}^{(i)}}$ of all elements that are similar to $\mathbf{c}^{(i)}$. The membership degree μ of $\mathbf{o}$ to this fuzzy set is the grade to which $\mathbf{o}$ and $\mathbf{c}^{(i)}$ are similar or *indistinguishable*:

$$\mu_{\mathbf{c}^{(i)}}(\mathbf{o}) = \widetilde{E}(\mathbf{c}^{(i)},\mathbf{o}) \; . \tag{4.20}$$

The fuzzy set $\mu_{\mathbf{c}^{(i)}}(\mathbf{o}) = \widetilde{E}(\mathbf{c}^{(i)},\mathbf{o})$ is called the *extensional hull* of $\mathbf{c}^{(i)}$ with respect to similarity relation $\widetilde{E}$.[4] Similarity relations are fuzzy generalizations of equivalence relations and clusters are *fuzzy equivalence classes* (4.21). We can therefore describe fuzzy clustering as the problem of constructing *fuzzy quotients* (4.22) with respect to similarity relations:

$$[\mathbf{c}^{(i)}]_{\widetilde{E}} = \left\{ \left(\mathbf{o}, \widetilde{E}(\mathbf{c}^{(i)},\mathbf{o})\right) \right\}, \tag{4.21}$$

$$\Xi/\widetilde{E} \doteq \left\{ [\mathbf{c}^{(i)}]_{\widetilde{E}} \right\} \; . \tag{4.22}$$

A similarity relation [Zad71, Wol98] on Ξ is a map whose values $\widetilde{E}(\mathbf{o},\mathbf{o}')$ are interpreted as the *degree* to which $\mathbf{o}$ and $\mathbf{o}'$ are equivalent. As with equivalence relations, $\widetilde{E}$ is required to satisfy three axioms interpreted as reflexivity, $\widetilde{E}(\mathbf{o},\mathbf{o}) = 1$, that is, each element is similar to itself to the degree 1; symmetry, $\widetilde{E}(\mathbf{o},\mathbf{o}') = \widetilde{E}(\mathbf{o}',\mathbf{o})$, *i.e.* $\mathbf{o}$ is similar to $\mathbf{o}'$ to the same degree as $\mathbf{o}'$ is similar to $\mathbf{o}$; and transitivity. In equivalence relations, transitivity describes the fact

$$E(\mathbf{o}_i,\mathbf{o}_j) \wedge E(\mathbf{o}_j,\mathbf{o}_k) \Rightarrow E(\mathbf{o}_i,\mathbf{o}_k) \; . \tag{4.23}$$

[4] In conventional set theory the subset consisting of all those elements which have a certain property is called the *extension* of the property. The 'extensionality' axiom for sets states that a set is completely determined just by specifying its elements. With respect to $\widetilde{E}$, the fuzzy set $\mu_{\mathbf{o}_0}$ represents the extensional hull of the set $\{\mathbf{o}_0\}$, *i.e.* it is the smallest extensional fuzzy set with $\mu_{\mathbf{o}_0}(\mathbf{o}_0) = 1$. In general, the extensionality of a fuzzy set A describes the compatibility of a t-norm $T(\cdot,\cdot)$, $\widetilde{E}$ and μ_A: $T(\mu_A(\mathbf{o}), \widetilde{E}(\mathbf{o},\mathbf{o}')) \le \mu_A(\mathbf{o}')$. That is, it is a multi-valued formulation of the statement "IF $\mathbf{o} \in A$, AND $\mathbf{o} \approx \mathbf{o}'$, THEN $\mathbf{o}' \in A$"

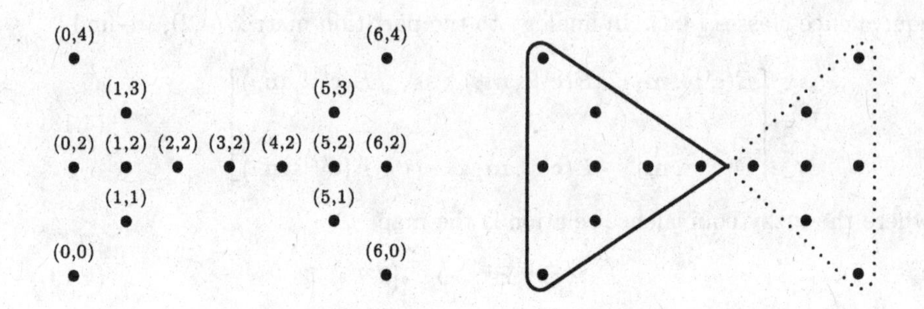

Fig. 4.4 The butterfly data set and hard-c-means result.

For similarity relations transitivity is generalized with the condition that there exists a continuous triangular norm, $T\colon \Xi \times \Xi \rightarrow [0,1]$, generalizing the triangle inequality (7.1), such that for all $\mathbf{o}_i, \mathbf{o}_j, \mathbf{o}_k \in \Xi$ the following inequality holds:

$$\widetilde{E}(\mathbf{o}_i, \mathbf{o}_k) \geq T\left(\widetilde{E}(\mathbf{o}_i, \mathbf{o}_j), \widetilde{E}(\mathbf{o}_j, \mathbf{o}_k)\right) \quad : \quad T\text{-transitivity.} \tag{4.24}$$

The triangular norm can be interpreted as the valuation function of a conjunction. For instance, let T be the Lukasiewicz norm (7.4). Then, for (4.24), we have

$$\widetilde{E}(\mathbf{o}_i, \mathbf{o}_k) \geq \widetilde{E}(\mathbf{o}_i, \mathbf{o}_j) + \widetilde{E}(\mathbf{o}_j, \mathbf{o}_k) - 1 \ ,$$

which states that if $\mathbf{o}_i$ and $\mathbf{o}_j$, as well as $\mathbf{o}_j$ and $\mathbf{o}_k$ are similar, then also $\mathbf{o}_i$ and $\mathbf{o}_k$ have to be similar. The origin of triangular norms, t-norms for short, is the so-called Poincaré paradox and Menger's probabilistic spaces (Section 7). The emergence of similarity relations from fundamental aspects of modelling and measurements is further discussed in Section 7. Section 4.6 presents a fuzzy rule-based system, based on similarity relations.

4.3 EXAMPLE: HARD VS. FUZZY CLUSTERING

The data set $\mathbf{M}$ consists of 15 points in the plane:

j	1	2	3	4	5	6	7	8
$\mathbf{m}_j$	(0,0)	(0,2)	(0,4)	(1,1)	(1,2)	(1,3)	(2,2)	(3,2)

j	9	10	11	12	13	14	15
$\mathbf{m}_j$	(4,2)	(5,1)	(5,2)	(5,3)	(6,0)	(6,2)	(6,4)

As depicted in Figure 4.4, the shape of the data set suggests a 'butterfly' which is frequently used to discuss hard vs. fuzzy clustering [Bez81]. Clustering these points by the hard-c-means algorithm yields two fuzzy clusters as shown in Figure 4.5. The plot shown on the right in Figure 4.4 shows the result after 2 iterations. We observe that, even though the butterfly is symmetric, the clusters in Figure 4.4 cannot be. This suggests that a fuzzy partitioning of the data into two clusters would be more natural. The result of the fuzzy-c-means algorithm 4.1 with $w = 1.25$ and $\delta = 0.01$ is shown in Figure 4.6. For $w = 1.25$, the algorithm stopped after 7 iterations. Figure 4.7 shows the fuzzy clusters for $w = 2$ obtained after 6 iterations and projected onto the x-axis. In Figure 4.8, the effect of the weighting factor w is demonstrated. For all results the following random initial partition matrix $\mathbf{U}^{(0)}$ was used:

$$\mathbf{U}^{(0)} = \begin{bmatrix} 0 & 0 & 1 & 1 & 1 & 1 & 1 & 0 & 0 & 0 & 0 & 1 & 1 & 0 & 1 \\ 1 & 1 & 0 & 0 & 0 & 0 & 0 & 1 & 1 & 1 & 1 & 0 & 0 & 1 & 0 \end{bmatrix} .$$

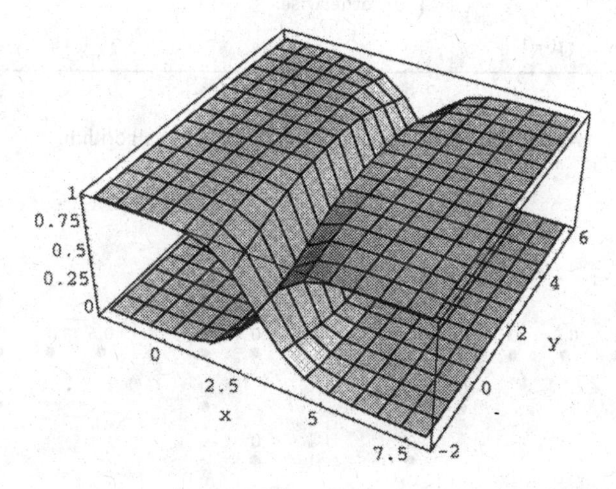

Fig. 4.5 Fuzzy sets obtained from fuzzy-c-means clustering of the Butterfly data set.

4.4 ORTHOGONAL TRANSFORMATION

As noted before, the distance measure employed in the objective function has considerable influence on how well the algorithm detects clusters (or shapes). To make the discussion in this section easier, let us consider a simple example of data in the plane similar to those in Figure 2.8 with $\Xi = X_1 \times X_2$. Let us shift our coordinate system of Ξ to bring the expected vector η (*i.e.* the cluster center $\mathbf{C}$) to the origin such that $\mathbf{o}$ denotes the new coordinate system: $\mathbf{o} \doteq \mathbf{o} - \eta$.

A Euclidean distance leads to the expression

$$\mathbf{o}^T \mathbf{I} \mathbf{o} ,$$

Fix c, $2 \leq c < d$, and choose any inner product norm metric for $\mathbb{R}^n$, the termination tolerance $\delta > 0$, e.g., between 0.01 and 0.001. Initialize $\mathbf{U}^{(0)} \in M_{hc}$, (e.g., randomly).

Repeat for $l = 1, 2, \ldots$:

Step 1: Calculate centers (centroids) of hard clusters, *i.e.* the c-mean vectors $\mathbf{c}^{(l)}$:

$$\mathbf{c}_l^{(i)} = \frac{\sum_{j=1}^{d} u_{ij}^{(l-1)} \cdot \mathbf{m}_j}{\sum_{j=1}^{d} u_{ij}^{(l-1)}}, \quad 1 \leq i \leq c .$$

Step 2: Update $\mathbf{U}^{(l)}$: Reallocate cluster memberships to minimize squared errors between the data and current cluster centers (prototypes):

$$u_{ij}^{(l)} = \begin{cases} 1 & \text{if } d(\mathbf{m}_j, \mathbf{c}_i^{(l)}) = \min_{1 \leq k \leq c} d(\mathbf{m}_j, \mathbf{c}_k^{(l)}) \\ 0 & \text{otherwise.} \end{cases}$$

Until $\left\| \mathbf{U}^{(l)} - \mathbf{U}^{(l-1)} \right\| < \delta$.

Algorithm 4.2 Hard-c-means (ISODATA) algorithm.

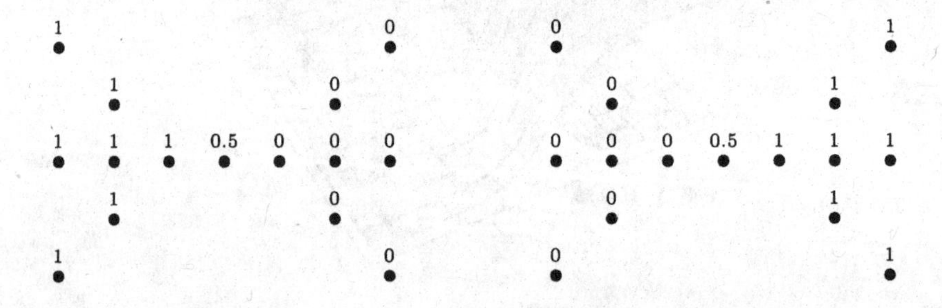

Fig. 4.6 Fuzzy-c-means clustering of the butterfly data set. $w = 1.25$

where $\mathbf{I}$ is the identity matrix, imposing a spherical structure onto the data space Ξ. In other words, the algorithm using the Euclidean norm should be particularly well suited to identify clusters of spherical shape. More often, however, we may find that clusters are shaped to what can be better described with an ellipsoid. What follows is a discussion on how such clusters can be characterized. In particular, we shall discuss the case where instead of the identity matrix $\mathbf{I}$ the covariance matrix $\mathbf{\Sigma}$ leads to what is called the Mahalonobis norm. Describing an ellipsoidal cluster by its two main axes is a task similar to principal component analysis. The motivation of principal component analysis is parsimony – dimension reduction. Two or more correlated variables are described by a single linear composite (the equation describing the longer axis of the ellipsoid or the regression line fitted to the

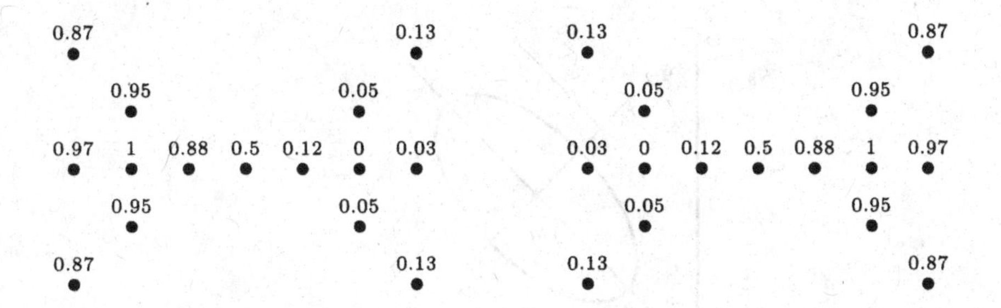

Fig. 4.7 Fuzzy-c-means clustering of the butterfly data set. $w = 2$

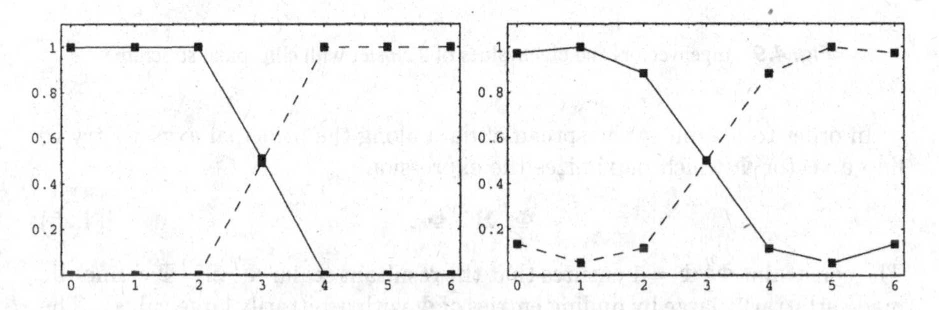

Fig. 4.8 FCM: $w = 1.25$ (left), $w = 2$ (right). Result after 7 iterations.

cluster). This linear composite ("principal component") attempts to account for as much as possible of the variation shared by the contributing variables. In other words, the data are maximally spread along the axis described by the linear composite. The axis is found by transforming the original coordinate system into a new one, maximizing the variance subject to the constraint $\mathbf{\Phi}^T\mathbf{\Phi} = 1$, where $\mathbf{\Phi}$ is one column of the *transformation matrix* $\mathbf{T}$ associated with the axis along which the data are spread with maximum variance (see Figure 4.9). This restriction on the length of $\mathbf{\Phi}$ will ensure that our transformation meets the unit length condition for rotation. The term 'orthogonal' refers to the square matrix $\mathbf{T}$ that exhibits the property

$$\mathbf{T}^T\mathbf{T} = \mathbf{T}\mathbf{T}^T = \mathbf{I}\,.$$

That is, any two column vectors or any two row vectors in the matrix $\mathbf{T}$ are mutually orthogonal and, furthermore, each vector is of unit length. Using such a transformation matrix on a set of points will preserve their angles, lengths, and interpoint distances, while at the same time referring them to a new, perhaps simpler, coordinate system.

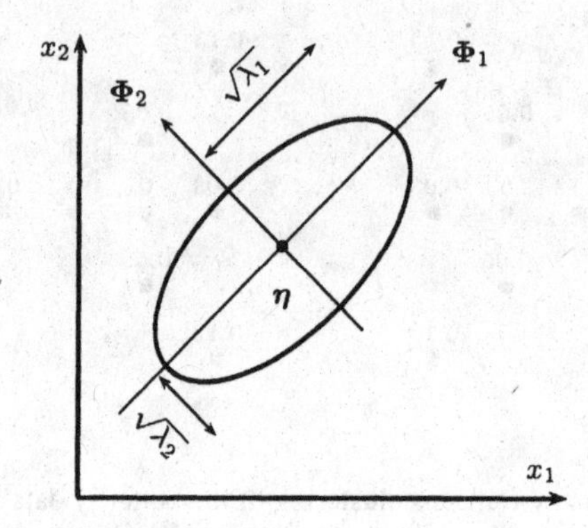

Fig. 4.9 Eigenvectors and eigenvalues of a cluster with ellipsoidal structure.

In order to maximize the spread of data along the principal axis, we try to find a vector $\mathbf{\Phi}$ which maximizes the expression

$$\mathbf{\Phi}^T \mathbf{\Sigma}^{-1} \mathbf{\Phi} . \tag{4.25}$$

The constraint $\mathbf{\Phi}^T \mathbf{\Phi} = 1$ ensures that the resultant scalar $\mathbf{\Phi}^T \mathbf{\Sigma}^{-1} \mathbf{\Phi}$ cannot be made arbitrarily large by finding entries of $\mathbf{\Phi}$ with arbitrarily large values. The expression is optimized by taking the partial derivatives w.r.t. to a Lagrange multiplier

$$\frac{\partial}{\partial \mathbf{\Phi}} \Big(\mathbf{\Phi}^T \mathbf{\Sigma}^{-1} \mathbf{\Phi} - \lambda (\mathbf{\Phi}^T \mathbf{\Phi} - 1) \Big) = 2 \mathbf{\Sigma} \mathbf{\Phi} - 2 \lambda \mathbf{\Phi} , \tag{4.26}$$

and setting this equal to zero

$$(\mathbf{\Sigma} - \lambda \mathbf{I}) \mathbf{\Phi} = 0 . \tag{4.27}$$

Note that the term $\partial / \partial \mathbf{\Phi}$ consists of n partial derivatives. Equation (4.27) resembles expression (1.25), $(\mathbf{F} - \lambda \mathbf{I})\mathbf{x} = 0$, which is often used to find the eigenstructure of system matrix $\mathbf{F}$ for dynamic systems. The only difference is that $\mathbf{\Sigma}$ is symmetric, and the eigenvectors $\mathbf{\Phi}$[5] to be normalized to unit length. If the matrix $\mathbf{\Sigma} - \lambda \mathbf{I}$, for a fixed λ, were nonsingular (*i.e.* possessed an inverse), it would always be the case that the only possible solution to the equation involves setting $\mathbf{\Phi}$ equal to the zero vector. Here we want the reverse, that is, to find a λ that will make $\mathbf{\Sigma} - \lambda \mathbf{I}$ singular. Recalling that singular matrices have determinants of zero, we want to find a value for λ that satisfies the *characteristic equation* of $\mathbf{\Sigma}$:

$$|\mathbf{\Sigma} - \lambda \mathbf{I}| = 0 .$$

[5] One should not confuse the notation of the eigenvector with the state-transition matrix.

In other words, we wish to find $\mathbf{\Phi}$ such that if premultiplied by $\mathbf{\Sigma}$, results in a vector $\lambda\mathbf{\Phi}$ whose components are proportional to those of $\mathbf{\Phi}$:

$$\mathbf{\Sigma}\mathbf{\Phi} = \lambda\mathbf{\Phi} .$$

Solving the characteristic equation yields n roots, but since we want to maximize the distances between points, (4.25), we look for the largest λ_i obtained from solving the characteristic equation. That is, we shall order the roots from large to small and choose that eigenvector $\mathbf{\Phi}$ corresponding to the largest λ_i. The eigenvectors corresponding to two different eigenvalues are orthogonal and hence form a new coordinate system. Collecting the eigenvectors in a transformation matrix $\mathbf{T}$

$$\mathbf{T} = [\mathbf{\Phi}_1, \ldots, \mathbf{\Phi}_n] ,$$

we can transform any given vector $\mathbf{o}$ into the new coordinate system by multiplying it with $\mathbf{T}$. In conclusion, the expression

$$(\mathbf{o} - \boldsymbol{\eta})^T \mathbf{\Sigma}^{-1} (\mathbf{o} - \boldsymbol{\eta}) = 1$$

using the covariance matrix $\mathbf{\Sigma}$, defines a (hyper)ellipsoid. Using the information of the covariance of the data we should therefore expect the clustering algorithm to perform better if the clusters are 'shaped'. We will explore this idea further when, in Section 5, we present the Gustafson-Kessel algorithm which does exactly that.

4.5 EXAMPLE: CLASSIFICATION

In this example we discuss fuzzy clustering in classifier design. *Discrimination* and *classification* are multi-variate techniques concerned with *separating* distinct sets of objects (or observations) and with *allocating* new objects (observations) to previously defined groups. The example and data we use in this section are taken from [JW98]. In addition to an introduction to classification we demonstrate the sensitivity of clustering to the scales of different variables. Subsequent sections will provide further examples of classifier design.

The admission officer of a business school has used an "index" of undergraduate grade point average (GPA) and graduate management aptitude test (GMAT) scores to help decide which applicants should be admitted to the school's graduate programs. In [JW98] pairs of values $\mathbf{x} = (x_1 \doteq \text{GPA}, x_2 \doteq \text{GMAT})$ for recent applicants have been categorized into three groups: C_1 – admit; C_2 – do not admit; and C_3 – borderline. The data are pictured in Figure 4.10. A decision will ultimately be to admit or not to admit a student and clustering the data into two groups would appear natural. The borderline group was presumably introduced to make the process fairer since there may

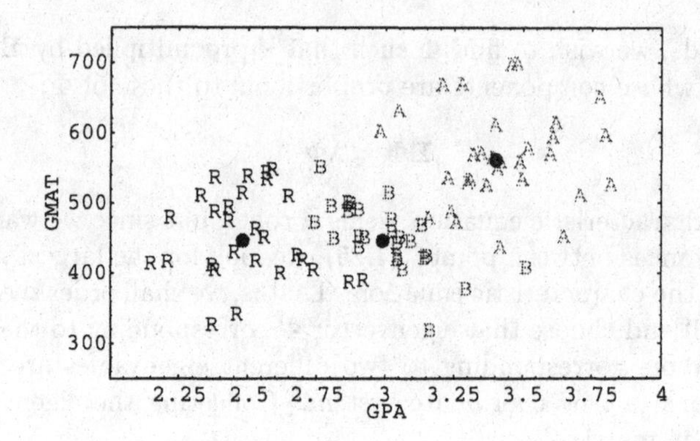

Fig. 4.10 Scatter plot of (x_1 = GPA, x_2 = GMAT) for applicants to a graduate school who have been classified as admit (A), reject application (R) and borderline (B).

be cases where two applicants with very similar scores may be classified into distinct groups.

Suppose a new applicant has an undergraduate GPA of $x_1 = 3.21$ and a GMAT score of $x_2 = 497$. Let us classify this applicant using the following allocatory rule with equal prior probabilities:

Assign $\mathbf{x}$ to the population C_i for which $-\frac{1}{2}d^2_{\Sigma_{\text{pooled}}}(\mathbf{x}, \mathbf{c}_i) + \ln p_i$ is largest.

In this rule the p_i are prior probabilities which, if unknown, are usually set to $p_1 = p_2 = \cdots = p_c = 1/c$. An observation is then assigned to the closest population where the distance of $\mathbf{x}$ to the sample mean vector $\mathbf{c}_i$ is calculated as

$$d^2_{\text{pooled}}(\mathbf{x}, \mathbf{c}_1) = (\mathbf{x} - \mathbf{c}_i)^T \, \mathbf{\Sigma}^{-1}_{\text{pooled}}(\mathbf{x} - \mathbf{c}_i) \, .$$

Matrix $\mathbf{\Sigma}$ is the *pooled* estimate of the covariance matrix

$$\mathbf{\Sigma}_{\text{pooled}} = \frac{1}{d_1 + d_2 + \cdots + d_c}\left((d_1 - 1)\mathbf{\Sigma}_1 + (d_2 - 1)\mathbf{\Sigma}_2 + \cdots + (d_c - 1)\mathbf{\Sigma}_c\right),$$

where d_i denotes the sample size and $\mathbf{\Sigma}_i$ the sample covariance matrix for population C_i. Obviously these values can only be calculated if a set of *labelled* training data exists, that is, a training set of correctly classified observations. We also note that for the probabilistic model, the populations do not actually exist, but are thought to be created artificially.

Using a probabilistic framework enables us to quantify the quality of our classification in terms of probabilities. However, as seen in this example we are required to make use of the following tricks and assumptions:

- Populations may be created artificially and assumed to be normal.

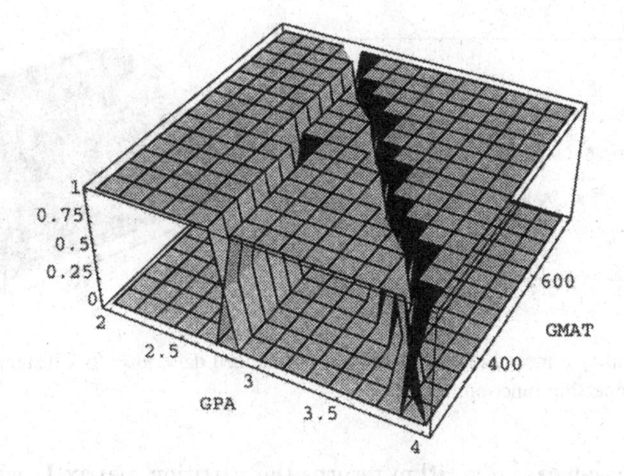

Fig. 4.11 Result of discriminant analysis for applicants to a graduate school who have been classified as admit (A), do not admit (R) and borderline (B).

- Prior probabilities need to be defined.
- A set of labelled training data is required.
- For data at the margins of a group an additional class is introduced.

For the training data shown in Figure 4.10 the following parameters were obtained using a statistical software package:

$$d_1 = 31 \qquad d_2 = 28 \qquad d_3 = 26$$

$$\mathbf{c}_1 = \begin{bmatrix} 3.40 \\ 561.23 \end{bmatrix} \qquad \mathbf{c}_2 = \begin{bmatrix} 2.48 \\ 447.07 \end{bmatrix} \qquad \mathbf{c}_3 = \begin{bmatrix} 2.99 \\ 446.23 \end{bmatrix}$$

$$\mathbf{c} = \begin{bmatrix} 2.97 \\ 488.45 \end{bmatrix} \qquad \Sigma_{\text{pooled}} = \begin{bmatrix} 0.0361 & -2.0188 \\ -2.0188 & 3655.9011 \end{bmatrix}$$

With $\mathbf{x} = [3.21, 497]^T$, the sample distances are calculated as

$$d^2_{\text{pooled}}(\mathbf{x}, \mathbf{c}_1) = 2.58 \quad d^2_{\text{pooled}}(\mathbf{x}, \mathbf{c}_2) = 17.10 \quad d^2_{\text{pooled}}(\mathbf{x}, \mathbf{c}_3) = 2.47 \,.$$

Since the distance from $\mathbf{x} = [3.21, 497]^T$ to the class mean $\mathbf{c}_3$ is smallest, the applicant is assigned to C_3, borderline. Figure 4.11 shows the *decision surface* for the whole space.

We now contrast the discrimination analysis and its result using fuzzy-c-means clustering. First, we consider three clusters, $c = 3$, as in the given example and set $w = 2$. After 14 iterations the fuzzy-c-means algorithm determined, from *unlabelled* data, the cluster centers shown on the left in Figure 4.12.

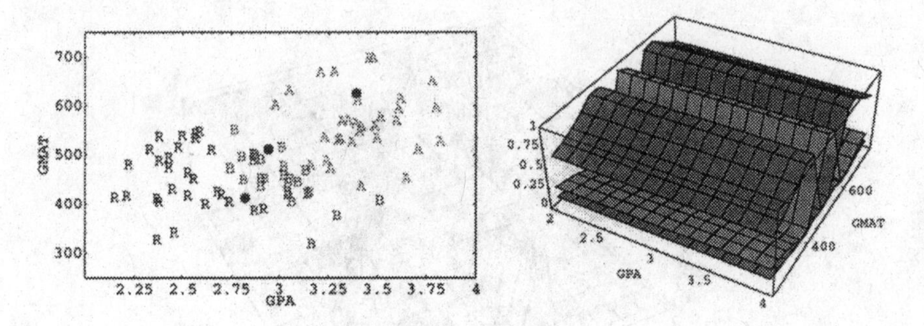

Fig. 4.12 Fuzzy-c-means clustering on non-normalized data, $w = 2$. Cluster centers (left) and class membership functions (right).

The fuzzy-c-means algorithm returns the partition matrix $\mathbf{U}$ which serves as the model for our classifier. The final cluster prototypes are obtained as

$$\mathbf{c}_i = \frac{\sum_{j=1}^{85} (u_{ij})^w \, \mathbf{m}_j}{\sum_{j=1}^{85} (u_{ij})^w}, \quad i = 1, 2 .$$

For any new applicant, from the pair $\mathbf{x} = [x_1 = \text{GPA}, x_2 = \text{GMAT}]^T$, the membership in one of the two classes is then calculated by

$$\mu_{C_i}(\mathbf{x}) \doteq \frac{1}{\left(\frac{d(\mathbf{x},\mathbf{c}_i)}{d(\mathbf{x},\mathbf{c}_1)}\right)^{\frac{2}{w-1}} + \left(\frac{d(\mathbf{x},\mathbf{c}_i)}{d(\mathbf{x},\mathbf{c}_2)}\right)^{\frac{2}{w-1}}} . \tag{4.28}$$

As can be seen in Figure 4.12, the cluster center for the 'reject' class is displaced leading to class membership functions which do not take account of the GPA scores. Distance measures are sensitive to variations in the numerical ranges. In our example, differences in GMAT scores can be around 200 while GPA score differences can be around 0.75. Since the membership evaluation in the iterative algorithm and the decision, (4.28), is based on a distance measure, the algorithm can generate misleading results. Clustering algorithms based on adaptive distance measures, introduced in Section 5, are less sensitive to data scaling. The normalization or scaling of data can help. If, in our example, we divide all values by the maximum value in the data set, the fuzzy-c-means algorithm identifies the three classes correctly, as shown in Figure 4.13.

In the design of the statistical classifier, it was recognized that two classes would lead to unfairness for the students near to the margins of any class. A 'borderline' class was introduced. Since a student will either be admitted or not, a further decision is required on the basis of the 'distance' to the two

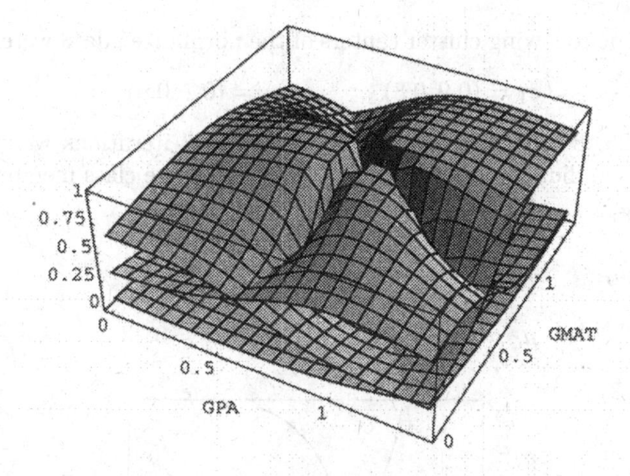

Fig. 4.13 Fuzzy-c-means clustering on normalized data. $w = 1.25$, 17 iterations.

decision regions ('accept', 'reject'). Let us now consider fuzzy clustering for only two classes, that is, $c = 2$. We use *unlabelled* but normalized data to classify applicants into two groups C_1 and C_2: admit and do not admit. The general rule applied is that if the GPA is low and the GMAT is low, then the student is likely to perform poorly and should be rejected. On the other hand, an applicant with a high GPA and a high GMAT value should be admitted. The data matrix is created from 85 applicants consisting of two rows and $d = 85$ columns:

$$\mathbf{M} = \begin{bmatrix} \cdots & \text{GPA} & \cdots \\ \cdots & \text{GMAT} & \cdots \end{bmatrix}.$$

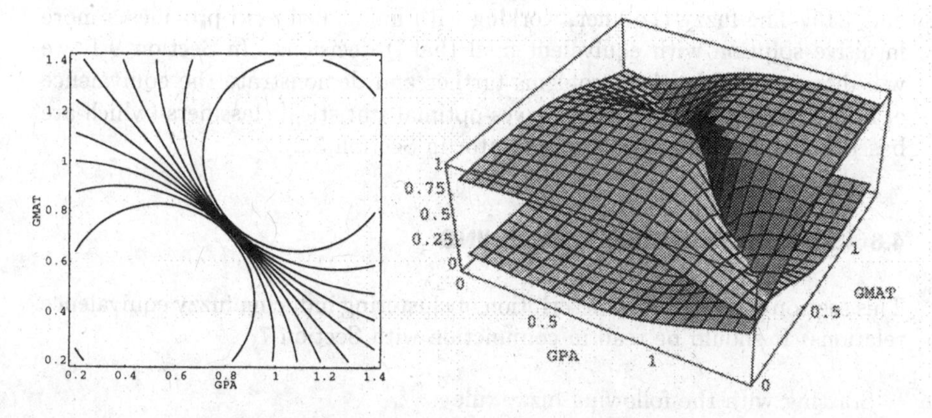

Fig. 4.14 Fuzzy-c-means classification for $w = 1.25$.

The weighting factor w reflects the fuzziness in the decision making (student most probably would refer to w as the (un)fairness factor). For both $w = 1.25$

and $w = 2$, the following cluster centers of the normalized data were obtained:

$$\mathbf{c}_1 = (0.9, 0.8) , \qquad \mathbf{c}_2 = (0.7, 0.6) .$$

For $w = 1.25$, 8 iterations and for $w = 2$, only 7 iterations were required. For the test candidate with scores, $\mathbf{x} = (3.21, 497)$, the class memberships for $w = 1.25$ are

$$\mu_{C_1}(\mathbf{x}) = 0.73 , \qquad \mu_{C_2}(\mathbf{x}) = 0.27$$

and for $w = 2$,

$$\mu_{C_1}(\mathbf{x}) = 0.67 , \qquad \mu_{C_2}(\mathbf{x}) = 0.33 .$$

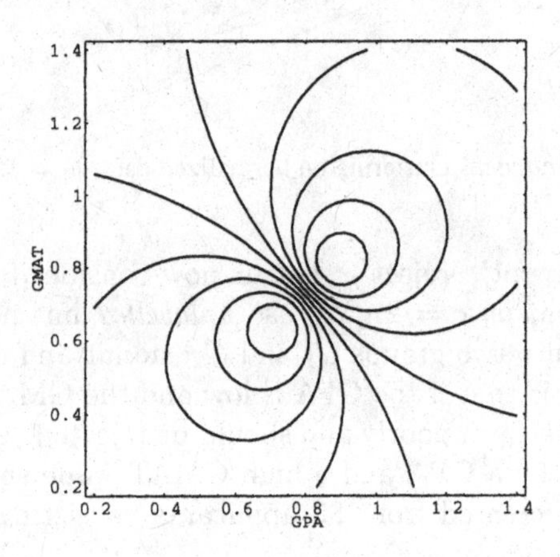

Fig. 4.15 Fuzzy-c-means classification for $w = 2$.

The decision surface and density contour plots are shown in Figures 4.14 and 4.15. The fuzzy classifier, working with unlabelled data provides a more intuitive solution with equivalent final (hard) decisions. In Section 9.1, we will discuss classification problems further and demonstrate the equivalence of fuzzy inference engines with Bayes-optimal statistical classifiers (which are based upon the kernel density estimator in Section 3.2).

4.6 SIMILARITY-BASED REASONING

This section follows up the description of clustering inducing fuzzy equivalence relations. It should be read in conjunction with Section 7.

Starting with the following fuzzy rule

$$R_i : \quad \text{IF } x_1 \text{ is } A_{i1} \text{ AND } x_2 \text{ is } A_{i2} \ldots \text{AND } x_r \text{ is } A_{ir}, \text{ THEN } y \text{ is } B_i \ . \quad (5.6)$$

In [KGK94], fuzzy modelling is described in terms of similarity relations in the sense that the fuzzy sets A_{ik}, B_i of model (5.6) correspond to the extensional hulls, (4.20), of single points $x_{k_0}^{(i)}$ and $y_0^{(i)}$, $k = 1, \ldots, r$. That is, $\mu_{x_{k_0}}^{(i)}(x_k) = \widetilde{E}_k^{(i)}(x_k, x_{k_0}^{(i)})$. The fuzzy rule can then be interpreted as a vague representation of the point $(x_{1_0}^{(i)}, \ldots, x_{r_0}^{(i)}, y_0^{(i)})$, $i = 1, 2, \ldots, n_R$, in $X_1 \times X_2 \times \cdots \times X_r \times Y$. For each rule, a multi-variate similarity relation on the product space is constructed by means of the similarity relations $\widetilde{E}_k$ defined on the subspaces (or vice versa). If, for instance, we require the similarity relation to be the coarsest similarity relation on the product space such that it does distinguish at least as well as each $\widetilde{E}_k$, for the antecedent variables on $X_1 \times \cdots \times X_r$, we have

$$\widetilde{E}_{\min\{\widetilde{E}_1, \ldots, \widetilde{E}_r\}} = \min\left\{\widetilde{E}_k(x_k, x_k')\right\}, \quad k = 1, 2, \ldots, r \,. \tag{4.29}$$

From the definition of the extensional hull of a single point, (4.20), the extensional hull of a set M_0 can be defined as the fuzzy set

$$\mu_{M_0}(\mathbf{o}) = \sup_{\mathbf{o}_0 \in M_0} \left\{\widetilde{E}(\mathbf{o}_0, \mathbf{o})\right\} . \tag{4.30}$$

μ_{M_0} is interpreted as the fuzzy set of elements that are similar to at least one of the elements of M_0. We can now construct the extensional hull of the set of points

$$M_0 = \left\{\left(x_{1_0}^{(i)}, \ldots, x_{r_0}^{(i)}, y_0^{(i)}\right)\right\}, \quad i = 1, 2, \ldots, n_R \,. \tag{4.31}$$

Combining equations (4.29) and (4.30), we have

$$\begin{aligned}
\mu_{M_0}(x_1, .., x_r, y) &= \max_i \left\{ \widetilde{E}^{(i)}_{\min\{\widetilde{E}_1, .., \widetilde{E}_r, \widetilde{E}\}} \left((x_1, .., x_r, y),\ (x_{1_0}^{(i)}, .., x_{r_0}^{(i)}, y_0^{(i)}) \right) \right\} \\
&= \max_i \left\{ \min \left\{ \min_{k=1,..,r} \left\{ \widetilde{E}_k(x_k, x_{k_0}^{(i)}),\ \widetilde{E}(y, y_0^{(i)}) \right\} \right\} \right\} \\
&= \max_i \left\{ \min \left\{ \min_{k=1,\ldots,r} \left\{ \mu_{x_{k_0}^{(i)}}(x_k) \right\},\ \mu_{y_0^{(i)}}(y) \right\} \right\} .
\end{aligned} \tag{4.32}$$

For a given tuple $(x_1, \ldots, x_r)$, for each possible response y the degree to which the tuple $(x_1, \ldots, x_r, y)$ is similar to at least one of the points in M_0, is computed by equation (4.32) replacing the sup by max. M_0 may be seen as (proto)typical points of the system model. The data space $\Xi = X_1 \times \cdots \times X_r \times Y$ is partitioned by the fuzzy sets induced by the typical points. The graph (1.8) of system $\mathfrak{S}$, is a fuzzy subset of Ξ, that is, it is the fuzzy graph

$$\widetilde{F} \doteq \left\{ \left((x_1, .., x_r), \bigvee_i \widetilde{E}^{(i)}_{\min\{\widetilde{E}_1, .., \widetilde{E}_r, \widetilde{E}\}} \left((x_1, .., x_r, y),\ (x_{1_0}^{(i)}, .., x_{r_0}^{(i)}, y_0^{(i)}) \right) \right) \right\} . \tag{4.33}$$

In [KK97], the set of typical points M_0, also referred to as a *partial mapping*, is identified with the vector of cluster prototypes $\mathbf{c}^{(i)}$, (4.17). Hence, the extensional fuzzy set A_i of Ξ is a *fuzzy point* since there exists an $\mathbf{c}^{(i)} \in \Xi$

such that the membership function of A_i equals the extensional hull of the (crisp) point $\{\mathbf{c}^{(i)}\}$ with

$$\mu_{\mathbf{c}^{(i)}}(\mathbf{o}) = \widetilde{E}\left(\mathbf{c}^{(i)}, \mathbf{o}\right) \tag{4.34}$$

and $\mu_{\mathbf{c}^{(i)}}(\mathbf{o})$ is ***normalized***, that is, there exists an $\mathbf{o}$ for which $\mu_{\mathbf{c}^{(i)}}(\mathbf{o}) = 1$. The definition of a fuzzy point implies

$$T\left(\mu_{A_i}(\mathbf{o}), \mu_{A_i}(\mathbf{o}')\right) \leq \widetilde{E}\left(\mathbf{o}, \mathbf{o}'\right) \ , \tag{4.35}$$

which is a multi-valued formulation of the statement "IF $\mathbf{o}$ is an element of A_i AND $\mathbf{o}'$ is an element of A_i, THEN $\mathbf{o}$ is similar to $\mathbf{o}'$". Then, if A_i is a fuzzy cluster in Ξ, we can use (4.35) to calculate $\widetilde{E}(\mathbf{o}, \mathbf{o}')$ for any $(\mathbf{o}, \mathbf{o}') \in \Xi$. The process of deriving an output fuzzy set or single output value in Y, given a fuzzy graph $\widetilde{F}$ and inputs, is based on the compositional rule of inference described on p. 158.

An open problem is how to extract some fuzzy logic rule-based system by exploiting the lattice structure of the quotient of Ξ by $\widetilde{E}$ defined by $\mathbf{U}$ as discussed in Section 4.7. Sections 7 and 8 describe various other fuzzy systems similar to (4.32) as part of an introduction to approximate reasoning. See also Section 11 for more details on fuzzy mappings and graphs.

4.7 THE STRUCTURE OF THE QUOTIENT INDUCED BY SIMILARITY RELATIONS

The fuzzy quotient induced by sampled data is a fuzzy partition of Ξ with respect to a similarity relation $\widetilde{E}$. By exploring the lattice structure of $\Xi/\widetilde{E}$ we aim to describe clusters as models of sets with respect to certain non-classical (fuzzy) logics, that is, maps with respect to T satisfying a list of logical axioms (cf. [FR94, NW97]). Fuzzy logic [NW97] means the extension of ordinary logic where the truth values consist of the two-element Boolean algebra to the case where they consist of the DeMorgan algebra $([0,1], \vee, \wedge, \neg, 0, 1)$ (*i.e.* a bounded distributive lattice where DeMorgan's laws hold). Such a reasoning system uses the connectives *and*, *or*, *not*, and *if-then*. Flexibility in the proposed methodology is ensured by a relatively large class of connectives forming the basis of a propositional logical system.

Höhle solved the quotient problem with reference to the ultra-metric approach to hierarchical clustering[6], via the definition of the category M-SET

[6]An overview of hierarchical clustering can be found in Appendix 13.3.

[Höh88]. Starting with an *integral commutative complete lattice monoid*[7]

$$M = (L, *, \preceq) \tag{4.36}$$

he defines *M-similarity* with the M-valued, symmetric, reflexive, fuzzy equivalence relation (4.19)

$$\widetilde{E}\colon \Xi \times \Xi \to L\ , \tag{4.37}$$

where transitivity is generalized by, cf. (4.24)

$$\widetilde{E}(\mathbf{o}_i, \mathbf{o}_k) \geq \widetilde{E}(\mathbf{o}_i, \mathbf{o}_j) * \widetilde{E}(\mathbf{o}_j, \mathbf{o}_k)\ . \tag{4.38}$$

As before, M is viewed as a set of truth values. If the partially ordered set $(L, \preceq)$ is a lattice, then it comes equipped with the two binary operations $\wedge$ and $\vee$ for every pair of elements in L:

$\alpha \vee \alpha' = \sup_L \{\alpha, \alpha'\}$: join (supremum - smallest upper bound of $\{\alpha, \alpha'\}$)

$\alpha \wedge \alpha' = \inf_L \{\alpha, \alpha'\}$: meet (infinimum - greatest lower bound of $\{\alpha, \alpha'\}$) .

The lattice is complete if every subset of L has a sup. If a lattice has an identity (*i.e.* it has a largest and smallest element) for $\wedge$ and $\vee$, then it is a bounded lattice. A lattice satisfying both distributive laws is called distributive. The lattice $([0,1], \preceq)$, used in Sections 4 and 7, is a bounded distributive lattice which however is not complemented, (*i.e.* there does not exist for every element in $[0,1]$ a complement). It is therefore not a Boolean lattice (Boolean algebra). For the unit-interval equipped with a t-norm and partial ordering $\preceq$, in case of $T_{\min}$, clusters are set structures with respect to the *intuistic logic* whereas for T_{Luk} clusters are models of sets with respect to the *Lukasiewicz logic*. On the other hand, (4.36) is also a *complete Heyting algebra* where $*$ is the meet $\wedge$.

Höhle showed that there exists a bijective mapping between the set of all M-similarities on Ξ and all *partition trees*, that is,

$$\mathcal{E}_{\widetilde{E}}(\alpha) = \text{partition induced by equiv. rel. } E_\alpha = \{(\mathbf{o}, \mathbf{o}')\ :\ E(\mathbf{o}, \mathbf{o}') \geq \alpha\}\ , \tag{4.39}$$

where $\alpha \in L$. $\widetilde{E}_\alpha \subseteq \Xi \times \Xi$ contains all pairs of points that are α-indistinguishable. For instance, from (7.6), $\widetilde{E}_\alpha = \{(\mathbf{o}, \mathbf{o}') \in \Xi \times \Xi\colon \|\mathbf{o} - \mathbf{o}'\| \leq \alpha\}$. As

[7] A ***groupoid*** is any abstract system consisting of a set on which a closed operation has been defined. If we throw away the group axiom requiring every element to have an inverse, we get a more general structure called a semi-group or monoid, *i.e.* $(L, \preceq)$ is a complete lattice, $(L, *)$ is a commutative monoid and $(L, *, \preceq)$ is integral iff the upper bound is unity w.r.t. $*$. A ***monoid*** is a semi-group with an identity element. A ***semi-group*** is an associative groupoid.

demonstrated in Section 7, E_α is not transitive for $\alpha \neq 1$, that is, E_α is not an equivalence relation.

If $\widetilde{E}(\mathbf{o}, \mathbf{o}')$ is interpreted as the degree of indistinguishability between $\mathbf{o}$ and $\mathbf{o}'$ in terms of an error or tolerance level α, a cluster with respect to a given $\widetilde{E}$ on Ξ is a subset of Ξ containing all those elements of Ξ whose similarity is sufficiently large. Using the category M-SET, Höhle defines the quotient w.r.t. $\widetilde{E}$ to be the object $(\Xi, E_\alpha)/\widetilde{E}$. With the definition of an equivalence class as the map

$$\xi_{E_\alpha} : \Xi \to \{0, 1\} \tag{4.40}$$

the relation

$$\widetilde{E}(\mathbf{o}, \mathbf{o}') = \sup_{\text{all clusters w.r.t. } \xi_{E_\alpha}} \xi(\mathbf{o}) * \xi(\mathbf{o}') \qquad \forall\, \mathbf{o}, \mathbf{o}' \in \Xi, \tag{4.41}$$

shows that $\widetilde{E} \to \{\text{clusters w.r.t. } \widetilde{E}\}$ is an injective map, that is, the 'chaining effect' in conventional hierarchical clustering does not occur.

An open question is how similar results can be obtained for partitional clustering; given fuzzy partition matrix $\mathbf{U}$. The investigation into the lattice structure of the fuzzy quotient, obtained from fuzzy clustering, would provide a rigorous basis for the engineering practice applying approximate reasoning and fuzzy modelling. The mathematical fusion of NARX modelling with a propositional fuzzy logic and fuzzy clustering would have a wide range of applications. For example, in time-series and multi-variate data analysis a propositional logical system would allow an interface of qualitative expert knowledge with the quantitative analysis.

To develop a propositional calculus based on fuzzy clustering in a multi-dimensional state-space a number of open problems need to be solved. By exploring the lattice structure of the quotient set, a (fuzzy) logic should form the basis for approximate reasoning on subsets of the product space used to represent a dynamic system. The main criticism of a propositional calculus, founded in bivalent logic, centers around the realism of modelling causation, that is, two-valued logical correlates of causal connections. In case of material implication, associating antecedent p with a cause and the consequent c with an effect, an absent cause may have an effect; things may be self-caused. To model causal links, relations between p and c are required to be non-reflexive (nothing is self-caused) and asymmetrical (irreversibility). Based on this requirement, conventional logic is inadequate for modelling cause-effect links. However, recent studies of fuzzy relations [FR94] offer a relational approach that provide appropriate mappings that are irreflexive, transitive (chain-like) and asymmetrical. At the same time the field of conditional event algebras [GMN97] advances formal links between logical and probabilistic descriptions and, in principle, it should therefore become possible to analyze dynamic systems in a measure-free, that is, (fuzzy) logical or (fuzzy) relational framework.

5

Fuzzy Systems and Identification

- ☐ *Fuzzy clustering provides an effective way to identify fuzzy rule-based models from data.*
- ☐ *Various fuzzy model structures exist and can be distinguished in terms of their simplicity, interpretability and suitability in diverse problems such as classification, prediction and control.*
- ☐ *Fuzzy systems can be shown to be equivalent to the basis function expansion model.*

> As the complexity of a system increases, our ability to make precise and yet significant statements about its behavior diminishes until a threshold is reached beyond which precision and significance (or relevance) become almost exclusive characteristics.
>
> —Lotfi Zadeh

In this section we follow up on the results of Section 4 to see how the fuzzy partition matrix $\mathbf{U}$ (4.18), can be used to describe *if-then* rule-based models, which were suggested in Section 3.5. The basic principle of identification by product space clustering is to approximate a nonlinear regression problem by decomposing it into several local linear subproblems described by *if-then* rules. The methods discussed here have found numerous applications in engineering. A comprehensive discussion, including the application in control, was provided by Babuska [Bab98].

Given a fixed number of clusters c, the weighting coefficient m, and an initial partition matrix, the fuzzy-c-means algorithm 4.1 returns, after some iterations, fuzzy partition matrix $\mathbf{U}$, describing the membership of the data samples in the clusters. Each cluster is characterized by its center and covariance matrix which represents the variance of the data in the cluster. The eigenstructure of the *cluster covariance matrix* $\mathbf{F}^{(i)}$, contains information about the shape and orientation of the ith cluster

$$\mathbf{F}^{(i)} = \frac{\sum_{j=1}^{d} (u_{ij})^{w} (\mathbf{m}_j - \mathbf{c}^{(i)})(\mathbf{m}_j - \mathbf{c}^{(i)})^T}{\sum_{j=1}^{d} (u_{ij})^{w}} . \tag{5.1}$$

Let λ_{ik} denote the kth eigenvalue of $\mathbf{F}^{(i)}$ and Φ_{ik} the kth unit eigenvector of $\mathbf{F}^{(i)}$. Let the eigenvalues be arranged in decreasing order as $\lambda_{i1} \geq \lambda_{i2} \geq \cdots \geq \lambda_{in}$. Then the eigenvectors Φ_{i1} to $\Phi_{i(n-1)}$ span the ith cluster's linear subspace and the nth eigenvector Φ_{in} is the normal to this linear subspace. The eigenvectors are labelled in the order of the eigenvalues, that is, if λ_{in} is the smallest eigenvalue, Φ_{in} is called smallest eigenvector. Whether the chosen model structure is correct can be evaluated using the proportions between the eigenvalues. Figure 5.1 illustrates an ellipsoid modelling the shape and orientation of a cluster by its eigenstructure.

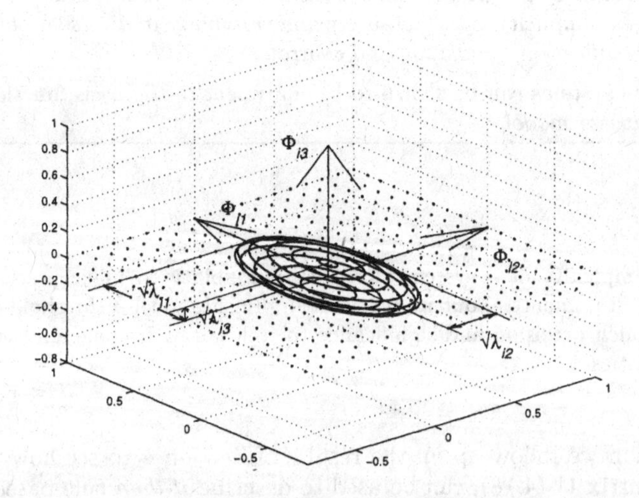

Fig. 5.1 Ellipsoid with eigenvectors and eigenvalues describing shape and orientation of a cluster.

For the basic fuzzy-c-means algorithm 4.1, where the distance d is simply the Euclidean norm, that is, $\mathbf{A}$ is the identity matrix, the fuzzy-c-means algorithm searches for spherical clusters of approximately the same size. By adapting the norm-inducing matrix $\mathbf{A}$ at each iteration using $\mathbf{F}^{(i)}$, ensures that the algorithm can detect hyperellipsoidal clusters of different shape and orientation. Then the equation $(\mathbf{o}-\mathbf{c})^T\mathbf{F}^{-1}(\mathbf{o}-\mathbf{c}) = 1$ defines a hyperellipsoid. The length of the kth axis of this hyperellipsoid is given by $\sqrt{\lambda_k}$ and its direction is spanned by Φ_k. A well-known implementation of this idea is the Gustafson-Kessel (GK) algorithm 5.1 [Bab98, Bez81]. For the GK-algorithm, each cluster has its own norm-inducing matrix $\mathbf{A}^{(i)}$. From (4.9), denote the inner-product norm by

$$d^2_{\mathbf{A}^{(i)}} = \left(\mathbf{c}_l^{(i)} - \mathbf{m}_j\right)^T \mathbf{A}^{(i)} \left(\mathbf{c}_l^{(i)} - \mathbf{m}_j\right) . \tag{5.2}$$

In order to be able to apply (4.16) and (4.17) iteratively with varying $\mathbf{A}^{(i)}$, its determinant (the cluster volume) is fixed, say $|\mathbf{A}^{(i)}| = 1$. To get a matrix $\mathbf{A}^{(i)\prime}$ from $\mathbf{A}^{(i)}$ with $|\mathbf{A}^{(i)}| = 1$, we have $\mathbf{A}^{(i)\prime} = (1/|\mathbf{A}^{(i)}|)^{1/(r+1)} \cdot \mathbf{A}^{(i)}$. Using the inverse of the covariance matrix, the $\mathbf{A}^{(i)}$ in equation (5.2) is calculated as

$$\mathbf{A}^{(i)} \doteq \left(|\mathbf{F}^{(i)}|\right)^{1/(r+1)} \cdot \left(\mathbf{F}^{(i)}\right)^{-1} . \tag{5.3}$$

In Figure 5.2, the GK-algorithm is applied to a data of a nonlinear first-order autoregressive process, $y(k) = f(\mathbf{x}; k)$, where $\mathbf{x} = [y(k-1)]$. For a first-order model, the data space Ξ is two-dimensional with $y(k)$ on the abscissa, values $y(k-1)$ on the ordinate and the ellipsoids, representing linear submodels, obtained by applying algorithm 5.1. In this example, the size of the cluster is about the same. In general, from the condition $|\mathbf{A}^{(i)}| = 1$ the GK-algorithm looks for clusters of equal volume. Hence, we have made an improvement to the FCM by considering shape and orientation but would require yet another refinement of the algorithm to take account of the cluster volume.

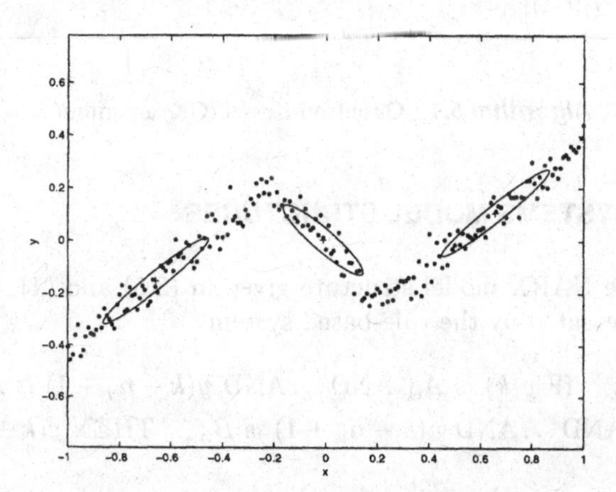

Fig. 5.2 First-order nonlinear autoregressive process and linear submodels identified by the GK-algorithm.

Fix c, $2 \le c < d$, and choose the termination tolerance $\delta > 0$, e.g., between 0.01 and 0.001, and fix w, $1 \le w < \infty$, e.g., 2. Initialize $\mathbf{U}^{(0)} \in M_{fc}$, (e.g., randomly).

Repeat for $l = 1, 2, \ldots$:

Step 1: Compute cluster prototypes (means) (4.17):

$$\mathbf{c}_l^{(i)} = \frac{\sum_{j=1}^{d} \left(u_{ij}^{(l-1)}\right)^w \mathbf{m}_j}{\sum_{j=1}^{d} \left(u_{ij}^{(l-1)}\right)^w} \quad 1 \le i \le c .$$

Step 2: Compute the cluster covariance matrices (5.1):

$$\mathbf{F}^{(i)} = \frac{\sum_{j=1}^{d} \left(u_{ij}^{(l-1)}\right)^w \left(\mathbf{m}_j - \mathbf{c}_l^{(i)}\right)\left(\mathbf{m}_j - \mathbf{c}_l^{(i)}\right)^T}{\sum_{j=1}^{d} \left(u_{ij}^{(l-1)}\right)^w} .$$

Step 3: Compute distances for $1 \le i \le c$ and $1 \le j \le d$:

$$d^2_{\mathbf{F}^{(i)}}\left(\mathbf{c}_l^{(i)}, \mathbf{m}_j\right) = \left(\mathbf{c}_l^{(i)} - \mathbf{m}_j\right)^T \left[|\mathbf{F}^{(i)}|^{\frac{1}{(r+1)}} \cdot \left(\mathbf{F}^{(i)}\right)^{-1}\right] \left(\mathbf{c}_l^{(i)} - \mathbf{m}_j\right)$$

Step 4: Update the partition matrix using (4.16):
If $d_{\mathbf{F}^{(i)}} > 0$ for $1 \le i \le c$, $1 \le j \le d$,

$$u_{ij}^{(l)} = \frac{1}{\sum_{k=1}^{c} \left(d_{\mathbf{F}^{(i)}} / d_{\mathbf{F}^{(i)}}\right)^{2/(w-1)}} ;$$

otherwise

$$u_{ij}^{(l)} = 0 \text{ if } d_{\mathbf{F}^{(i)}}\left(\mathbf{c}^{(j)}, \mathbf{m}_j\right) > 0, \text{ and } u_{ij}^{(l)} \in [0, 1] \text{ with } \sum_{i=1}^{c} u_{ij}^{(l)} = 1 .$$

Until $\|\mathbf{U}^{(l)} - \mathbf{U}^{(l-1)}\| < \delta$.

Algorithm 5.1 Gustafson-Kessel (GK) algorithm.

5.1 FUZZY SYSTEMS MODEL STRUCTURES

Considering the NARX model structure given in (3.5) and (4), the mapping $X \to Y$ is represented by the rule-based system

$$R_i: \quad \text{IF } y(k) \text{ is } A_{i1} \text{ AND} \ldots \text{AND } y(k - n_y + 1) \text{ is } A_{in_y} \text{ AND}$$
$$u(k) \text{ is } B_{i1} \text{ AND} \ldots \text{AND } u(k - n_u + 1) \text{ is } B_{in_u}, \text{ THEN } y(k+1) \text{ is } C_i , \tag{5.4}$$

where the conjunctive operator 'AND' is represented by a t-norm and we wish to find the fuzzy sets A_{ik}, B_{ik} and C_i from the fuzzy partition matrix $\mathbf{U}$. This can be achieved by pointwise *projection* of the multi-dimensional fuzzy sets, defined pointwise, in the rows of partition matrix $\mathbf{U}$, onto the subspaces referring to individual variables in the antecedent and/or consequent part of the rule. The *projection* of the partition matrix onto the regressors should result in a semantically interpretable partition with unimodal fuzzy sets.

In general, we have the following rule-based structure of the *Linguistic Model*:

$$R_i : \quad \text{IF } \mathbf{x} \text{ is } A_i, \text{ THEN } y \text{ is } B_i, \qquad i = 1, 2, \ldots, n_R \;, \tag{5.5}$$

where $\mathbf{x}$ is the *antecedent variable* and y the *consequent variable*. Note that in general antecedent and consequent variables can be *fuzzy variables* which have fuzzy sets as their values, $\mathbf{x} \in \mathcal{F}(X)$, $y \in \mathcal{F}(Y)$. For $\mathbf{x} \in \mathbb{R}^r$, A_i is a fuzzy set defined by a multi-variate membership function $\mu_{A_i}(\mathbf{x})\colon X_1 \times \cdots \times X_r \to [0,1]$. A_i and B_i may then be viewed as *fuzzy restrictions* on X and Y as they partition the input and output space. In order to achieve a semantically interpretable model, it is desirable to have fuzzy sets associated with each of the regressors, leading to the *conjunctive form*

$$R_i : \quad \text{IF } x_1 \text{ is } A_{i1} \text{ AND } x_2 \text{ is } A_{i2} \ldots \text{AND } x_r \text{ is } A_{ir}, \text{ THEN } y \text{ is } B_i \tag{5.6}$$

with the *degree of fulfillment* β_i of the rule given by the conjunction

$$\beta_i(\mathbf{x}) \doteq \mu_{A_{i1}}(x_1) \wedge \mu_{A_{i2}}(x_2) \wedge \cdots \wedge \mu_{A_{ir}}(x_r) \;, \tag{5.7}$$

where $\wedge$ is a suitable t-norm. As illustrated in Figure 5.3, for $r = 2$, we can obtain fuzzy subsets A_{i1} and A_{i2} for the conjunctive form from projections of the two-dimensional fuzzy set A_i.

In the *Takagi-Sugeno (TS) Model*, the rule consequents are functions of the model inputs:

$$R_i : \quad \text{IF } \mathbf{x} \text{ is } A_i, \text{ THEN } y_i = f_i(\mathbf{x}), \qquad i = 1, 2, \ldots, n_R \;. \tag{5.8}$$

As before, the antecedent proposition "$\mathbf{x}$ is A_i", defined by the multi-variate membership function $\mu_{A_i}(\mathbf{x})\colon \mathbb{R}^r \to [0,1]$, can be decomposed into a conjunctive combination of regressor variables and univariate fuzzy sets:

$$R_i : \quad \text{IF } x_1 \text{ is } A_{i1} \text{ AND } x_2 \text{ is } A_{i2} \ldots \text{AND } x_r \text{ is } A_{ir}, \text{ THEN } y_i = f_i(\mathbf{x}) \;. \tag{5.9}$$

A commonly used parameterization of y leads to the *affine linear TS-model*:

$$y_i = \mathbf{a}_i^T \mathbf{x} + b_i \;, \tag{5.10}$$

where $\mathbf{a}_i$ is a parameter vector and b_i is a scalar offset. The consequents of the affine TS-model are hyperplanes (r-dimensional linear subspaces) in $\mathbb{R}^{r+1}$, whereas the if-part of the rule partitions the input space and determines the validity of the n_R locally linear models for different regions of the antecedent space.

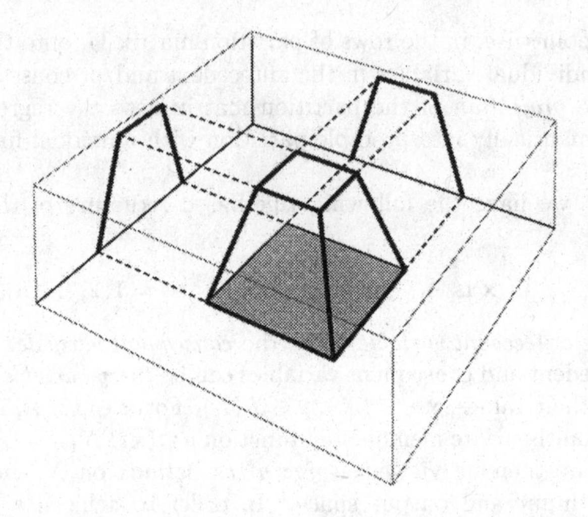

Fig. 5.3 Two-dimensional fuzzy set A_i and its projections A_{i1} and A_{i2}.

5.2 IDENTIFICATION OF ANTECEDENT FUZZY SETS

Antecedent membership functions for the 'Linguistic' and 'Takagi-Sugeno' models can be obtained by projection of the multi-dimensional fuzzy set defined pointwise in the rows of partition matrix **U** (4.18). Consider first the model structures given in (5.5) and (5.8), that is,

$$R_i : \quad \text{IF } \mathbf{x} \text{ is } A_i, \text{ THEN } \ldots, \qquad i = 1, 2, \ldots, n_R \, .$$

Considering one rule per cluster, $c = n_R$, for the partition matrix we use indices

$$u_{ij} \in \mathbf{U}, \qquad i = 1, 2, \ldots, c \, , \qquad j = 1, 2, \ldots, d \, .$$

The regressor vector (antecedent variables) is denoted by

$$\mathbf{x} \doteq [x_1, \ldots, x_k, \ldots, x_r]^T, \qquad k = 1, 2, \ldots, r \, .$$

The multi-dimensional fuzzy sets A_i define fuzzy regions in the antecedent space for which the ith consequent proposition is valid. In general, the membership function is defined by $\mu_{A_i}(\mathbf{x})$, but since **U** is finite, we have a pointwise definition for

$$\mathbf{x}_j \doteq [m_{1j}, \ldots, m_{kj}, \ldots, m_{rj}]^T \, , \tag{5.11}$$

vectors of sampled values for antecedent variables $x_1, \ldots, x_r$, $m_{kj} \in \mathbf{M}$ and **M** is a $(r+1) \times d$ matrix. The projection of the $(r+1)$-dimensional fuzzy set, defined by the ith row in **U**, onto the r-dimensional antecedent subspace is an r-dimensional fuzzy set A_i, defined pointwise for $\mathbf{x}_j$ and membership function

$$\forall \, (i, j) \qquad \mu_{A_i}(\mathbf{x}_j) = \max_{j'=1,\ldots,d} \{u_{ij'} \in \mathbf{U} : \; \mathbf{x}_{j'} = \mathbf{x}_j\} \, . \tag{5.12}$$

For model structures (5.6) and (5.9) in conjunctive form,

R_i : IF x_1 is A_{i1} AND ... AND x_k is A_{ik} AND ... AND x_r is A_{ir}, THEN...

the membership functions for each antecedent variable $x_k \in \mathbf{x}$ are obtained by projecting each of the c rows of the partition matrix onto each of the subspaces corresponding to the r antecedent variables; for all (i, j, k):

$$\mu_{A_{ik}}(x_{kj}) = \max_{m_{k'j'} \in \mathbf{M}} \{u_{ij'} \in \mathbf{U} \,:\, m_{kj} = m_{k'j'}, j' = 1, \ldots, d, k' = 1, \ldots, r\} \,. \tag{5.13}$$

In other words, $\mu_{A_{ik}}(x_{kj})$ is the projection of fuzzy cluster i onto the kth coordinate space, defined pointwise for all values m_{kj}. In order to extend this discrete fuzzy set to the set of real numbers, one usually computes the convex hull or fits a parametric function. Figure 5.4 gives an example for a pointwise projection onto a subspace and subsequent fitting of a parametric function. In general, the degree of membership for a value x to the kth projection of fuzzy cluster i, is the supremum over the membership degrees of all vectors $\mathbf{m}_j$ with x_k as the kth component to the fuzzy cluster i, that is,

$$\mu_{A_{ik}}(x) = \max \{u_{ij} \,:\, \mathbf{x}_j = (x_1, \ldots, x_k, \ldots, x_r) \in \mathbb{R}^r\} \,. \tag{5.14}$$

The axis-orthogonal projection onto the axes of the antecedent variables has the disadvantage that for clusters which are shaped and not orthogonal to the axis, the projection would imply a loss of information. A transformation of the antecedent variables, that is, the axis, by means of eigenvector projection, using the r largest eigenvectors of the cluster covariance matrices (5.1) would compensate for such problems. This however compromises the interpretability of the rule-based structure.

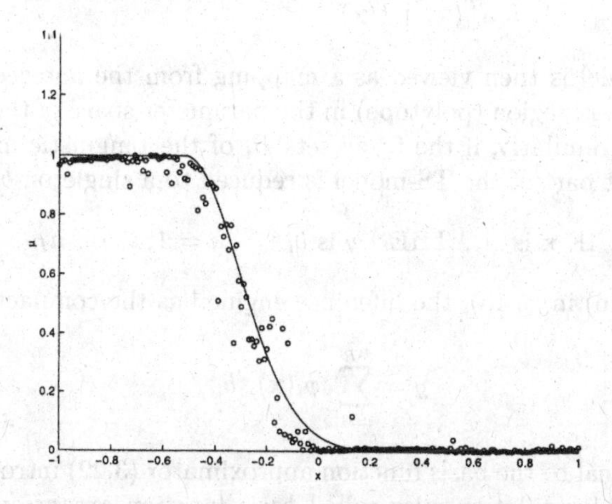

Fig. 5.4 Pointwise projection of a cluster and fitted parametric function.

5.3 PARAMETER IDENTIFICATION IN THE TAKAGI-SUGENO MODEL

Inference in fuzzy rule-based systems is the process of deriving an output fuzzy set from given rules (the 'knowledge base') and the inputs. The inference in the Linguistic model is based on the *compositional rule of inference* described in Section 7.2. In the Takagi-Sugeno fuzzy model, for a given input $\mathbf{x}$ the degree of fulfillment for the antecedent part of the rule is either calculated by $\beta_i = \mu_{A_i}(\mathbf{x})$, referring to model (5.8), or using (5.7) with respect to model (5.9). Then the inference is reduced to

$$y = \frac{\sum_{i=1}^{n_R} \beta_i(\mathbf{x}) \cdot (\mathbf{a}_i^T \mathbf{x} + b_i)}{\sum_{i=1}^{n_R} \beta_i(\mathbf{x})} . \tag{5.15}$$

An advantage of the TS-model structure is that it facilitates the analysis of stability and approximation capabilities in the framework of 'polytopic systems' [Bab98]. By denoting the normalized degree of fulfillment

$$\phi_i(\mathbf{x}) \doteq \frac{\beta_i(\mathbf{x})}{\sum_{k=1}^{n_R} \beta_k(\mathbf{x})} , \tag{5.16}$$

the affine Takagi-Sugeno model can be expressed as a quasi-linear model with input-dependent parameters:

$$\begin{aligned} y &= \left(\sum_{i=1}^{c} \phi_i(\mathbf{x}) \cdot \mathbf{a}_i^T \right) \mathbf{x} + \sum_{i=1}^{c} \phi_i(\mathbf{x}) \cdot b_i \\ &\doteq \mathbf{a}^T(\mathbf{x}) + b(\mathbf{x}) . \end{aligned} \tag{5.17}$$

The fuzzy model is then viewed as a mapping from the antecedent or input space to a convex region (polytope) in the parameter space of the quasi-linear model (5.17). Similarly, if the fuzzy sets B_i of the Linguistic model (5.5) or the consequent part of the TS-model is reduced to a singleton b_i, we have

$$R_i : \quad \text{IF } \mathbf{x} \text{ is } A_i, \text{ THEN } y \text{ is } b_i, \qquad i = 1, 2, \ldots, n_R . \tag{5.18}$$

Replacing (5.16) in (5.15), the inference engine has the compact form

$$y = \sum_{i=1}^{n_R} \phi_i(\mathbf{x}) \cdot b_i , \tag{5.19}$$

which is identical to the basis function approximator (3.22) introduced in Section 3.2! This formulation, also called *basis function expansion*, describes a general class of nonlinear function approximators including Radial Basis Function (RBF) networks, multi-variate adaptive regression splines and neurofuzzy spline networks. In the singleton model (5.18), the basis functions $\phi_i(\mathbf{x})$ are given by the normalized degrees of fulfillment of the rule antecedents.

What remains to complete the description of TS-models is a method to estimate parameters $\mathbf{a}_i$ and b_i for the affine model (5.10). Assuming each cluster represents a local linear model, the consequent parameters $\mathbf{a}_i$ and b_i can be identified from $i = 1, \ldots, c$ independent ordinary least-squares solutions. To separate data into their cluster, the membership degrees u_{ij} of the partition matrix $\mathbf{U}$, (4.18), are used as weights for each data pair $(\mathbf{x}_j, y_j)$. We arrange the identification data as in (2.27) with weighting matrix $\mathbf{W}_i$,

$$\mathbf{X} = \begin{bmatrix} \mathbf{x}_1^T \\ \mathbf{x}_2^T \\ \vdots \\ \mathbf{x}_d^T \end{bmatrix}, \qquad \mathbf{Y} = \begin{bmatrix} y_1 \\ y_2 \\ \vdots \\ y_d \end{bmatrix}, \qquad \mathbf{W}_i = \begin{bmatrix} u_{i1} & 0 & \cdots & 0 \\ 0 & u_{i2} & \cdots & 0 \\ \vdots & \vdots & \ddots & \vdots \\ 0 & 0 & \cdots & u_{id} \end{bmatrix}. \tag{5.20}$$

Hence, the consequent parameters for each rule are concatenated into a single parameter vector

$$\boldsymbol{\theta}_i = \left[\mathbf{a}_i^T, b_i\right]^T. \tag{5.21}$$

Appending a unitary column to $\mathbf{X}$, (2.27) gives the extended regressor matrix

$$\mathbf{X}_e = [\mathbf{X}, \mathbf{1}]. \tag{5.22}$$

We find the weighted least-squares solution of

$$\mathbf{Y} = \mathbf{X}_e \boldsymbol{\theta} + \varepsilon \tag{5.23}$$

by replacing $\mathbf{X}$ by $\mathbf{X}_e$ and $\mathbf{W}$ by $\mathbf{W}_i$ in (2.32):

$$\hat{\boldsymbol{\theta}}_i = \left[\mathbf{X}_e^T \mathbf{W}_i \mathbf{X}_e\right]^{-1} \mathbf{X}_e^T \mathbf{W}_i \mathbf{Y}. \tag{5.24}$$

The parameters $\mathbf{a}_i$ and b_i are combined in a single vector

$$\boldsymbol{\theta}_i = \left[\theta_{i1}, \theta_{i2}, \ldots, \theta_{ir}, \theta_{i(r+1)}\right], \tag{5.25}$$

where

$$\begin{aligned} \mathbf{a}_i &= [a_{i1}, a_{i2}, \ldots, a_{ir}] = [\theta_{i1}, \theta_{i2}, \ldots, \theta_{ir}] \\ b_i &= \theta_{r+1}. \end{aligned}$$

5.4 EXAMPLE: TS-MODELLING AND IDENTIFICATION

The following example is taken from one of the seminal papers that introduced TS-models [TS85].

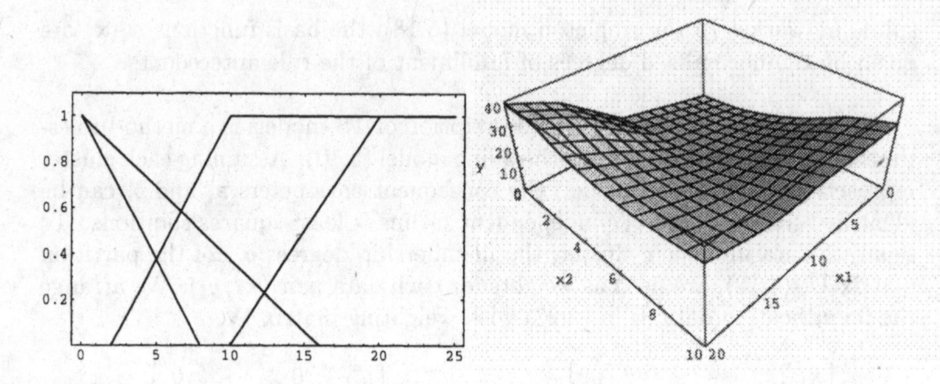

Fig. 5.5 Fuzzy sets and input-output relationship of the TS-model in the first example.

In TS-models, (5.9), the premise of an implication (IF-part of a rule) is the description of fuzzy spaces of inputs and its consequence is a linear input-output relation. Suppose that we have the following three rules (implications):

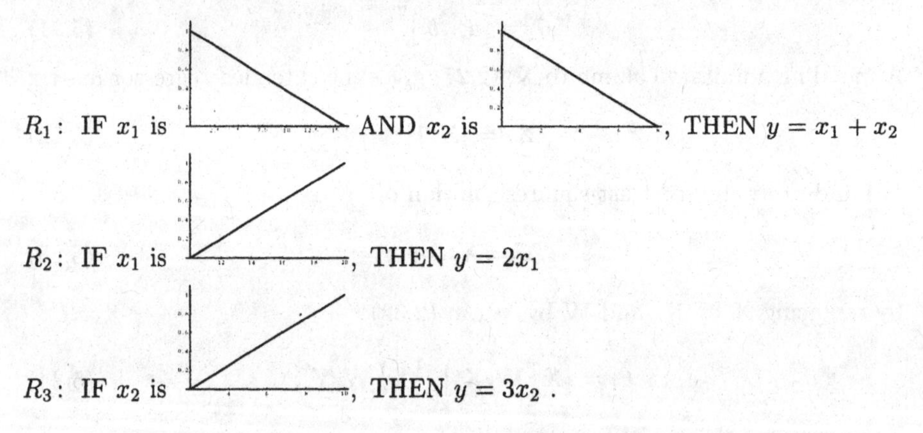

R_1: IF x_1 is [fuzzy set], AND x_2 is [fuzzy set], THEN $y = x_1 + x_2$

R_2: IF x_1 is [fuzzy set], THEN $y = 2x_1$

R_3: IF x_2 is [fuzzy set], THEN $y = 3x_2$.

The output for the conjunctive TS-model structure is given by

$$y = \frac{\sum_{i=1}^{n_R} \left(\mu_{A_{i1}}(x_1) \wedge \cdots \wedge \mu_{A_{ir}}(x_r)\right) \cdot y_i}{\sum_{i=1}^{n_R} \left(\mu_{A_{i1}}(x_1) \wedge \cdots \wedge \mu_{A_{ir}}(x_r)\right)} . \tag{5.26}$$

If we are given $x_1 = 12$, $x_2 = 5$, the value inferred by the rules is

$$y = \frac{0.25 \cdot 17 + 0.2 \cdot 24 + 0.375 \cdot 15}{0.25 + 0.2 + 0.375} \simeq 17.8 .$$

Figure 5.5 shows the antecedent fuzzy sets on the left and the input-output relationship obtained from (5.26) on the right. For a second example, suppose that we have the following two implications (rules):

$$R_1: \quad \text{IF } x \text{ is } A_{11}, \text{ THEN } y = 2 + 0.6x$$
$$R_2: \quad \text{IF } x \text{ is } A_{21}, \text{ THEN } y = 9 + 0.2x\ .$$

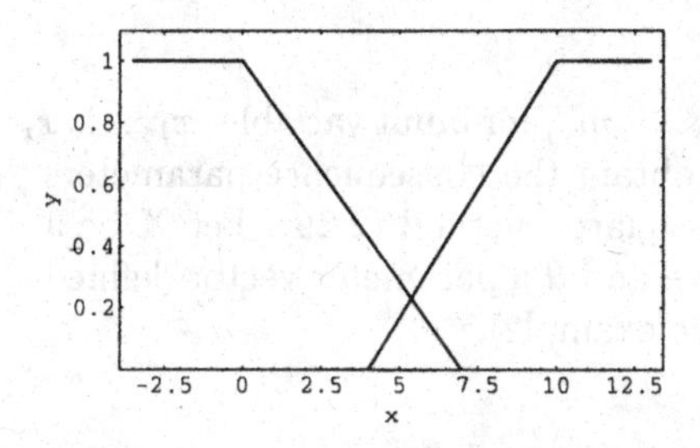

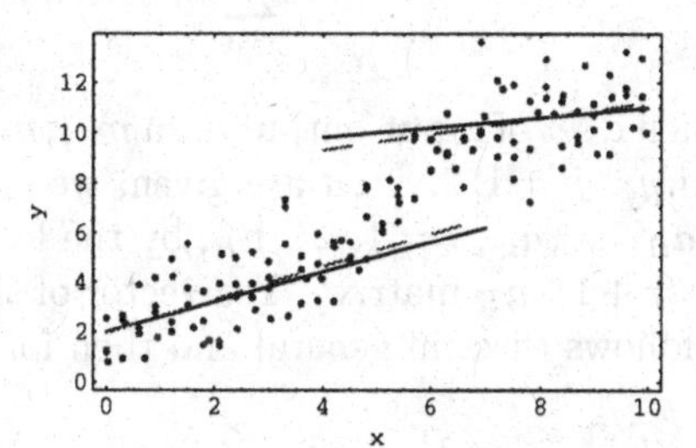

Fig. 5.6 Antecedent fuzzy sets, input-output data and identified model of the second example.

On the left in Figure 5.6, the antecedent fuzzy sets A_{11} (R_1, the fuzzy set on the left) and A_{21} (R_2, The fuzzy set on the right) are shown. The equation in the consequence can be interpreted to represent a law that holds in the fuzzy subspace defined in the premise. On the right in Figure 5.6, the dashed lines are the original equations in the model above. Note that the fuzzy partitioning of the input space enables us to connect equations smoothly. In fact, with hard clustering, an additional third relation would be required to connect both equations. The output y for the input x is obtained from (5.26) as

$$y = \frac{\sum_{i=1}^{n_R} (\mu_{A_{i1}}(x_1) \wedge \cdots \wedge \mu_{A_{ir}}(x_r)) \cdot (b_i + a_{i1}x_1 + \cdots + a_{ir}x_l)}{\sum_{i=1}^{n_R} (\mu_{A_{i1}}(x_1) \wedge \cdots \wedge \mu_{A_{ir}}(x_r))}$$

$$= \frac{\sum_{i=1}^{2} \mu_{A_{i1}}(x) \cdot (b_i + a_{i1}x)}{\sum_{i=1}^{2} \mu_{A_{i1}}(x)}\ .$$

Let ϕ_i be

$$\phi_i = \frac{\mu_{A_{i1}}(x_1) \wedge \cdots \wedge \mu_{A_{ir}}(x_r)}{\sum_{k=1}^{n_R} (\mu_{A_{k1}}(x_1) \wedge \cdots \wedge \mu_{A_{kr}}(x_r))}$$

$$= \frac{\mu_{A_{i1}}(x)}{\sum_{k=1}^{2} \mu_{A_{k1}}(x)}\ .$$

Then

$$y = \sum_{i=1}^{n_R} \phi_i \cdot (b_i + a_{i1}x_1 + \cdots + a_{ir}x_r)$$
$$= \sum_{i=1}^{2} \phi_i \cdot (b_i + a_{i1}x) \ .$$

When a set of input-output data $m_{1j}, m_{2j}, \ldots, m_{rj}$ for input variables $x_1, \ldots, x_r$ and y_j, $j = 1, 2, \ldots, d$ are given, we can obtain the consequence parameters $b_i, a_{i1}, \ldots, a_{ir}$, $i = 1, \ldots, n_R$, by the least-squares method (2.29). Let $\mathbf{X}$ be a $d \times (r+1) \cdot n_R$ matrix[1], $\mathbf{Y}$ a vector of size d and $\boldsymbol{\theta}$ a parameter vector defined as follows (first in general and then for the example):

$$\mathbf{Y} = [y_1, \cdots, y_d]^T$$
$$\boldsymbol{\theta} = [b_1, \cdots, b_{n_R}, a_{11}, \cdots, a_{n_R 1}, \cdots, a_{1r}, \cdots, a_{n_R r}]^T$$
$$\mathbf{X} = \begin{bmatrix} \phi_{11} & \phi_{21} \cdots & \phi_{n_R 1} & m_{11}\phi_{11} \cdots & m_{11}\phi_{11} \cdots & m_{r1}\phi_{11} \cdots & m_{r1}\phi_{n_R 1} \\ \phi_{12} & \phi_{22} \cdots & \phi_{n_R 2} & m_{12}\phi_{12} \cdots & m_{12}\phi_{22} \cdots & m_{r2}\phi_{12} \cdots & m_{r2}\phi_{n_R 2} \\ \vdots & \vdots & \vdots & \vdots & \vdots & \vdots & \vdots \\ \phi_{1d} & \phi_{2d} \cdots & \phi_{n_R d} & m_{1d}\phi_{1d} \cdots & m_{1d}\phi_{2d} \cdots & m_{rd}\phi_{1d} \cdots & m_{rd}\phi_{n_R d} \end{bmatrix},$$

where

$$\phi_{ij} \doteq \frac{\mu_{A_{i1}}(x_{1j}) \wedge \cdots \wedge \mu_{A_{ir}}(x_{rj})}{\sum_k \mu_{A_{k1}}(x_{kj}) \wedge \cdots \wedge \mu_{A_{kr}}(x_{rj})} \ .$$

Then the parameter vector $\boldsymbol{\theta}$ is calculated by (2.29),

$$\hat{\boldsymbol{\theta}} = \left[\mathbf{X}^T\mathbf{X}\right]^{-1}\mathbf{X}^T\mathbf{Y} \ .$$

In Figure 5.6, the identified two models are plotted as solid lines, while the (noise-free) models R_1 and R_2 are represented by dashed lines.

5.5 EXAMPLE: PREDICTION OF A CHAOTIC TIME-SERIES

In this example the fuzzy-c-means algorithm is used to identify the model of a the (chaotic) Mackey-Glass time-series. The time-series is generated by the following delay differential equation:

$$\frac{\mathrm{d}x(t)}{\mathrm{d}t} = \frac{0.2 \cdot x(t-\tau)}{1 + x^{10}(t-\tau)} - 0.1x(t) \ . \tag{5.27}$$

For $\tau > 17$, (5.27) displays chaotic behavior. Here, we chose $\tau = 25$, and Figure 5.7 shows a section of the sampled time-series.

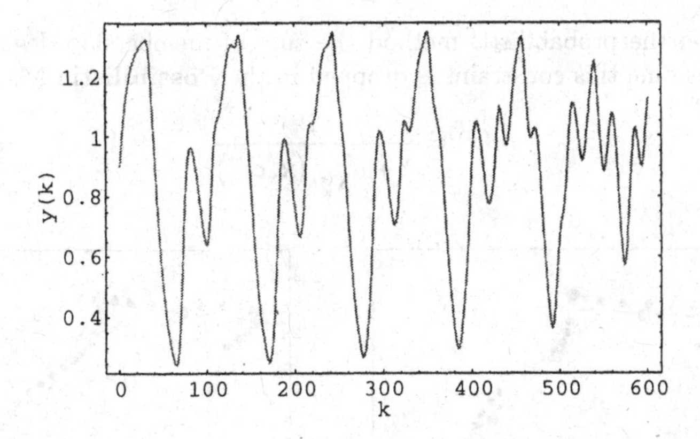

Fig. 5.7 A section of the Mackey-Glass time-series for $\tau = 25$. 60 values from $k = 200$ to 260 were used for training.

In time-series prediction we are not particularly interested in individual antecedent fuzzy sets and the interpretability of the rule. The aim is to have accurate predictions and one may therefore consider multi-dimensional antecedent membership functions. The membership degree $\mu_{A_i}(\mathbf{x}) \doteq \beta_i(\mathbf{x})$ of the regressor vector $\mathbf{x}$ is calculated directly using the inverse of the distance from the cluster prototype in subspace X. Let $\mathbf{F_x}$ denote the part of the cluster covariance matrix related to the regressor $\mathbf{x}$. The matrix is therefore the same as (5.1), $\mathbf{F}^{(i)}$, but with the last column dropped. Since we consider subspace X of $X \times Y$, the norm-inducing matrix is calculated as

$$\mathbf{A}_{\mathbf{x}}^{(i)} \doteq \left(|\mathbf{F}^{(i)}|\right)^{1/r} \cdot \left(\mathbf{F}^{(i)}\right)^{-1} . \tag{5.28}$$

Similarly, let $\mathbf{c}_{\mathbf{x}}^{(i)}$ denote the projection of cluster prototype $\mathbf{c}^{(i)}$ (4.17), onto X such that the inner-product norm $d^2_{\mathbf{A}_{\mathbf{x}}^{(i)}}$ measures the distance of an input vector $\mathbf{x}$ to the prototype $\mathbf{c}_{\mathbf{x}}^{(i)}$:

$$d^2_{\mathbf{A}_{\mathbf{x}}^{(i)}}\left(\mathbf{x}, \mathbf{c}_{\mathbf{x}}^{(i)}\right) = \left(\mathbf{c}_{\mathbf{x}}^{(i)} - \mathbf{m}_j\right)^T \mathbf{A}_{\mathbf{x}}^{(i)} \left(\mathbf{c}_{\mathbf{x}}^{(i)} - \mathbf{m}_j\right) . \tag{5.29}$$

The membership to a cluster (rule) is somehow inverse proportional to the distance and the following methods have been suggested.

Probabilistic Method:

$$\beta_i(\mathbf{x}) = \frac{1}{\sum_{k=1}^{c} \left(d^2_{\mathbf{A}_{\mathbf{x}}^{(i)}}\left(\mathbf{x}, \mathbf{c}_{\mathbf{x}}^{(i)}\right) / d^2_{\mathbf{A}_{\mathbf{x}}^{(i)}}\left(\mathbf{x}, \mathbf{c}_{\mathbf{x}}^{(i)}\right) \right)^{1/(w-1)}} . \tag{5.30}$$

[1]The matrix $\mathbf{X}$ given in the original paper by Takagi and Sugeno is not correct.

Whereas in the probabilistic method the sum of membership degrees of all the rules is one, this constraint is dropped in the **Possibilistic Method:**

$$\beta_i(\mathbf{x}) = \frac{1}{1 + d^2_{\mathbf{A}_{\mathbf{x}}^{(i)}}\left(\mathbf{x}, \mathbf{c}_{\mathbf{x}}^{(i)}\right)} . \tag{5.31}$$

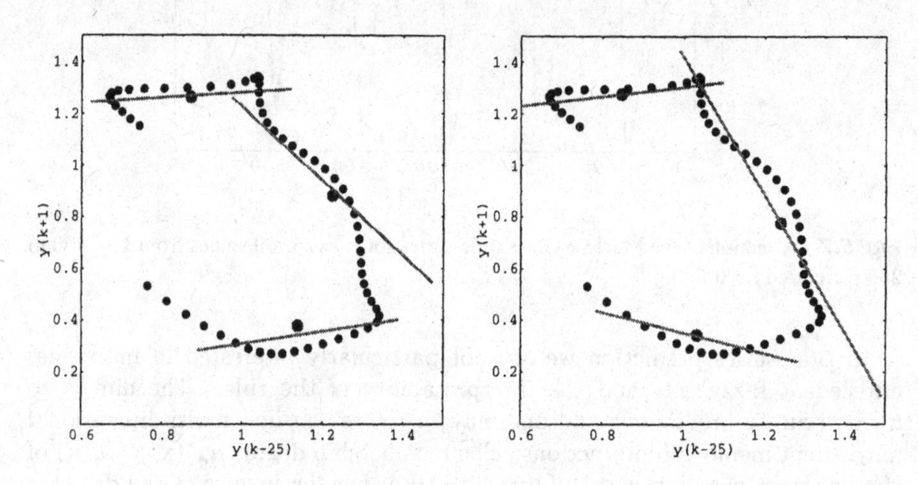

Fig. 5.8 Result of fuzzy-c-means (left) and GK-clustering (right).

The effectiveness of these two methods and others is discussed in [Bab98]. Here, we compare both methods for fuzzy-c-means and the GK-algorithm for 60 training data taken from $k = 200$ to $k = 260$. In all cases, we chose $w = 1.5$, $c = 3$ and the model structure $y(k+1) = f\big(y(k-25)\big)$ or equivalently $y(k+26) = f\big(y(k)\big)$. First, considering the results of the fuzzy-c-means, Figure 5.8 shows the training data in the product space, the identified cluster centers and the linear models identified using weighted least-squares.

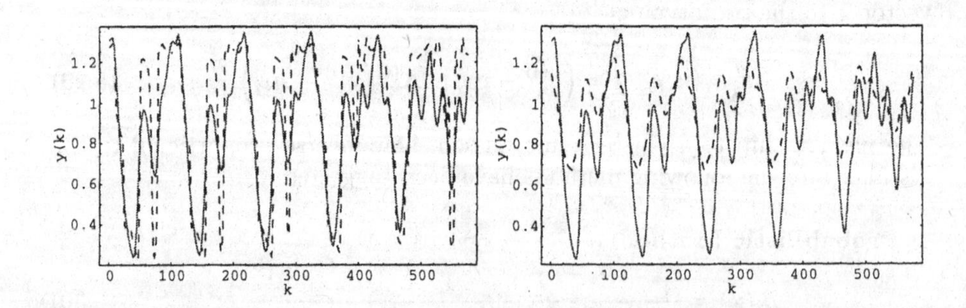

Fig. 5.9 Fuzzy-c-means: probabilistic method (left) vs. possibilistic method (right).

For the Gustafsson-Kessel algorithm, the clustering result is shown in Figure 5.8. Though both models, the possibilistic method in particular, produce better forecasts, the algorithm is computationally much more expensive and with only 60 values for training the GK-algorithm is sensitive to the initial partition matrix. The forecast results are shown in Figure 5.10.

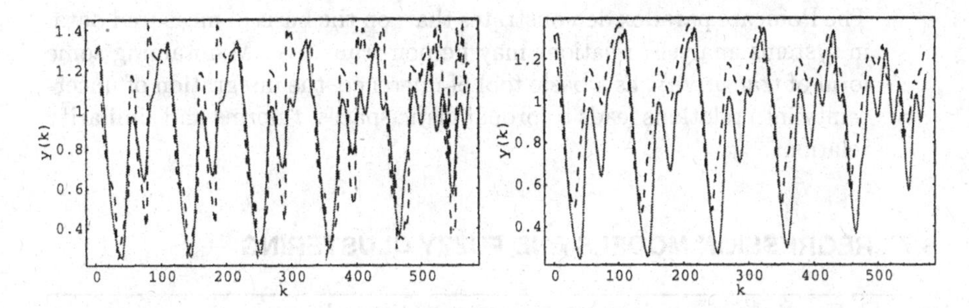

Fig. 5.10 GK-algorithm: probabilistic method (left) vs. possibilistic method (right).

5.6 DISCUSSION: PRODUCT SPACE CLUSTERING FOR SYSTEM IDENTIFICATION

You can only find truth with logic if you have already found truth without it.
—G.K. Chesterton

The input-output variables of a system and their past values specify a multi-dimensional space which translates the dynamics into a static regression problem. The main problems of clustering for system identification are:

▷ *Structure selection*, that is, to determine the 'relevant' input and output variables with respect to the aim of the modelling exercise.

▷ When identifying dynamic systems, the structure and the order (n_u, n_y) of the model dynamics must be chosen.

In many practical situations this may be very difficult from prior knowledge alone. Also, the discussed algorithms assume the number of cluster is known or fixed. Once the structure is fixed, fuzzy clustering has proven successful in systems modelling and pattern recognition.

From the foregoing discussion, there are at least three important reasons why fuzzy relations are relevant in systems analysis:

1. In hard clustering, objects somewhere "in between" two clusters, are 'forced' to be associated with exactly one cluster though partial membership may be more realistic.

2. Continuous degrees of membership allow analytical optimal solutions to the optimization problems in clustering, simplifying the computation as well as providing more realistic results with overlapping clusters.

3. The Poincaré paradox demonstrates that, on the basis of measured data, in systems analysis, relations may be non-transitive. Maintaining some form of transitivity, as a basic tool of inference, the integration of uncertainty into relations leads to probabilistic spaces, t-norms and similarity relations.

5.7 REGRESSION MODELS AND FUZZY CLUSTERING

- ☐ *Fuzzy clustering does not make assumptions about the randomness but can be related to regression analysis.*
- ☐ *Fuzzy-c-regression models yield simultaneous estimates for the parameters of c-regression models.*

Oft expectation fails and most oft there
Where most it promises, and oft it hits
Where hope is coldest and despair most fits
—William Shakespeare, "All's Well That Ends Well"

In 1993, Hathaway and Bezdek [HB93] introduced a family of objective functions, called fuzzy-c-regression models which can be used to fit switching regression models to certain types of mixed data. In close analogy to the fuzzy-c-means algorithm, minimization of the objective functions yields simultaneous estimates for the parameters of c-regression models, together with a fuzzy-c-partitioning of the data[2]. The notation in this section is identical to the concepts defined in Section 4.

Instead of assuming that a single regression model (2.15), can account for the d pairs of data $(\mathbf{x}_j, y_j)$, the *switching regression model* from which the data are assumed to be drawn, is specified by

$$y = f_i(\mathbf{x}; \boldsymbol{\theta}_i) + \varepsilon_i\ , \qquad 1 \leq i \leq c\ , \tag{5.32}$$

where the the objective is to determine the parameter vectors $\boldsymbol{\theta}_i \in \Theta \subset \mathbb{R}^l$ for c different models. The data $\mathbf{m}_j = (\mathbf{x}_j, y_j)$ are *unlabelled*, that is, it is not known which model from (5.32) applies[3].

[2] An alternative approach is to first cluster the data, for instance with either the hard-c-means or fuzzy-c-means algorithms, and then apply linear model identification to the data in each of the obtained clusters.

[3] As data are unlabelled, *i.e.* it is unknown to which *class* (cluster) they belong, fuzzy-c-means clustering is also called *unsupervised learning*.

In a statistical framework the definition of an optimal estimate of $\boldsymbol{\theta}$ depends on assumptions made about the distribution of random vectors ε_i and the set of feasible parameters. Commonly, the ε_i are assumed to be independently generated from some probability distribution function $p(\varepsilon;\eta,\sigma)$ such as the Gaussian distribution with mean 0 and unknown standard deviation σ_i,

$$p(\varepsilon;\eta,\sigma)=\frac{1}{\sigma\sqrt{2\pi}}\,e^{-\frac{(\varepsilon-\eta)^2}{2\sigma^2}}\,. \tag{2.57}$$

For example, let $c=2$ and $l=2$. Then, from (5.32),

$$\begin{aligned} y &= f_1(\mathbf{x};\boldsymbol{\theta}_1)+\varepsilon_1 \\ y &= f_2(\mathbf{x};\boldsymbol{\theta}_2)+\varepsilon_2\,. \end{aligned} \tag{5.33}$$

Each data sample $\mathbf{m}_j$ is assumed to be generated with probability $Pr(i)$ from model i such that $\sum_{i=1}^{c} Pr(i)=1$. The log-likelihood function (2.55) of the data is

$$\mathcal{L}(\boldsymbol{\theta};\mathbf{M}) \doteq \ln \ell(\boldsymbol{\theta};\mathbf{M}) = \ln Pr(\mathbf{M}|\boldsymbol{\theta}) \tag{2.55}$$

$$= \sum_{j=1}^{d} \ln \sum_{i=1}^{c} Pr(i)\cdot p(\mathbf{m}_j|i,\boldsymbol{\theta}_i) \tag{3.25}$$

$$\begin{aligned} &= \sum_{j=1}^{d} \ln\big[Pr(i)\cdot p(y_j-\theta_{11}x_j-\theta_{12}|0,\sigma_1) \\ &\quad + \big((1-Pr(i))\cdot p(y_j-\theta_{21}x_j-\theta_{22}|0,\sigma_2)\big)\big]\,. \end{aligned} \tag{5.34}$$

To this point, the formulation of the problem of the mixture density model discussed in Section 3.3 and the EM-algorithm can be used to iteratively optimize $\mathcal{L}$ as detailed in Section 3.4. The motivation to use the EM-algorithm comes from the fact that the observed data are *incomplete* in the sense that they are *unlabelled* - we do not know for any $\mathbf{m}_j$ which model i generated it.

If we could partition $\mathbf{M}$ into c subsets corresponding to the models (5.32), we would have complete information and then estimators for $\boldsymbol{\theta}_i$ could be obtained by regression. For instance, we can first use the hard-c-means algorithm to identify a crisp-c-partition of $\mathbf{M}$ such that $\mathbf{M}=\bigcup \mathbf{M}_i$, $\mathbf{M}_i\cap\mathbf{M}_j=\emptyset$ for $i\neq j$, and then solve c separate single models using standard least-squares, (2.29).

Alternatively, Hathaway and Bezdek suggested to (fuzzy) partition $\mathbf{M}$ and estimate $\boldsymbol{\theta}_1,\ldots,\boldsymbol{\theta}_c$ simultaneously by modifying the fuzzy-c-means algorithm. Clustering in $\mathbf{M}$ assigns (fuzzy) label vectors $u_{ij}\in\mathbf{U}$, (4.18), to each $\mathbf{m}_j$ describing the membership of the object represented by $\mathbf{m}_j$ in the ith class or submodel. If the u_{ij} came from maximum likelihood estimation in mixture density decomposition, it would describe a posterior probability that the object described by u_{ij} (j fixed) came from class i. For the switching regression problem, u_{ij} is considered a weight describing the extent to which the model value $f_i(\mathbf{x}_j; \boldsymbol{\theta}_i)$ matches $\mathbf{y}_j$. The quality of the model is quantified with the loss $Q(\mathbf{m}, \boldsymbol{\theta})$, that is, the expectation $E_{ij}[\boldsymbol{\theta}_i]$ of the error in $f_i(\mathbf{x}_j; \boldsymbol{\theta}_i)$ approximating $\mathbf{y}_j$. For example, let

$$E_{ij}[\boldsymbol{\theta}_i] = \|\mathbf{y}_j - f_i(\mathbf{x}_j; \boldsymbol{\theta}_i)\|^2 . \tag{5.35}$$

With the general definition, E_{ij}, the family of fuzzy-c-regression model objective functions for $\mathbf{U} \in M_{fc}$ and $u_{ij} \in \mathbf{U}$ is defined by

$$E_w[\mathbf{U}, \{\boldsymbol{\theta}_i\}] = \sum_{i=1}^{c} \sum_{j=1}^{d} u_{ij}^w \cdot E_{ij}[\boldsymbol{\theta}_i] , \tag{5.36}$$

where $w > 1$ is fixed as in (4.11). The difference with the fuzzy objective functions (4.11) is that the fit of the regression models to each y_j replaces the distance of $\mathbf{m}_j$ to some prototype. With respect to (5.35) the objective function $E_w[\mathbf{U}, \{\boldsymbol{\theta}_1, \ldots, \boldsymbol{\theta}_l\}]$ is a fuzzy, multi-model extension of the least-squares criterion. If the regression functions $f_i(\mathbf{x}; \boldsymbol{\theta}_i)$ are linear in the parameters $\boldsymbol{\theta}_i$, the parameters can be obtained as a solution of the weighted least-squares problem specified by (5.36). Using the same approach as in (5.20), where the membership degrees of the fuzzy partition matrix $\mathbf{U}$ serve as the weights in $\mathbf{W}$, and with (5.22), the optimal parameters $\boldsymbol{\theta}_i$ are computed as

$$\boldsymbol{\theta}_i = \left[\mathbf{X}_e^T \mathbf{W}_i \mathbf{X}_e\right]^{-1} \mathbf{X}_e^T \mathbf{W}_i \mathbf{Y} . \tag{5.24}$$

The procedure is summarized in algorithm 5.2. Note that in the mixture density approach for the EM-algorithm, the probability that component i generated data point j is denoted by p_{ij}, (3.28), and in the fuzzy-c-regression model it is denoted by u_{ij}. For our example (5.33), we would have in the ML formulation

$$p_{ij} = \frac{Pr(i) \cdot p(y_j - \theta_{i1} x_j - \theta_{i2} \mid 0, \sigma_i)}{\sum_{k=1}^{c} Pr(k) \cdot p(y_j - \theta_{k1} x_j - \theta_{k2} \mid 0, \sigma_k)} .$$

5.8 EXAMPLE: PH NEUTRALIZATION PROCESS

In this example, we apply the fuzzy regression model, discussed in the previous section, to data obtained from a highly nonlinear pH neutralization process[4] [Bab98]. The system consists of a neutralization tank to which a base Q is added and the pH level is the output variable to be controlled. The identification data set, consisting of $d = 1250$ values is shown in Figure 5.11.

[4]The pH model was described in [MHL72].

Fix c, $2 \leq c < d$, and choose the termination tolerance $\delta > 0$ and fix w, $1 \leq w < \infty$. Initialize $\mathbf{U}^{(0)} \in M_{fc}$ randomly.

Repeat for $l = 1, 2, \ldots$:

Step 1: Using (5.24), calculate model parameters $\boldsymbol{\theta}_i^{(l)}$ to globally minimize (5.36).

Step 2: Update the partition matrix with $E_{ij} = E_{ij}[\boldsymbol{\theta}_i^{(l-1)}]$ to satisfy

$$u_{ij}^{(l)} = \frac{1}{\sum_{k=1}^{c} \left(\frac{E_{ij}}{E_{kj}} \right)^{\frac{2}{w-1}}}$$

if $E_{ij} > 0$ for $1 \leq i \leq c$, and otherwise $u_{ij} = 0$ if $E_{ij} > 0$, and $u_{ij} \in [0, 1]$ with $(u_{1j} + \cdots + u_{cj}) = 1$.

Until $\|\mathbf{U}^{(l)} - \mathbf{U}^{(l-1)}\| < \delta$.

Algorithm 5.2 Fuzzy-c-regression models.

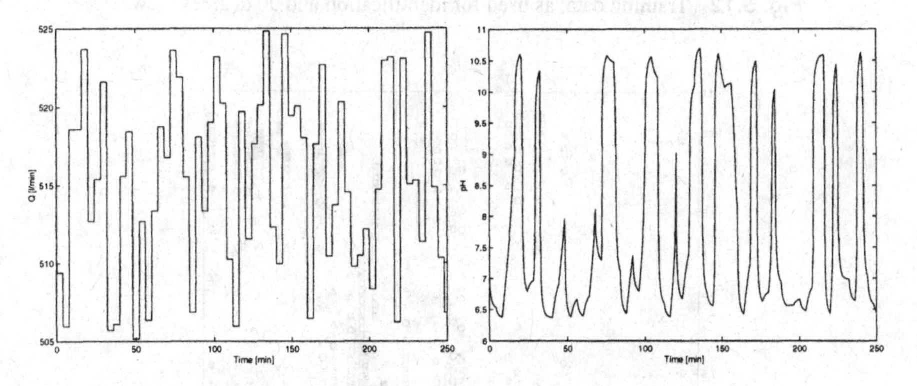

Fig. 5.11 Input and output data used for identification.

Here we represent the process as a first-order discrete-time NARX model:

$$pH(k+1) = f\left(pH(k), Q(k)\right) ,$$

where k denotes the sampling instant and f is the unknown relationship we wish to identify. In Figure 5.12, the data are plotted in the data space and it can be seen that clusters are not immediately apparent. The s-shaped area with a higher density of data corresponds to the equilibrium of the system. For $c = 3$, $w = 2$ and the stopping criterion set equal to $\delta = 10^{-10}$, the fuzzy-$\dot{c}$-regression model leads to the model shown in Figure 5.13. Note that the switching regression model does not produce compact sets. Instead, it finds the dynamically similar regions on the pH-Q space. This means the points marked by squares are mainly equilibrium points, while the other two clusters cover two different off-equilibrium regions. Hence, the method is useful for extracting "fuzzy" knowledge about the possible dynamic regions of the system, but because of the non-compact fuzzy sets one would obtain from the fuzzy clusters, it is difficult to transform the result into a rule base.

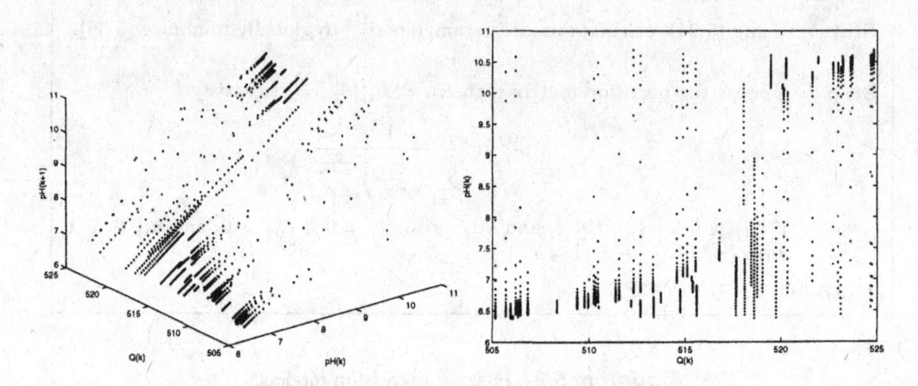

Fig. 5.12 Training data, as used for identification and 90 degrees view.

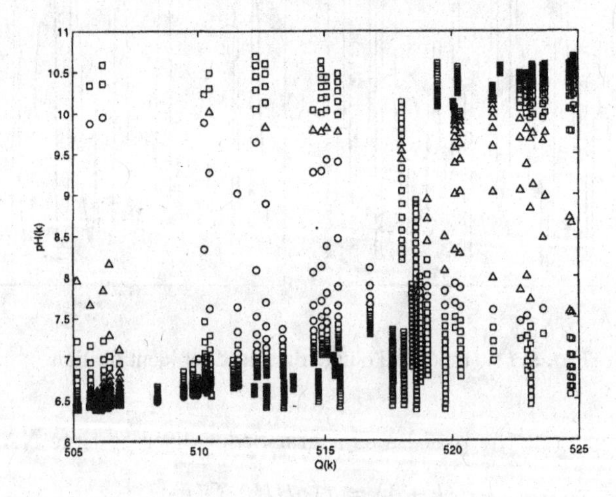

Fig. 5.13 Fuzzy switching regression model. 90 degrees view where points are assigned to the cluster in which they have the maximum membership.

6

Random-Set Modelling and Identification

- *Random sets can be viewed as multi-valued maps or random variables.*
- *For very small data sets, local uncertainty models (random subsets) can be used to generalize information in the data space.*
- *The estimation of coverage functions of random sets yields possibility distributions.*
- *While probabilities describe whether or not an events occurs, possibility describes the degree of confidence or feasibility to which some condition exists.*
- *A random set approach to identification leads to qualitative predictions.*
- *Qualitative predictions are fuzzy restrictions and may therefore provide a mechanism to combine quantitative data analysis with rule-based systems describing qualitative expert knowledge.*

Prediction is difficult, especially if it concerns the future.

—Mark Twain

In this section, we describe a random-set approach to time-series analysis. The principle idea is that, based on only a small data set **M**, we are unable to choose a distribution function for data in Ξ and therefore a specified number of nearest neighbors $\mathbf{m}_j$ in Ξ induce random subsets - representing a local uncertainty model. The motivation is that random-set theory provides us with a mechanism to integrate statistical objects into rule-based systems on the basis of generalized (fuzzy) sets. More specifically, predictions of the models developed in this section are determined from a possibility distribution which can be processed in a rule-based system as described in Section 7.2. The possibility distribution itself is an estimate of the generalized distribution function of projected random subsets. Before introducing random-sets, we review the more familiar concept of random variables and point-valued maps.

6.1 RANDOM VARIABLES, POINT-VALUED MAPS

The possible outcomes of an experiment, determined in part by random factors form the space of elementary events Ω. Other events, expressed as qualitative concepts such as whether the die produces an even or odd number, are formed by combinations of elementary events, that is, subsets of Ω. These events may thus be organized in a structure of subsets σ-algebra on Ω, denoted by σ_Ω. This structure ensures (by imposing certain conditions to the subsets) a predictable behavior in operations with subsets such as complement, union, intersection, and so on, and then allows us to measure how likely an event is by introducing a probability measure Pr_Ω in the measurable space (Ω, σ_Ω). The choice of σ_Ω depends on the kind of experiment under consideration. The tuple $(\Omega, \sigma_\Omega, Pr_\Omega)$ is called a probability space and it summarizes the experiment or process.

A random variable is a rule that associates for each elementary event $\omega \in \Omega$ an element $\mathbf{x}(\omega)$ in a space X, in which the elements are organized by a σ-algebra, σ_X . The aim of using a random variable is to generate a probability measure on (X, σ_X) such that the probability space $(X, \sigma_X, Pr_\mathbf{x})$ is the mathematical description of the experiment as well as the original probability space $(\Omega, \sigma_\Omega, Pr_\Omega)$. The benefit arises when (X, σ_X) is a well-known measurable space where mathematical tools such as integration are established. One of the most commonly used measurable space is $(\mathbb{R}, \mathcal{B})$, where $\mathcal{B}$ is the σ-algebra of Borel (Borel algebra), which is generated from the topological space of the open subsets of $\mathbb{R}$.

For the formal definition of a random variable, let $(\Omega, \sigma_\Omega, Pr_\Omega)$ be a probability space and (X, σ_X) a measurable space. Every $(\sigma_\Omega\text{-}\sigma_X)$-measurable mapping

$$\mathbf{x} : \Omega \to X \tag{6.1}$$

is called a random variable. The mapping $\mathbf{x}$ is said to be σ_Ω-σ_X measurable iff $\forall A \in \sigma_X$, $\mathbf{x}^{-1}(A) \in \sigma_\Omega$. Measurability ensures equivalent representations of the experiment in both probability spaces $(\Omega, \sigma_\Omega, Pr_\Omega)$ and (X, σ_X, Pr_X) where Pr_X is explained below. If $X = \mathbb{R}$ and $\sigma_X = \mathcal{B}$, it is called a numerical random variable. Since every elementary outcome is 'mapped' into a 'point', a random variable is also referred to as a *point-valued map*.

The distribution or probability law of $\mathbf{x}$ is defined as $Pr_{\mathbf{x}} = Pr_{\Omega} \circ \mathbf{x}^{-1}$. This means that an event $A \in \sigma_X$ has the probability

$$\begin{aligned} Pr_{\mathbf{x}}(A) &= Pr_{\Omega} \circ \mathbf{x}^{-1}(A) \\ &= Pr_{\Omega}\left(\mathbf{x}^{-1}(A)\right) \\ &= Pr_{\Omega}\{\omega : \mathbf{x}(\omega) \in A\}. \end{aligned} \tag{6.2}$$

From (6.2), for any event A, given a probability law defined on Ω, a measure of the probability of that event in X can be obtained using the inverse image $\mathbf{x}^{-1}(A)$ of A. The inverse map is defined by

$$\begin{aligned} \mathbf{x}^{-1} : \mathcal{B}_X &\to \sigma_{\Omega} \\ A &\mapsto \mathbf{x}^{-1}(A) = \{\omega : \mathbf{x}(\omega) \in A\} . \end{aligned} \tag{6.3}$$

Depending on whether we deal with a discrete or continuous random variables, we have

$$Pr_{\mathbf{x}}(A) = \sum_{\omega \in \mathbf{x}^{-1}(A)} Pr(\{\omega\}) \qquad \text{(discrete case)}, \tag{6.4}$$

$$= \int_{\mathbf{x}^{-1}(A)} \mathrm{d}Pr \qquad \text{(continuous case)}. \tag{6.5}$$

As illustrated in Figure 6.1, the event A is defined by a crisp subset. A set of elements can be described by listing its elements or by the definition of its characteristic function (Chapter 2). In terms of the expectation operator $E[\cdot]$, we have equivalently:

$$\begin{aligned} Pr_{\mathbf{x}}(A) &= \int_{\mathbf{x}^{-1}(A)} \mathrm{d}Pr_{\Omega} = \int_{\Omega} \zeta_{\mathbf{x}^{-1}(A)}(\omega)\, \mathrm{d}Pr_{\Omega} \\ &= \int_A \mathrm{d}Pr_{\mathbf{x}} = \int_X \zeta_X(x)\, \mathrm{d}Pr_{\mathbf{x}} \\ &= E[\zeta_A(x)] . \end{aligned} \tag{2.2}$$

The concept of a random variable $\mathbf{x}$ can be summarized in the following commutative diagram:

$$\begin{array}{ccc} \mathcal{B}_X & \xrightarrow{\mathbf{x}^{-1}} & \sigma_{\Omega} \\ & \searrow^{Pr_{\mathbf{x}} = Pr_{\Omega} \circ \mathbf{x}^{-1}} & \downarrow Pr_{\Omega} \\ & & [0,1] \end{array}$$

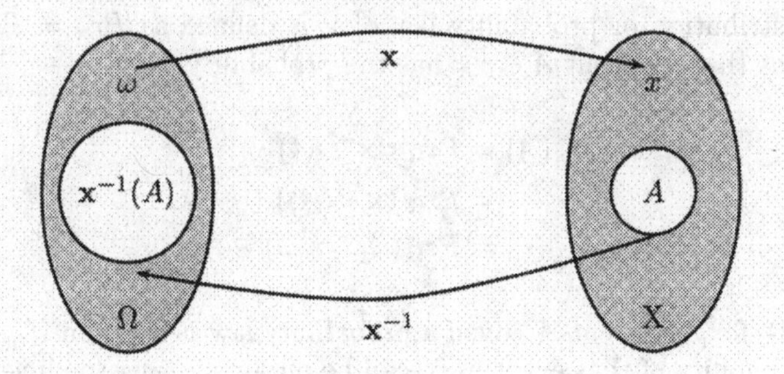

Fig. 6.1 A random variable as a point-valued mapping.

6.2 RANDOM-SETS, MULTI-VALUED MAPS

It has been argued[1] that in many cases the situation described by ordinary random variables is somewhat ideal and, in fact, we may not know into which x any particular ω maps. Such considerations lead to a multi-valued map where any $\omega \in \Omega$ maps into a subset of X. In other words, the only difference to a random variable is that it associates every elementary event $\omega \in \Omega$ with an event of $\mathcal{U}$ which is a set of subsets of X. That is, we associate with each $\omega \in \Omega$ a subset of X. Associated with a probability measure, a multi-valued map therefore describes a *random-set* [GMN97, Wol01]. As can be seen from the illustration of a random-set in Figure 6.2, in case of multi-valued maps, the inverse image is not unambiguous and subsequently the mathematics of the spaces and measures involved is more complicated. However, we will find that random-set theory provides a convenient interface between probability theory and possibility theory.

Let $(\Omega, \sigma_\Omega, Pr_\Omega)$ be a probability space and let $(\mathcal{U}, \sigma_\mathcal{U})$ be a measurable space where $\mathcal{U}$ is a set of subsets of X, that is, $\mathcal{U} \subseteq \mathcal{P}(X)$, where $\mathcal{P}(\Xi)$ is the power class of Ξ and $\sigma_\mathcal{U}$ is a σ-algebra defined on $\mathcal{U}$. The *power class* $\mathcal{P}(X)$ is defined as the set of sets $\mathcal{P}(X) = \{C: C \subseteq X\}$. Then a random-set is defined as a $(\sigma_\Omega - \sigma_\mathcal{U})$-measurable mapping $\Gamma : \Omega \to \mathcal{U}$. The mapping associates elementary events of Ω with elements of $\mathcal{U}$, so really it is a random variable between the probability space $(\Omega, \sigma_\Omega, Pr_\Omega)$ and the measurable space $(\mathcal{U}, \sigma_\mathcal{U})$. Then in some way, as we define the distribution or the probability law of a random variable, we define the distribution of a random-set by $Pr_\Gamma =$

[1] The introduction of multi-valued maps in statistics is attributed to A. Dempster, who showed that the multi-valued mapping from Ω to X carries a probability measure, Pr_Ω, defined over subsets of Ω into a system of upper and lower probabilities defined over subsets of X. Upper and lower probabilities were then later described by G. Shafer as subjective probabilities modelling ***belief*** and ***plausibility***, in what is referred to as *evidence theory*.

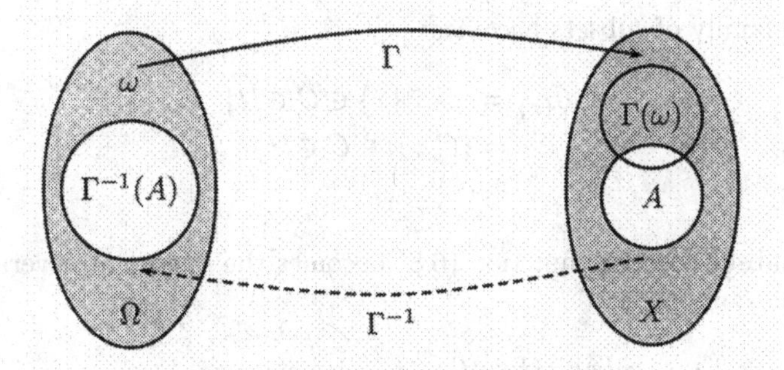

Fig. 6.2 Random-set: Set-valued map.

$Pr_\Omega \circ \Gamma^{-1}$ in analogy with random variables:

$$\begin{aligned} Pr_\Gamma(\mathcal{A}) &= Pr_\Omega \circ \Gamma^{-1}(\mathcal{A}) \\ &= Pr_\Omega\{\omega : \Gamma(\omega) \in \mathcal{A}\} \quad \forall \mathcal{A} \in \sigma_{\mathcal{U}}. \end{aligned} \tag{6.6}$$

Note that a random-set can be seen as a random variable from Ω to $\mathcal{U}$ or as a multi-valued mapping from Ω to X since $\Gamma(\omega) \in \mathcal{U} \Rightarrow \Gamma(\omega) \subseteq X$. The probability space $(\mathcal{U}, \sigma_{\mathcal{U}}, Pr_\Gamma)$ is the mathematical model that is used to represent and explain the experiment.

In order to be able to associate (probability) measure with events in $\mathcal{P}(X)$, we need to define measurable spaces. $\mathcal{U}$ is a set of subsets included in X and is chosen according to the type of process that is being studied. The σ-algebra on $\mathcal{U}$, $\sigma_{\mathcal{U}}$, being a set of sets of subsets of X, implies a rather complicated structure of subsets to identify and work with. Note also that $\sigma_{\mathcal{U}} \subsetneq \mathcal{P}(\mathcal{U}) \subseteq \mathcal{P}(\mathcal{P}(X))$ since $\mathcal{U} \subseteq \mathcal{P}(X)$. $(\mathcal{U}, \sigma_{\mathcal{U}})$ is termed a hypermeasurable space.

For any $A \in \mathcal{U}$, that is,, $A \subset X$, we distinguish the following family of subsets:

$$\mathcal{C}_A = \{C : A \subseteq C \in \mathcal{U}\} . \tag{6.7}$$

Let us suppose that $(\mathcal{U}, \sigma_{\mathcal{U}})$ is a measurable space such as $\forall A \in \mathcal{U}$, $\mathcal{C}_A \in \sigma_{\mathcal{U}}$. Then (6.7) determines the following measure on X such that $\forall A \subseteq X$:

$$\begin{aligned} c_\Gamma(A) &\doteq Pr_\Gamma(\mathcal{C}_A) \\ &= Pr_\Omega(\Gamma(\omega) \in \mathcal{C}_A) \\ &= Pr_\Omega\{\omega : A \subseteq \Gamma(\omega)\} . \end{aligned} \tag{6.8}$$

The function $c_\Gamma(\cdot)$ in (6.8) is called subset coverage function. We now focus our attention on the special case of the family of subsets $\mathcal{C}_A$ when $A = \{x\}$.

Then the family of subsets becomes

$$\mathcal{C}_{\{x\}} = \{C : \{x\} \in C \in \mathcal{U}\} \tag{6.9}$$
$$= \{C : x \in C \in \mathcal{U}\} \tag{6.10}$$
$$= \mathcal{C}_x$$

and the subset coverage function (6.8) becomes the one-point coverage function of the random-set Γ

$$\begin{aligned} c_\Gamma(x) &= Pr_\Gamma(\mathcal{C}_x) \\ &= Pr_\Omega \circ \Gamma^{-1}(\mathcal{C}_x) \\ &= Pr_\Omega\{\omega : \Gamma(\omega) \subset \mathcal{C}_x\} \\ &= Pr_\Omega\{\omega : x \in \Gamma(\omega)\} \quad \forall x \in X. \end{aligned} \tag{6.11}$$

From (6.11), we can define yet another distribution for subsets of X:

$$\Pi_\Gamma(A) = \sup_{x \in A}\{c_\Gamma(x)\} \qquad \forall A \subset X \ . \tag{6.12}$$

Both (6.11) and (6.12) are mappings into the unit interval. Note that $c_\Gamma : X \to [0,1]$ therefore defines a fuzzy restriction on X. (6.12) is called a possibility measure and was already introduced in Section 2.

Let the outcomes of a process be in the form of subsets of X and let $C_1, ..., C_d$ be a sequence of random subsets obtained from n realizations of the process. We assume that $C_1, ..., C_d$ is a random-set sample. Then set-valued statistics are required to estimate the one-point coverage function and their properties. An estimator of the one-point coverage function is

$$\hat{c}_\Gamma(x) = \frac{1}{d}\sum_{k=1}^{d} \zeta_{C_k}(x), \quad \forall x \in X \ , \tag{6.13}$$

where ζ_{C_k} is the indicator or characteristic function of the crisp set C_k, $k \in \{1, ..., d\}$. The difference between point-valued and set-valued statistics is illustrated in Figure 6.3.

Problems subject to uncertainty, imposed by subjective and imprecise information, are present in many real-world problems and are more and more often considered in engineering applications. Fuzzy set theory has been successfully applied to those problems where a lack of precision exists in the outcome of an experiment, for instance, the definition of a concept by a group of experts, where everyone has a own idea of the outcome. It is called a fuzzy concept because the borderlines are not clearly defined. Fuzzy mathematics gives good descriptions for those concepts by using fuzzy set membership functions.

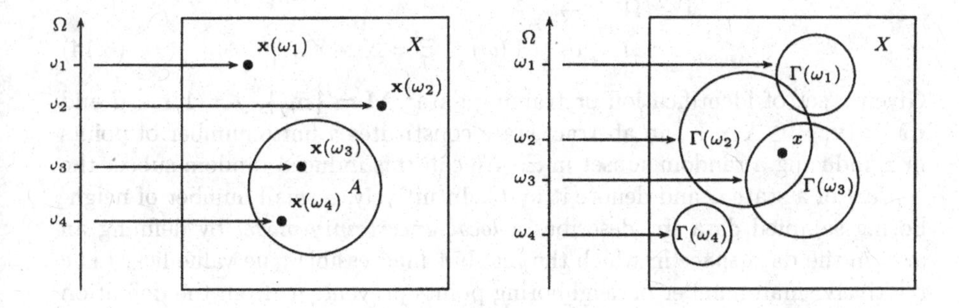

Fig. 6.3 Point-valued statistics (left) vs. set-valued statistics (right).

Set-valued statistics based on random-set theory introduces a practical method to set up these membership functions from a sample of outcomes such as some different opinions about a concept. This allows us to study the process by applying fuzzy mathematics, which has become a useful extension to probability theory and statistics.

In the previous section we introduced the definition of a random-set and its distribution by drawing on the similarity with the definition of a random variable. This similarity allows us to study processes governed by a random-set (i.e. processes with subsets of a space as possible outcomes) by applying probability theory on the $(\mathcal{U}, \sigma_{\mathcal{U}})$ hypermeasurable space.

6.3 A RANDOM-SET APPROACH TO SYSTEM IDENTIFICATION

Based upon previous definitions regarding random sets and possibility distributions, this section we introduces a random-set model for multi-variate data and dynamic systems in particular. The principle idea is to let sampled data induce random-sets (local uncertainty models) in the data space. Thus, the estimation of the coverage function describes a fuzzy set. In case of $\Xi = X \times Y$ where we consider an autoregressive structure for $y = f(\mathbf{x})$, the projections of random subsets onto space Y leads to a possibility distribution as a qualitative forecast, expressing explicitly the uncertainty (confidence) of the model w.r.t. any potential candidate $y(k+1)$.

We begin with a simple example of the idea using a first-order autoregressive model structure and square-shaped subsets. Let Ω be a set of abstract states and Γ the observable

$$\begin{aligned} \Gamma : \Omega &\to \mathcal{U} \\ \omega &\mapsto \Gamma(\omega) \subset \Xi = X \times Y . \end{aligned} \tag{6.14}$$

Given a set of identification or training data[2] $\mathbf{M} = \{\mathbf{m}_j\}$, $j = 1, \ldots, d$ and $\mathbf{m}_j = (\mathbf{x}, y) \in X \times Y$, an abstract state constitutes a finite number of points in Ξ inducing a random subset in Ξ. We call this induced random subset the *context* of a state ω and denote it by C. Intuitively, a small number of neighboring sampled data $\mathbf{m}_j$ describes a *local uncertainty model* by defining an area in the data space in which the real but inaccessible true value lies. (The relatively small number of neighboring points prevents us from the definition of a probability density.) One possibility to create a context is for all $\mathbf{m}_j \in \mathbf{M}$ to use a metric between two points $\mathbf{m}_j$ and $\mathbf{m}_i$ to define the neighborhood of $\mathbf{m}_j \in \mathbf{M}$ as the closed set (hypercube) whose boundaries are defined by $\mathbf{m}_j$ and its n_n nearest neighbors:

$$C_j = [\min x_1^i, \max x_1^i] \times \cdots \times [\min x_r^i, \max x_r^i] \times [\min y^i, \max y^i] . \tag{6.15}$$

The point $\mathbf{m}_j$ and its n_n nearest neighbors are indexed by i. The data set $\mathbf{M}$ thus generates a sample of Γ, denoted

$$\mathcal{C} = \{C_j\} \qquad \text{where} \quad C_j \subseteq X \times Y,\ j = 1, \ldots, d .$$

The set of hypercubes, $\mathcal{C}$, essentially comprises our identified model on which we shall base our decision making. We notice that the identification itself therefore, initially, does not involve any optimization. Given a set of training data, the identified model is a coverage function of a random-set (6.13). Since the one-point coverage function (6.11) is also a fuzzy restriction (11.1), we may consider the random-set model as a fuzzy model.

Now, consider the situation in which the data space is defined in terms of the nonlinear autoregressive model

$$y(k+1) = f(\mathbf{x}) ,$$

where

$$\mathbf{x} = [y(k), \ldots, y(k - n_y + 1), u(k), \ldots, u(k - n_u + 1)]^T$$

such that each element in vector $\mathbf{x}$ defines an axis in the product space $X = X_1 \times \cdots \times X_r$. A prediction model is then built from the projections of random subsets $C_j \subset \Xi$ onto Y, the space of candidates for $y(k+1)$. Formally, for any $\mathbf{x} \in X$, extended into $X \times Y$, the projected random intervals on Y are defined by

$$C_i^{\mathbf{x}} = \{y : \mathbf{x}_{\text{ext}} \cap C_i \neq \emptyset\} . \tag{6.16}$$

[2]Note that '$\mathbf{x}$' denoted a random variable in the previous section, while here it denotes the regression vector as introduced in Section 2.

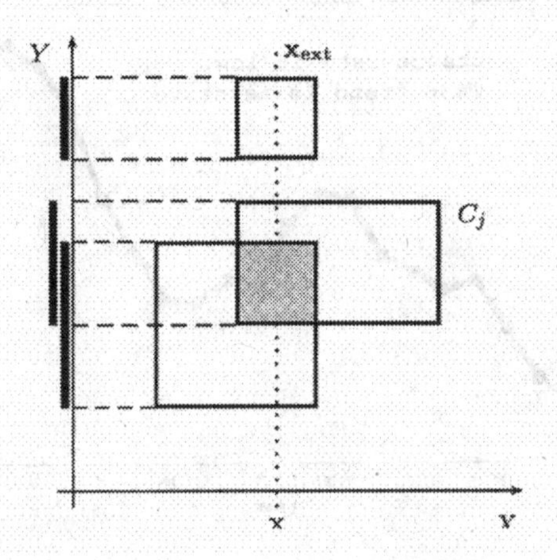

Fig. 6.4 Projection of random subsets in $X \times Y$ onto Y.

Let $i = 1, \ldots, n_p$ be the number of random intervals for which $\mathbf{x}_{ext}$ is covered by any C_j. Figure 6.4 illustrates the concept of a realization of random-set Γ and projections onto Y. Then, $\{C_i^{\mathbf{x}}\}$ is a set-valued sample for $\mathbf{x} \in X$ and from (6.13), we obtain a conditional possibility distribution

$$\hat{\pi}(y|\mathbf{x}) = \frac{1}{n_p} \sum_{i=1}^{n_p} \zeta_{C_i^{\mathbf{x}}}(y) \quad \text{where} \quad \zeta_{C_i^{\mathbf{x}}}(y) = \begin{cases} 1 & \text{if } y \in C_i^{\mathbf{x}}, \\ 0 & \text{otherwise.} \end{cases} \tag{6.17}$$

In (6.17), for any $y' \in Y$, $\hat{\pi}(y'|\mathbf{x})$ determines the possibility of $y(k+1) = y'$. In other words, each $\pi(y)$ specifies the degree of feasibility, model confidence, that y is the next output at time $k+1$. The measure is conditional on the *experience* we have from set of training data $\mathbf{M}$. Consequently, regions in which training data are sparse will produce low degrees of confidence. In fact, if for any given $\mathbf{x}$, $\mathbf{x}_{ext}$ extends into regions of the data space in which no random subsets exist, the model confidence will be zero. (We can still make a prediction by extrapolating as in conventional regression analysis).

The difference with a probability distribution or density is important. Statistical laws (established either empirically or on the basis of hypothetical models) are not used in prediction of the probability of occurrence of *individual* events. Their chief importance, as far as prediction is concerned, lies in that they help in forecasting collective properties, that is, properties of large collections of entities that are similar in some respects. Here, however, we consider time-series with small numbers of data which are non-repeatable or only one realization is available for analysis. Hence, we are unable to associate with a forecast a probability distribution which for any $y \in Y$ would tell us whether or not, on average, we could expect y to be the value $y(k+1)$. We contend

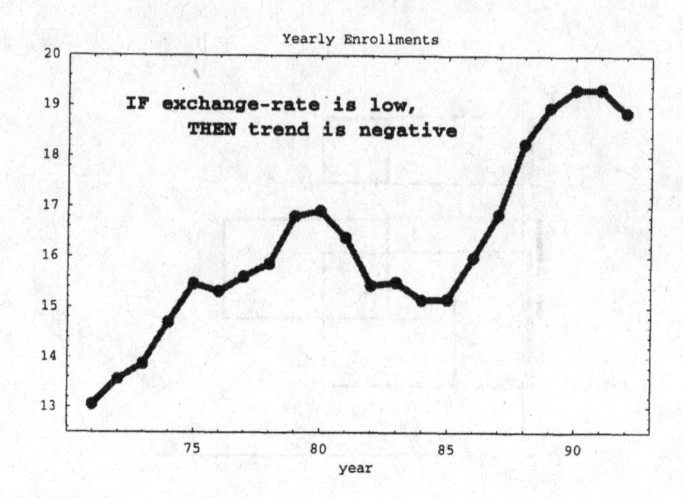

Fig. 6.5 Student numbers (in thousands) enrolling between 1971 and 1992.

that reporting a single number for a prediction (or parameter estimate) is almost always inadequate. Quantification of the *uncertainty* associated with a single number, while often challenging, is critical for subsequent decision making. Here, instead of generating first a single-valued prediction and then building an uncertainty model, we produce a qualitative forecast, $\pi(y)$ and then extract a single number if required. For instance, we obtain a numerical prediction of a value in Y by finding a representative value from the possibility distribution, for example,

$$\hat{y}(k+1) = \frac{\int_Y y \cdot \hat{\pi}(y|\mathbf{x}) \, \mathrm{d}y}{\int_Y \hat{\pi}(y|\mathbf{x}) \, \mathrm{d}y} \ . \tag{6.18}$$

Since, in general, we do not have $\forall y \in Y$ that there exists a y' for which $\pi(y') = 1$, we normalize (6.17) to the unit-interval. Note that this does not influence the numerical value predicted in any way (cf. Eq. (6.18)) but provides an intuitive interpretation of π (see Figure 6.12).

The random-set model itself can be described by a fuzzy graph (cf. (4.33))

$$\widetilde{F} \doteq \{(\mathbf{x}, c_\Gamma(\cdot, y))\} \ . \tag{6.19}$$

The model can therefore be generalized to fuzzy inputs. Thus, the formulation of the fuzzy system and inference described here is equivalent to the concepts of approximate reasoning (Section 7.2). The more important point, however, is that the information provided by π (from *quantitative* analysis) can be combined with vague or fuzzy information in a *qualitative* knowledge base as described in Section 7.2. The motivation for such *diagnostic signal analysis* is as follows: We started off with the assumption that only a small training data set M is available, the experiment may not be repeatable and that in the case of a time-series the process is nonlinear and non-stationary. As an example, consider the problem of forecasting the number of students at a particular university. Assuming we have records for 22 years, as shown in Figure 6.5, we wish to make a forecast for the upcoming year. Surely, we can extract cycles and trends from past data but *context-dependent knowledge* should help us to improve our forecast. In our example, knowledge of economical and political events provide additional information which is not buried in historical data. For instance, following the currency crisis in Asia, UK universities with a large number of Asian students could expect a negative trend in enrollments. It is this type of *qualitative* context-dependent knowledge which is conveniently represented by fuzzy if-then rules, and the random-set model suggested here allows us to combine both quantitative analysis with qualitative information. In Algorithm 6.1 the process of fuzzy random set modelling and forecasting is summarized. Note that modelling essentially comprises the storage of local uncertainty models (such as hypercubes or ellipsoids) but does not involve any numerical optimization.

1. Modelling:

1.1 Fix n_n, the number of nearest neighbors forming the local uncertainty models.

1.2 Calculate local uncertainty model $\mathcal{C} = \{C_j\}$, $j = 1, 2, \ldots, d$, where d is the number of training data, $\mathbf{m}_j \in \mathbf{M}$, $\mathbf{m}_j = [\mathbf{x}_j^T, y_j]^T$.

2. Forecasting:

2.1 Obtain n_p random subsets $C_i^{\mathbf{x}}$, where $\mathbf{x}$ extended into Ξ is covered by C_j. From this set, determine projections $C_i^{\mathbf{x}}$, $i = 1, 2, \ldots, n_p$ onto Y.

2.2 Estimate possibility distribution (qualitative forecast quantifying model uncertainty):

$$\hat{\pi}(y|\mathbf{x}) = \frac{1}{n_p} \sum_{i=1}^{n_p} \zeta_{C_i^p}(y) .$$

2.3 For a single-valued numerical forecast, 'defuzzify' $\pi(\cdot)$ calculating for instance the mean value.

Algorithm 6.1 Fuzzy random-set modelling and forecasting.

6.4 EXAMPLE 1: NONLINEAR AR PROCESS

The nonlinear first-order autoregressive process used here has been used to test other fuzzy models elsewhere [Bab98]. As before, it is assumed that the system identification problem is transformed into a static approximation problem: $y \approx f(\mathbf{x})$, where $\mathbf{x} = (x_1, x_2, \ldots, x_r)$, $\mathbf{x} \in (X_1 \times X_2 \times \cdots \times X_r) \subset \mathbb{R}^r$ and $y \in Y \subset \mathbb{R}$. The considered nonlinear AR(1) dynamic system is simulated

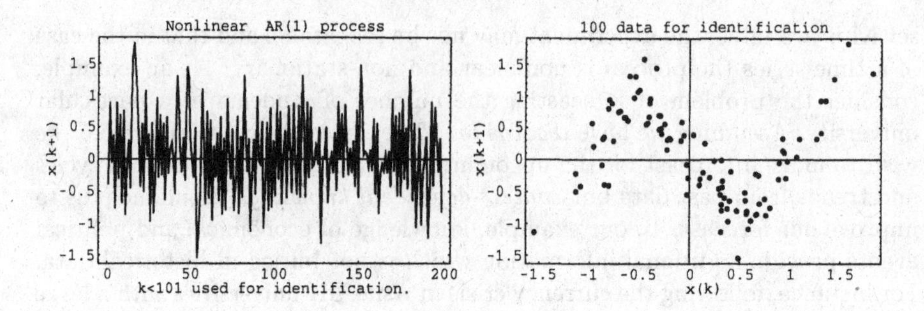

Fig. 6.6 Ex.1: Nonlinear AR(1) process: Time series and data space.

as follows:

$$x(k+1) = f\big(x(k)\big) + \varepsilon(k), \quad f(x) = \begin{cases} 2x - 2, & 0.5 \leq x, \\ -2x, & -0.5 < x < 0.5 \\ 2x + 2, & x \leq -0.5\ , \end{cases} \tag{6.20}$$

where $\varepsilon(k) \sim N(0, \sigma^2)$ with $\sigma = 0.3$. $x(0) = 0.1$ and $0 \leq k \leq 200$ of which the first hundred values are used for identification and the rest for model validation. Note that we keep the notation as in [Bab98], that is, $x(k+1)$ takes its values in Y.

The only model assumptions are that the data were generated by some nonlinear autoregressive system:

$$x(k+1) = f\big(x(k), x(k-1), \ldots, x(k-r+1)\big), \tag{6.21}$$

where r is the system order. We choose $r = 1$, leading to the *planar* case for the random set model. Figure 6.6 shows the signal on the left and observations $\mathbf{m}_j \in \mathbf{M}$, used for identification, on the right.

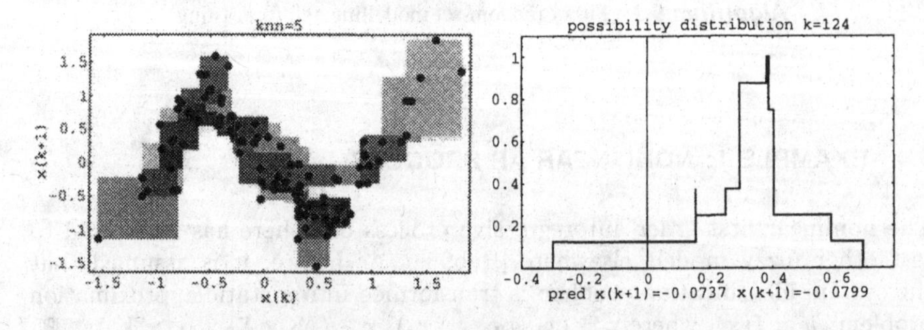

Fig. 6.7 Ex.1: Random-set model and forecast at $k = 124$, $n_n = 5$.

Figure 6.7 shows the random-set model obtained when considering five nearest neighbors. At $k = 124$, the *qualitative* forecast is a possibility distribution which for a value $y \in Y$, specifies the degree of confidence that this value will correspond to $x(k+1)$.

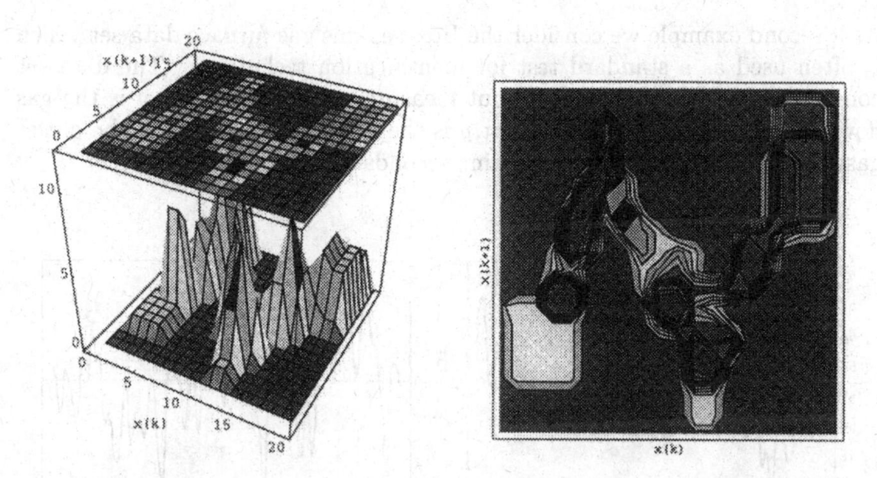

Fig. 6.8 Ex.1: Estimated fuzzy model $c_\Gamma(\mathbf{x}, y)$ and its contour plot.

Figure 6.8 shows the (non-normalized) fuzzy model and the contour plot calculated for a discretized product space $X \times Y$. From the contour plot, we can see that the random set model only conveys information about the regions covered by the identification data. Events $(\mathbf{x}, y)$ which have not been 'experienced' before will have only little or zero reliability assigned. Little extrapolation occurs outside regions covered by identification data, and the model explicitly quantifies the confidence of the model. This may be seen as an advantage but implies that the model may not provide any prediction for some input values.

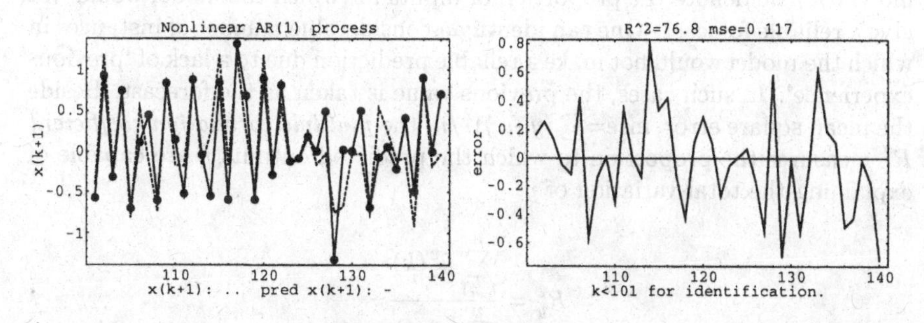

Fig. 6.9 Ex.1: Model validation.

For the five nearest neighbor model, Figure 6.9 shows a sequence of forecasts and the error. For better visibility only a section of the sequence is shown.

6.5 EXAMPLE 2: BOX-JENKINS GAS-FURNACE DATA

As a second example we consider the Box-Jenkins gas furnace data set which is often used as a standard test for identification techniques. The data set consists of 296 pairs of input-output measurements. The input u is the gas flow rate into a furnace, the output y is the CO_2 concentration in the outlet gases with a sampling interval of nine seconds.

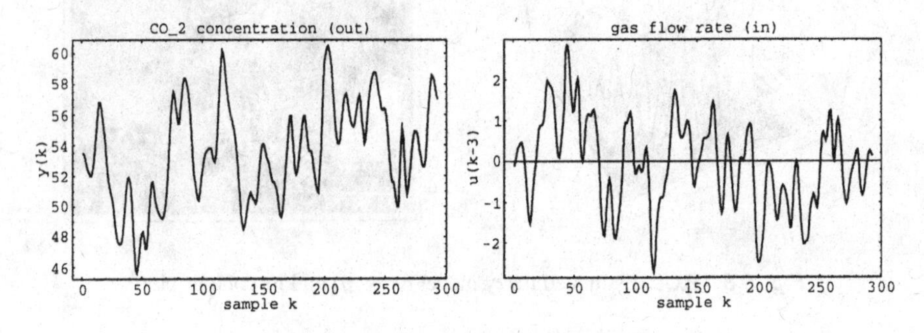

Fig. 6.10 Ex.2: Box-Jenkins gas furnace data.

The model structure chosen is $y(k) = f(y(k-1), u(k-3))$. The two data sets are plotted in Figure 6.10. Only the first 50 values are used for identification. Figure 6.11 shows the random set model for two nearest neighbor sets (cubes). An increase in the number of nearest neighbors, forming a random set, increases the area of the data space into which the model generalizes/extrapolates but also increases the model complexity.

Figure 6.12 illustrates the validation of a 5-nearest neighbor model. With md (in %) we denote the proportion of inputs for which the model would not give a reliable forecast. One can identify a considerable number of instances in which the model would not make a reliable prediction due to a lack of 'previous experience'. In such cases, the previous value is taken as the forecast. Beside the mean square error, mse=$\sum (y-\hat{y})^2/n$, the ***multiple correlation coefficient*** R_y^2 measures the proportion to which the predicted outputs $\hat{y}$ are capable of explaining the total variation of y:

$$R_y^2 = \frac{\sum_{k=1}^{n} \hat{y}_n^2(k)}{\sum_{k=1}^{n} y^2(k)} .$$

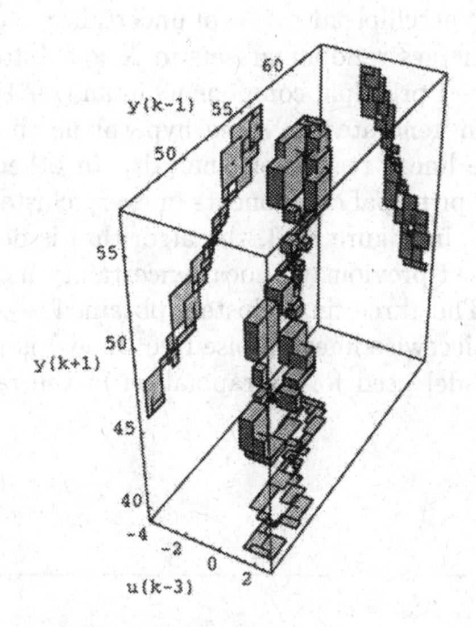

Fig. 6.11 Ex.2: Random set model. n_n=2.

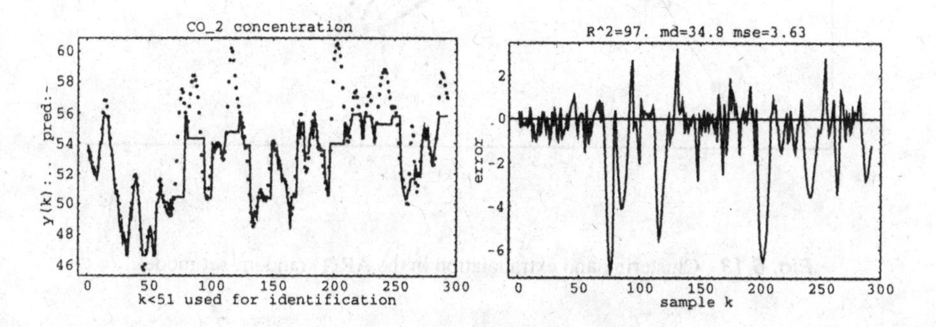

Fig. 6.12 Ex.2: Model validation. n_n=5.

In Figure 6.12, if the model confidence was zero, the current value was assumed as the forecast. As can be seen from the plot, this leaves to a rather large error at times. As suggested before, in these cases we would have to extrapolate. Using hyperellipsoids as local uncertainty models, a merging algorithm iteratively merges random subsets in $X \times Y$ into a small number of clusters. The $(r+1)-1$ principal components in an $(r+1)$-dimensional space of a set of data then generate the same hyperplane that we would obtain by applying multiple linear regression analysis. In other words, the hyperplane generated by r principal components in every cluster are used as model to make predictions. In Figure 6.13, the algorithm is demonstrated for the AR(1) model discussed previously. Local uncertainty models are formed by random ellipsoids. The three final clusters obtained are marked by thicker lines. The original piecewise-linear (noise-free) model is plotted as a dashed line, whereas the model used for extrapolation in the random-set model is shown as a solid line.

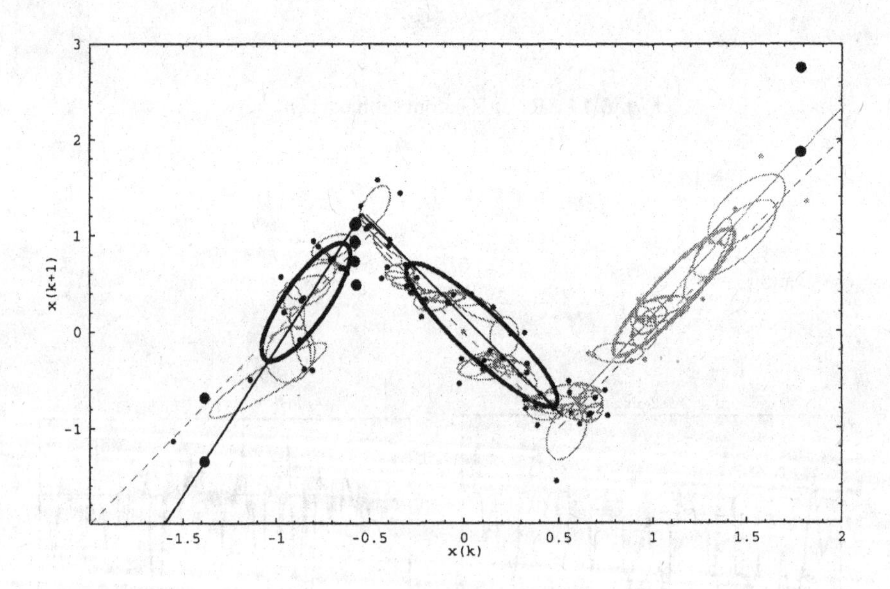

Fig. 6.13 Clustering and extrapolation in the AR(1) random-set model.

7

Certain Uncertainty

- ☐ *Scientific investigation relies on two principal concepts: comparing and reasoning.*
- ☐ *Mathematical formulations of distance and transitivity are at the core of the modelling problem.*
- ☐ *The Poincaré paradox describes the indistinguishability of individual elements in non-mathematical continua and hence proves that uncertainty is certain.*
- ☐ *Taking account of uncertainty leads to similarity (fuzzy) relations.*
- ☐ *Fuzzy concepts therefore occur 'naturally' from fundamental analysis.*
- ☐ *Fuzzy relations motivate approximate reasoning.*
- ☐ *Approximate reasoning is a concept to capture qualitative (context-dependent) expert knowledge.*

> The physical laws, in their observable consequences, have a finite limit of precision.
>
> —Kurt Gödel

If we are to reduce the process of a scientific investigation to two concepts, it would be *comparing* and *reasoning*. We use sets and operations on sets or, equivalently, relations in order to group and hence compare objects, variables, and so forth. In Section 1 we saw that an analysis of a system by means of observables inevitably induces equivalence relations. They play a fundamental role which we shall discuss further in this section. More general and philosophical consequences are discussed in a summary at the end of this section and in Section 12.2.

Let Ξ be a set, then a (crisp) relation R on Ξ will be a subset of the Cartesian product, $R \subset \Xi \times \Xi$. If $(\mathbf{o}, \mathbf{o}') \in R$, we shall write $R(\mathbf{o}, \mathbf{o}') = 1$ (or just $R(\mathbf{o}, \mathbf{o}')$ for short) to state that u is *related* to $\mathbf{o}'$ via R. A relation R in Ξ is an *equivalence relation*, denoted E if it satisfies the following conditions:

$$\begin{aligned} E(\mathbf{o}, \mathbf{o}) &= 1 \qquad \forall\, \mathbf{o} \in \Xi & &\text{(reflexive)}, \\ E(\mathbf{o}, \mathbf{o}') &= 1 \;\Rightarrow\; E(\mathbf{o}', \mathbf{o}) = 1 & &\text{(symmetry)}, \\ E(\mathbf{o}, \mathbf{o}') = 1 \;\wedge\; E(\mathbf{o}', \mathbf{o}'') &= 1 \;\Rightarrow\; E(\mathbf{o}, \mathbf{o}'') = 1 & &\text{(transitivity)}\,. \end{aligned}$$

Intuitively, an equivalence relation is a generalization of equality (which itself is an equivalence relation). We have therefore already heavily relied on two simple equivalence relations, equality ($=$) and elementhood ($\in$) for the purpose of comparison. With respect to equality, it is obvious that it satisfies the three conditions for an equivalence relation:

$$\begin{aligned} a = a \qquad &\text{holds} & &\text{(reflexivity)} \\ a = b \;\Rightarrow\; &b = a & &\text{(symmetry)} \\ a = b \;\wedge\; b = c \;\Rightarrow\; &a = c & &\text{(transitivity)}\,. \end{aligned}$$

From this example, we can see that transitivity enables us to infer something new about the relationship of two variables given two pieces of information. The concept plays consequently a fundamental role in reasoning. We can illustrate transitivity with another important tool for comparison: the concept of a metric.

The function $d(\cdot, \cdot)$ defines a *distance* between elements of Ξ. Let for any $\mathbf{o}_i, \mathbf{o}_j, \mathbf{o}_k$ in Ξ:

$$\begin{aligned} d(\mathbf{o}_i, \mathbf{o}_j) &= 0 & &\text{iff} \quad \mathbf{o}_i = \mathbf{o}_j \\ d(\mathbf{o}_i, \mathbf{o}_j) &> 0 & &\text{iff} \quad \mathbf{o}_i \neq \mathbf{o}_j \\ d(\mathbf{o}_i, \mathbf{o}_j) &= d(\mathbf{o}_j, \mathbf{o}_i) & &\text{symmetry}\,. \end{aligned}$$

A distance is called *metric* iff $\forall \mathbf{o}_i, \mathbf{o}_j, \mathbf{o}_k \in X$, it is transitive:

$$d(\mathbf{o}_i, \mathbf{o}_k) \leq d(\mathbf{o}_i, \mathbf{o}_j) + d(\mathbf{o}_j, \mathbf{o}_k)\,. \tag{7.1}$$

This inequality is called the *triangle inequality*:

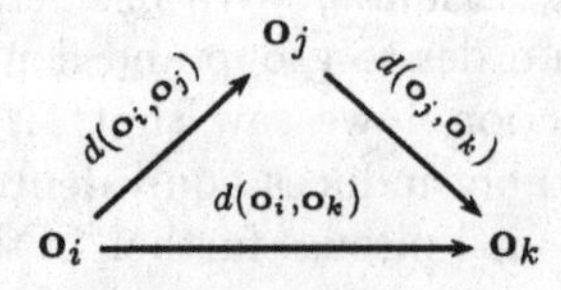

A simple example for a metric is the absolute value of the difference:

$$d(x, x') = |x - x'| .$$

Having established the basic tools for comparison and reasoning it remains to define a mechanism of *order*. Examples of relations which establish an order are the 'greater than' and 'subsethood' relations. Again these relations are transitive:

"greater or equal" $\geq$: $\mathbf{o}_i > \mathbf{o}_j \quad \wedge \quad \mathbf{o}_j > \mathbf{o}_k \quad \Rightarrow \quad \mathbf{o}_i > \mathbf{o}_k$

"set-inclusion" $\subseteq$: $A \subseteq B \quad \wedge \quad B \subseteq C \quad \Rightarrow \quad A \subseteq C$

Formally, these relations are establishing a *partial order* on Ξ, making Ξ a partially ordered set (poset). A *partial ordering* (or semi-ordering) on Ξ is a binary relation $\preceq$ on Ξ such that the relation is

reflexive, *i.e.* $x \preceq x$,
antisymmetric, *i.e.* $x \preceq x'$ and $x' \preceq x$ implies $x = x'$,
transitive, *i.e.* $x \preceq x'$ and $x' \preceq x''$ implies $x \preceq x''$.

In Section 1, for the analysis of a system we were required to establish the quality of states, that is, we were testing values of observables for equality. If such an analysis is to be implemented in a computer or validated with measured data, we may find it impossible to establish theoretical equality for real numbers. We are therefore forced to take some *imprecision* into account. Matching mathematical idealism with physical reality, however, has unfortunate consequences illustrated by the Poincaré paradox. In short, the Poincaré paradox describes the *indistinguishability* of individual elements in non-mathematical continua. More specifically, for three points $\mathbf{o}_i$, $\mathbf{o}_j$ and $\mathbf{o}_k$, let δ denote a threshold (tolerance, significance level). Then, two elements $\mathbf{o}$ and $\mathbf{o}'$ are indistinguishable for $d(\mathbf{o}, \mathbf{o}') \leq \delta$, where $d(\cdot, \cdot)$ denotes a proximity measure such as a metric. Therefore, it turns out that physical equality is not transitive, that is, (7.1) does not apply if we study physical systems by means of observations. The Poincaré paradox demonstrates that an element, $\mathbf{o}_j$, may be indistinguishable from two others, $\mathbf{o}_i$ and $\mathbf{o}_k$ that can be distinguished from one another. Let us consider the following measurements in $\mathbb{R}$: $\mathbf{o}_i = 1.5$, $\mathbf{o}_j = 2$, $\mathbf{o}_k = 2.2$, and $\delta = 0.6$, and let us use the metric $d(\mathbf{o}, \mathbf{o}) = |\mathbf{o} - \mathbf{o}'|$, w.r.t. a threshold, accuracy or error bound δ, to identify observations. Our analysis is based on the *theoretical model* that if observations $\mathbf{o}_i$ and $\mathbf{o}_j$ are similar, as well as $\mathbf{o}_j$ and $\mathbf{o}_k$ are similar, then so should be $\mathbf{o}_i$ and $\mathbf{o}_k$, in other words,

$$\mathbf{o}_i = \mathbf{o}_j \wedge \mathbf{o}_j = \mathbf{o}_k \Rightarrow \mathbf{o}_i = \mathbf{o}_k .$$

Now, considering *actual data*,

$$
\begin{aligned}
& |\mathbf{o}_i - \mathbf{o}_j| = 0.5 < \delta \quad \Rightarrow \quad \mathbf{o}_i = \mathbf{o}_j \\
& |\mathbf{o}_j - \mathbf{o}_k| = 0.2 < \delta \quad \Rightarrow \quad \mathbf{o}_j = \mathbf{o}_k \\
\text{but} \quad & |\mathbf{o}_i - \mathbf{o}_k| = 0.7 > \delta \quad \Rightarrow \quad \mathbf{o}_i \neq \mathbf{o}_k\,.
\end{aligned} \tag{7.2}
$$

A solution to this dilemma leads us directly to fuzzy relations which have already played an important role in clustering in Section 4. To bridge the mathematical idealization with physical reality, Karl Menger suggested retaining the transitive relation but introducing a measure between 0 and 1, probabilities, to quantify uncertainty. More specifically, we associate $d(\mathbf{o}, \mathbf{o}')$ by a (cumulative) distribution function $F_{\mathbf{o},\mathbf{o}'}$ whose value $F_{\mathbf{o},\mathbf{o}'}(a)$ for any a is interpreted as the probability that the distance between $\mathbf{o}$ and $\mathbf{o}'$ is less than a.' As a result, metric spaces become ***probabilistic metric spaces*** [SS83]. The most important fact, however, is that the triangle inequality given in (7.1) has no unique generalization:

$$
F_{\mathbf{o}_i\mathbf{o}_k}(a+b) \geq T\left(F_{\mathbf{o}_i\mathbf{o}_j}(a), F_{\mathbf{o}_j\mathbf{o}_k}(b)\right)\,, \tag{7.3}
$$

with a *choice* of functions $T(\cdot,\cdot)$. This means that the model under consideration is a formal model which cannot model reality adequately and hence assumptions are in a sense arbitrary, that is, the model builder can freely decide which model characteristics he or she chooses. We can arrive at inequality (7.3) by starting with the triangle inequality (7.1), which implies the logical proposition

$$
d(\mathbf{o}_i, \mathbf{o}_j) < a \quad \wedge \quad d(\mathbf{o}_j, \mathbf{o}_k) < b \Rightarrow d(\mathbf{o}_i, \mathbf{o}_k) < a + b\,.
$$

Since $A \Rightarrow B$ implies that $Pr(A) \leq Pr(B)$, we get

$$
Pr\big(d(\mathbf{o}_i, \mathbf{o}_j) < a \wedge d(\mathbf{o}_j, \mathbf{o}_k) < b\big) \leq Pr\big(d(\mathbf{o}_i, \mathbf{o}_k) < a + b\big) = F_{\mathbf{o}_i\mathbf{o}_k}(a+b).
$$

Thus, if T is such that $T(Pr(A), Pr(B)) \leq Pr(A \wedge B)$ for any two propositions A and B, the desired inequality (7.3) will follow. The function T is a mapping $[0,1] \times [0,1] \rightarrow [0,1]$. For example,

$$
\begin{aligned}
T_{\min}(a,b) &= \min(a,b) && \text{(minimum operator)}, \\
T_{\text{Luk}}(a,b) &= \max(a+b-1, 0) && \text{(Lukasiewicz norm)}, \\
T_{\text{pro}}(a,b) &= a \cdot b && \text{(algebraic product)}.
\end{aligned} \tag{7.4}
$$

Let $T = a \cdot b$. Then (7.3) becomes

$$
F_{\mathbf{o}_i\mathbf{o}_k}(a+b) \geq F_{\mathbf{o}_i\mathbf{o}_j}(a) \cdot F_{\mathbf{o}_j\mathbf{o}_k}(b)\,, \tag{7.5}
$$

which states that the probability of the distance between $\mathbf{o}_i$ and $\mathbf{o}_k$ being smaller than $a+b$ is at least the joint probability of the independent occurrence of the distance between $\mathbf{o}_i$ and $\mathbf{o}_j$ being smaller than a and the distance between $\mathbf{o}_j$ and $\mathbf{o}_k$ being smaller than b. In other words,

$$Pr\big(d(\mathbf{o}_i, \mathbf{o}_k) < a + b\big) \geq Pr\big(d(\mathbf{o}_i, \mathbf{o}_j) < a,\ d(\mathbf{o}_j, \mathbf{o}_k) < b\big) \ .$$

So much for probabilistic uncertainty and the introduction to triangular norms. We now show that a metric induces a similarity (fuzzy) relation for which transitivity is generalized in the form of inequality (7.3). What we look for is a transitive relation that defines a degree of indistinguishability for values being very close, $\mathbf{o} \approx \mathbf{o}'$. A *fuzzy equivalence relation* or *similarity relation* [Zad71], $\widetilde{E}$, is a fuzzy relation which is reflexive, symmetric, and transitive. It defines a function $\widetilde{E}\colon \Xi \times \Xi \to [0,1]$ that satisfies the conditions (cf. (4.24)):

$$\begin{aligned}
\widetilde{E}(\mathbf{o},\mathbf{o}) &= 1 \qquad \forall\, \mathbf{o} \in \Xi && \text{(reflexive)},\\
\widetilde{E}(\mathbf{o},\mathbf{o}') &= \widetilde{E}(\mathbf{o}',\mathbf{o}) && \text{(symmetric)},\\
\widetilde{E}(\mathbf{o},\mathbf{o}'') &\geq T\big(\widetilde{E}(\mathbf{o},\mathbf{o}'), \widetilde{E}(\mathbf{o}',\mathbf{o}'')\big) && \text{(transitive)} .
\end{aligned}$$

Transitivity for fuzzy relations is therefore defined in analogy to Menger's inequality, (7.3), for probabilistic metric spaces. In this context, the triangular norm T, extends the domain of logical conjunction from the set $\{0,1\}$ to the interval $[0,1]$. Using the min-operator, we speak of min-transitivity as a natural extension of the equivalence relation above. The equivalence classes partition U into sets containing elements that are all similar to each other to degree at least δ.

For a bounded metric space (Ξ, d) there exists a non-negative value $\delta \in \mathbb{R}^+$ such that $d(\mathbf{o},\mathbf{o}') \leq \delta$ for all $\mathbf{o}$ in Ξ. The distance between values of factors on objects then induces a fuzzy relation over Ξ:

$$\widetilde{E}(\mathbf{o},\mathbf{o}') = 1 - \frac{1}{\delta} d(\mathbf{o},\mathbf{o}') \ .$$

The bound δ allows scaling such that the distance between any two values in Ξ lies in the unit-interval $[0,1]$. The correspondence of transitivity for a distance function and transitivity of fuzzy relations depends on the T-norm employed. The metric equivalent of the Lukasiewicz norm (7.4), is the triangle inequality, (7.1); the metric equivalent of product transitivity is the inequality

$$d(\mathbf{o}_i, \mathbf{o}_k) \leq d(\mathbf{o}_i, \mathbf{o}_j) + d(\mathbf{o}_j, \mathbf{o}_k) - d(\mathbf{o}_i, \mathbf{o}_j) d(\mathbf{o}_j, \mathbf{o}_k)$$

related to (7.5) w.r.t. probabilistic uncertainty. If a fuzzy equivalence relation is min-transitive the distance satisfies the more restrictive ultrametric inequality:

$$d(\mathbf{o}_i, \mathbf{o}_k) \leq \max\big(d(\mathbf{o}_i, \mathbf{o}_j), d(\mathbf{o}_j, \mathbf{o}_k)\big) \ .$$

The Lukasiewicz norm turns out to be the least restrictive one. For the comparison of factors on objects u it would usually be reasonable to assume that to objects are similar in their contribution to the model if $|\mathbf{o}-\mathbf{o}'| \leq \delta$, where δ is a number representing our "indifference" w.r.t. to the measurement process.

We saw that if we are to use equivalence relations, in practical situations, we may allow for a tolerance to identify two objects as the same (as having the same observable consequence). The inequality $|\mathbf{o}-\mathbf{o}'| \leq \delta$ describes a subset (relation) $R_\delta \subset \Xi \times \Xi$

$$R_\delta = \{(\mathbf{o},\mathbf{o}') \in \Xi \times \Xi : \ |\mathbf{o}-\mathbf{o}'| \leq \delta\} \ .$$

The Poincaré paradox (7.2) demonstrated that this relation is not an equivalence relation, that is, it is not a transitive relation. We therefore could not study the quotient set induced by this relation. Kruse et al. [KGK94][1] showed, however, that we can define a mapping $\widetilde{E}_\delta$ such that $\widetilde{E}_\delta(\mathbf{o},\mathbf{o}')$ is greater than $1-\delta$ if and only if $\mathbf{o}$ and $\mathbf{o}'$ are indistinguishable with respect to the tolerance δ:

$$(\mathbf{o},\mathbf{o}') \in R_\delta \quad \text{if and only if} \quad \widetilde{E}_\delta(\mathbf{o},\mathbf{o}') \geq 1-\delta \ ,$$

where

$$\begin{aligned} \widetilde{E}_\delta : \quad \Xi \times \Xi &\to [0,1] \\ (\mathbf{o},\mathbf{o}') &\mapsto 1 - \inf\{\delta \in [0,1] : \ (\mathbf{o},\mathbf{o}) \in R_\delta\} \end{aligned}$$

with $\delta \in [0,1]$ and if there is no δ for which the relation holds, we define $\inf \emptyset \doteq 1$. $\widetilde{E}_\delta$ is a fuzzy equivalence relation w.r.t. T_{Luk}. The value $\widetilde{E}_\delta(\mathbf{o},\mathbf{o}') = 1-\min\{|\mathbf{o}-\mathbf{o}'|,1\}$ describes the degree to which two objects $\mathbf{o}$ and $\mathbf{o}'$ have similar observable consequences and transitivity of this relation implies that if $\mathbf{o}$ and $\mathbf{o}'$ are similar and $\mathbf{o}'$ and $\mathbf{o}''$ are similar in their values in Ξ, then $\mathbf{o}$ is similar to $\mathbf{o}''$. To arrive at the inequality (4.24) defining transitivity for similarity relations, we show that any (pseudo)metric (in the unit-interval) induces a similarity relation and vice versa [Zad71]:

$$\widetilde{E}(\mathbf{o},\mathbf{o}') = 1 - \inf\big(d(\mathbf{o},\mathbf{o}'),1\big) \ . \tag{7.6}$$

Let $d'(\mathbf{o},\mathbf{o}') = 1-\inf(d(\mathbf{o},\mathbf{o}'),1)$ be a pseudo-metric. Inserted in the transitivity law for fuzzy relations (4.24) with $T_{\min}$, the greatest t-norm, gives

$$1 - d'(\mathbf{o}_i,\mathbf{o}_k) \geq \min\big(1-d'(\mathbf{o}_i,\mathbf{o}_j),\ 1-d'(\mathbf{o}_j,\mathbf{o}_k)\big)$$

[1]The book [KGK94] and various papers by R. Kruse and F. Klawonn provide an extensive treatment of equivalence relations and how rule-based systems can be build from them (see Section 4.6). They also generalize the case described here with $\delta \in [0,1]$ and $\widetilde{E}_\delta(\mathbf{o},\mathbf{o}') = 1-\min\{|\mathbf{o}-\mathbf{o}'|,1\}$ to any unit in Ξ by means of a scaling factor $s > 0$, $\widetilde{E}_\delta(\mathbf{o},\mathbf{o}') = 1-\min\{|s\cdot\mathbf{o}-s\cdot\mathbf{o}'|,1\}$.

and since $T(A(\mathbf{o}), B(\mathbf{o})) = 1 - S(1 - A(\mathbf{o}),\ 1 - B(\mathbf{o}))$, the t-conorm,

$$1 - d'(\mathbf{o}_i, \mathbf{o}_k) \geq 1 - \max(d'(\mathbf{o}_i, \mathbf{o}_j),\ d'(\mathbf{o}_j, \mathbf{o}_k))$$
$$d'(\mathbf{o}_i, \mathbf{o}_k) \leq \max(d'(\mathbf{o}_i, \mathbf{o}_j),\ d'(\mathbf{o}_j, \mathbf{o}_k)),$$

which is the ultra-metric, implying the triangle inequality (7.1). Then from (7.6), with respect to (7.3), $\widetilde{E}(\mathbf{o}, \mathbf{o}')$ can also be interpreted as the probability that the distance between $\mathbf{o}$ and $\mathbf{o}'$ is equal to 0.

This section has focused on equivalence relations and transitivity as two fundamental concepts in any scientific investigation that makes use of a formal model. Transitivity is not only of importance in science and engineering. 'Non-technical' decision making is based on prioritization: In pairwise comparisons the importance (dominance, preference) is relative. For example, we may consider q three times more important than p and r twice as important than p. To be consistent, we need a rule, such as, if q is more important than p and r is more important than q then also r should be more important than p. More formally, if we denote by $i(\cdot, \cdot)$ the relative importance, to ensure consistency we require the transitive relation

$$\text{IF } i(q,p) = x \text{ AND } i(r,q) = y, \text{ THEN } i(r,p) = x \cdot y$$

to hold[2]. We have seen that comparisons, using set operations, norms and metrics quantifies (dis)similarity by inducing formal relations. Consistent reasoning requires the use of some rule, that is, relations are expected to be transitive. Both the definition of functions to evaluate comparisons and the form of the transitive law are *defined* - choice, *a priori*, *ad hoc*, whether it is formal, rigorous or not.

7.1 UNCERTAINTY IN SYSTEMS ANALYSIS

The sense of the world must lie outside the world... What we cannot speak about we must remain silent about.

—Ludwig Wittgenstein, Tractatus Logico-Philosophicus

To this point our study of system models and the analysis of data in the presence of uncertainty suggest a number of conclusions about the encoding of natural systems through formal systems:

▷ As Karl Popper demonstrated, scientific theories deal with concepts not reality. Formula and theories are so formulated as to correspond in

[2]The 'product-transitive' relation is also the basis for *Analytical Hierarchy Processes* (AHP); a decision making and prioritization technique developed by L. Saaty.

some 'useful' way to the real world. However, this is an approximate correspondence. Mathematical forms say by *themselves* nothing about material reality. Any objective content lies entirely in the (biological, physical, ..) meaning attached *ad hoc* to the symbols appearing in mathematical formulations. There is no wrong theory or model. Instead, one may be more useful or convenient than another. The quest for precision is analogous to the quest for certainty and both precision and certainty are impossible to attain. It is therefore important to be precise about uncertainty, not to ignore it but to incorporate it in our models and theories. In systems engineering, uncertainty appears in various distinct forms, illustrated in Figure 7.1. A successful methodology to analyze complex systems will have to embrace all types of uncertainty including those induced by human experts or operators. Data alone, without knowledge of the context in which they were generated, will not be sufficient. (More details of the concepts presented in Figure 7.1 can be in found in [Wol98].)

▷ Triangular norms, which in many ways are at the root of fuzzy mathematics and fuzzy logic, are motivated by an analysis of the fundamental mechanism we use in the analysis of data.

▷ Similarity relations arise from metrics employed to quantify the similarity of objects (7.6). As for conventional equivalence relations, the definition of transitivity is important. In general, transitivity for some fuzzy relation R, is commonly denoted $R \supseteq R \circ R$ and is defined in terms of membership functions by

$$\mu_R(x,z) \geq \sup_{y \in Y} T\big(\mu_R(x,y), \mu_R(y,z)\big) .$$

Or, equivalently, expressed in terms of fuzzy equivalence relations $\widetilde{E}$ we obtain (4.24), which resembles Menger's inequality (7.3). In Section 4.6, we describe an approach to approximate reasoning based on similarity relations, developed by R. Kruse and F. Klawonn [KGK94]. The principle idea of approximate reasoning is introduced in the following section.

More philosophical consequences from these considerations and the arguments put forward throughout the book are further discussed in Section 12.2.

7.2 A FUZZY PROPOSITIONAL CALCULUS

Inferences of Science and Common Sense differ from those of deductive logic and mathematics in a very important respect, namely, when the premises are true and the reasoning correct, the conclusion is only *probable*.

—Bertrand Russell

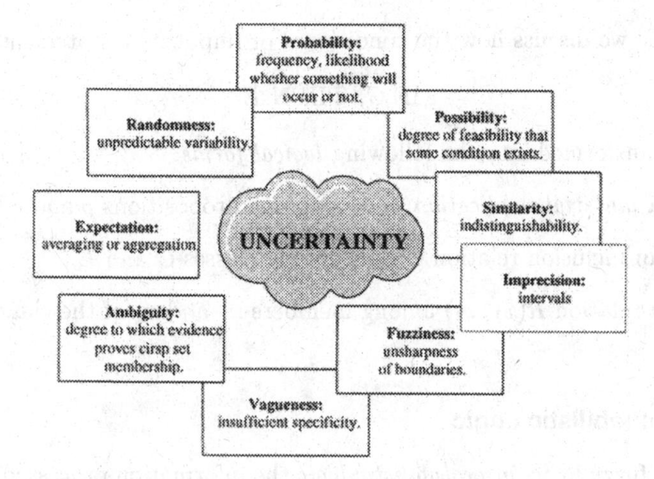

Fig. 7.1 Forms and representations of uncertainty.

In subsequent sections, we will describe the problem of how to capture qualitative, context-dependent expert or operator knowledge in form of fuzzy logic, if-then rules. The models developed in Sections 4 and 5 take a very similar, if not identical, form but there the fuzzy graph $\widetilde{F}$ is interpreted as a set of rules or statements about local dependencies among variables. In Section 6, we developed a time-series model in which forecasts are formulated as possibility distributions, which, in turn, can be processed in a fuzzy logic rule-based system as will be discussed in Section 7.2. In Section 4 we then suggested to interpret the cluster structure *directly* in terms of a multi-valued *logic*. In this section, we discuss the question how logic or which form of logical relations are suitable for modelling dynamics, states, and dependencies.

Though the concepts and ideas introduced in previous sections, lead to models which are sets of rules, statements about local dependencies among variables, we acknowledge that the *causal problem*[3] is an ontological, not a logical question, it cannot be *reduced* to logical terms but it can be analyzed with the help of logic. Reducing a causal problem to its logical aspects, the terms 'cause' and 'effect' are associated with *truth values* as follows: Event C is the cause of event E, and event E the effect of event C, is translated into three logical propositions, u, p, c, such that u is a true universal law, p describes c, and c is the logical consequence of u and p. p and c are statements about concrete objects - events, processes, conditions and so on. In the following

[3] As a definition of a 'causal law', which is not strictly bound to any specific philosophical perspective, we shall understand by a 'causal dependency' a general proposition by virtue of which it is possible to *infer* the existence of an event from the existence of another.

subsections we discuss how the conditional or implicative statement

IF C, THEN E

may be transformed into the following *logical forms*:

- As a *material implication*, $p \Rightarrow c$ among propositions p and c.
- As an inclusion relation $C \subset E$ among classes C and E.
- As a relation $R(x_1, x_2)$ among members x_1 and x_2 of the classes C and E.

7.2.1 Probabilistic Logic

Similar to fuzzy logic, in *probabilistic logic* the information processed is uncertain. The statements contain quantifiers such as "many", "almost" instead of just $\exists, \forall$ in first-order logic. Viewing degrees of beliefs in the truth of propositions as subjective probabilities, one would like to use standard calculus of probabilities to implement a 'logic' of uncertain information [NW97]. Formally, the objective is to identify logical formulations with set-theoretic ones - propositions are viewed as elements of a Boolean algebra that is a lattice. This is accomplished with a partial order given by $p \preceq c$ if $p \wedge c = p$ and the connectives $\wedge$ (meet) and $\vee$ (join) are defined in terms of $\preceq$ by $p \vee c = \sup\{p, c\}$ and $p \wedge c = \inf\{p, c\}$. The difficulties of a synergy of set theory and logic becomes obvious when comparing the conditional probability model for "c given p" - $Pr(c|p) = Pr(c \text{ and } p)/Pr(p)$ and the two-valued logical equivalent of it, $Pr(p \Rightarrow c)$, determined as the truth of "c or not p". We may therefore have $Pr(p \Rightarrow c) > Pr(c|p)$. This problem of defining an algebraic synthesis of the foundations of logic and probabilities was the original program of George Boole [Boo58], which is now discussed in the context of *conditional event algebras* [GMN97].

7.2.2 Classical Two-Valued Logic

The basis of *propositional calculus* is a set of formal entities, that is, simple statements - primitive propositions, often called *variables* of the logic. Such variables are combined using basic logical connectives, $\vee$ (*or*), $\wedge$ (*and*), $\neg$ (*not*), to build expressions or *formulas*. The mapping from the set of all expressions into the set of truth values is called *truth evaluation*. In two-valued logic the 'truth' of a proposition can take the values 0 ("false") and 1 ("true") only. The formula $p \Rightarrow c$, modelling "IF p THEN c" or "p implies c" is called *material implication* and is defined by

$$p \Rightarrow c \doteq \neg p \vee c \tag{7.7}$$

$$= (p \vee c) \vee \neg p \, , \tag{7.8}$$

that is, it is defined in terms of the three basic connectives. From the table above it becomes apparent that two-valued logic is insufficient to deal with all those cases for which we might employ rule-based knowledge. In particular, the truth values of p and $p \Rightarrow c$ cannot be chosen independently and it is not possible to quantify gradual changes in the antecedent and consequent of a rule. In fact, if the antecedent p were interpreted as the cause and the consequent c as the effect, the material implication would mean that an absent cause entails any effect. Further, every proposition implies itself as $p \Rightarrow p$, meaning everything is self-caused.

A propositional calculus is a logic of atomic propositions which cannot be broken down. The validity of arguments does not depend on the meaning of these atomic propositions, but rather on the form of the argument. If we consider propositions of the form "all *as* are *b*" which involves the ***quantifier*** "all" and the ***predicate*** b, then the validity of an argument should depend on the relationship between parts of the statement as well as the form of the statement. In order to reason with this type of proposition, propositional calculus is extended to ***predicate calculus***. A predicate on a set is a relation. Generalizing the Boolean modus ponens, we can either redefine the implication or allow fuzzy concepts (fuzzy sets) in the premise and conclusion parts of the rule. This leads to what is known as ***approximate reasoning***.

Drawing conclusions from hypotheses, approximate reasoning is an approach which models IF-THEN types of knowledge representations by generalizing the Boolean (two-valued) logic ***modus ponens*** ("forward reasoning"):

	Approximate Reasoning	Two-valued logic
Implication	IF $\mathbf{x}$ is A, THEN $\mathbf{y}$ is B.	$p \Rightarrow c$
Premise	$\mathbf{x}$ is A'.	p
Conclusion	$\mathbf{y}$ is B'.	c

In fault or change detection, the truth of a proposition is interpreted as the ***feasibility*** or ***possibility*** to which some practical condition exists. The rule IF C, THEN E, describes the truth value or feasibility of E as a function of the truth value of event C. In other words, the rule serves as a description, explanation:

$$\text{IF } p, \text{ THEN } c$$
$$\text{or IF } C \text{ is true, THEN } E \text{ is true },$$

where $p =$ "C is true" essentially is a test of hypothesis. On the other hand, in prediction we identify the succession of events

$$\text{IF } \omega(k), \text{ THEN } \omega(k+1) \text{ is } \omega' ,$$

where again in the premise part of the rule we test a hypothesis for the possibility or feasibility that the systems is in a particular state or condition. The uncertainty is with the truth evaluation or measurement of the current state, condition the system is in. In conclusion, it seems that a relational framework would be most promising in modelling and analyzing local dependencies among variables. The next section will introduce fuzzy sets and relations as a basis for approximate reasoning.

7.2.3 Approximate Reasoning

> Everything is vague to a degree you do not realize till you have tried to make it precise.
>
> —Bertrand Russell

To introduce the idea of approximate reasoning consider the set of "people", denoted P. Let $A \subset P$ be the subset of all "tall people" and let $h(p)$ be a function describing the height for any $p \in P$, $h\colon P \to \mathbb{R}$ such that $h(p_1) \geq h(p_2)$ means person p_1 is at least as tall as p_2. Suppose $P \neq \emptyset$, there is at least one tall person. (This assumption should be reasonable by experience: the author of this book is 2.11 m tall and has not met anyone considering him as "not tall".) We can formulate the following inductive rule:

IF p_1 is tall, AND p_1 is indistinguishable from p_2, THEN p_2 is tall.

Starting with three persons, p_1 known to be tall, p_2 only 1 cm taller than p_1 and p_3 being 1cm taller than p_2, we find that the rule is true for p_1 and p_2, p_2 is concluded to be a tall person. Since p_3 is only just as much taller than p_2 as p_2 is taller than p_1 we also conclude p_3 is tall. Declaring one person as tall, it turns out from a finite iteration of modus ponens that all people are tall – which clearly does not conform with the author's experience. The problem lies with the application of the *law of the excluded middle*, $A \cap A^c = \emptyset$, applied to vague concepts such as "tall people". Using a fuzzy set to represent the set of tall people avoids this paradoxical situation. Consider the map

$$A : P \to [0,1] .$$

Then A should be some function of the height of a person, $A(p) = f\big(h(p)\big)$. Function f should be continuous, monotonically increasing and for at least one value of h it should be equal to one. The inductive premise is now reformulated as a relation $R(p_1, p_2)$ such that (see also p. 145)

$$R(p_1, p_2) = 1 \quad \text{given} \quad (p_1, p_2) \in \{(p_1, p_2) \,:\, 0 \leq |h(p_1) - h(p_2)| \leq \delta\} .$$

A simple way of deducing the truth value $A(p_2)$ of "p_2 is tall" from the truth value $A(p_1)$ is by means of the following relation representing the concept of "relative height":

$$R(p_1, p_2) = \frac{A(p_2)}{A(p_1)} ,$$

from which we obtain the basic deductive relationship

$$A(p_2) \geq R(p_1, p_2) \cdot A(p_1) \qquad \forall (p_1, p_2) .$$

If both persons are of similar height, $R(p_1, p_2)$ is close to one and it turns out that for finite iterations of modus ponens, as above, the validity of the deductive process becomes smaller and smaller. It seems that fuzzy sets and fuzzy relations provide a methodology to capture *approximate reasoning*.

In approximate reasoning, classical propositions p, c, which can be either true or false, are replaced by fuzzy propositions such as "**x** is A" where **x** is a fuzzy variable and the fuzzy concept A is represented by a fuzzy subset. A given fact "**x** is A'", conjunctively combined with the prior knowledge of the implication rule, leads to gradual truth values taking values on the unit-interval. In standard logic the emphasis is on formal validity and truth is to be preserved under any and every interpretation. In contrast, in approximate reasoning one tries to preserve information within the situation (context) in which the reasoning takes place. In general, we identify the triple $(\neg, \wedge, \vee)$ with $(^c, \cap, \cup)$ and here in particular with $(1-\mu, T, S)$. Here, a t-norm is a binary function that extends the domain of logical conjunction from the set $\{0, 1\}$ to the interval $[0, 1]$. Similarly, S models disjunctive operations.

A *proposition* takes the form "**x** is A" with fuzzy variable **x** taking values in X and A modelled by a fuzzy set defined on the *universe of discourse* X by *membership function* $\mu\colon X \to [0, 1]$. A *compound statement*, "**x** is A AND **y** is B", is taken as a fuzzy set $A \cap B$ in $X \times Y$ with

$$\mu_{A \cap B}(x, y) = T\big(\mu_A(x), \mu_B(y)\big) .$$

For the sake of simplicity, we consider a single rule of type

$$\text{IF } \mathbf{x} \text{ is } A, \text{ THEN } \mathbf{y} \text{ is } B ,$$

which can be regarded as a fuzzy relation

$$\begin{aligned} R : X \times Y &\to [0, 1] \\ (x, y) &\mapsto R(x, y) , \end{aligned}$$

where $R(x, y)$ is interpreted as the strength of relation between x and y. Viewed as a fuzzy set, with $\mu_R(x, y) = R(x, y)$ denoting the degree of membership in the (fuzzy) subset R, $\mu_R(x, y)$ is computed by means of a *fuzzy implication*. Replacing the negation $\neg$ in (7.7) with the basic fuzzy complement $1 - \mu$, and the disjunction $\vee$ with the fuzzy union max-operator, we obtain the so-called *Dienes-Rescher implication*

$$\mu_R(x, y) = \max\big(1 - \mu_A(x), \mu_B(y)\big) . \tag{7.9}$$

From (7.8), replacing negation by the fuzzy complement, disjunction by the max-operator and conjunction by the min-operator, we obtain the *Zadeh implication*

$$\mu_R(x,y) = \max\left(\min\left(\mu_A(x), \mu_B(y)\right), 1 - \mu_A(x)\right) \ . \tag{7.10}$$

Other possibilities are

$$\mu_R(x,y) = \min\left(1, 1 - \mu_A(x) + \mu_B(y)\right) \qquad : \text{Lukasiewicz implication,} \tag{7.11}$$

$$\mu_R(x,y) = \begin{cases} 1 & \text{if } \mu_A(x) \le \mu_B(y), \\ \mu_B(y) & \text{otherwise.} \end{cases} \qquad : \text{Gödel implication,} \tag{7.12}$$

or defined using t-norms

$$\mu_R(x,y) = \min\left(\mu_A(x), \mu_B(y)\right) \qquad : \text{Minimum implication,} \tag{7.13}$$

$$\mu_R(x,y) = \mu_A(x) \cdot \mu_B(y) \qquad : \text{Product implication.} \tag{7.14}$$

Finally, given some 'input data', the generalized *modus ponens* provides a mechanism for inference on the basis of some input:

Implication:	IF **x** is A, THEN **y** is B.
Premise:	**x** is A'.
Conclusion:	**y** is B'.

In terms of fuzzy relations the output fuzzy set B' is obtained as the relational sup-t composition, $B' = A' \circ R$. The computation of the conclusion $\mu_{B'}(y)$ is realized on the basis of what is called the *compositional rule of inference*: Given $\mu_{A'}(x)$, and $\mu_R(x,y)$, $\mu_{B'}(y)$ is found by generalizing the 'crisp' rule

$$\text{IF } \mathbf{x} = a \text{ AND } \mathbf{y} = f(\mathbf{x}), \text{ THEN } y = f(a) \ .$$

The inference can be described in three steps as illustrated in Figure 7.2:

1. Extension of A' to $X \times Y$, *i.e.* $\mu_{A'_{\text{ext}}}(x,y) = \mu_{A'}(x)$.

2. Intersection of A'_{ext} with R, *i.e.*

$$\mu_{A'_{\text{ext}} \cap R}(x,y) = T\left(\mu_{A'_{\text{ext}}}(x,y), \mu_R(x,y)\right) \qquad \forall\ (x,y) \ .$$

3. Projection of $A'_{\text{ext}} \cap R$ on Y, *i.e.* the compositional rule of inference is defined by

$$\begin{aligned} \mu_{B'}(y) &= \sup_{x \in X} \mu_{A'_{\text{ext}} \cap R}(x,y) \\ &= \sup_{x \in X} T\left(\mu_{A'_{\text{ext}}}(x,y),\ \mu_R(x,y)\right) \ , \end{aligned} \tag{7.15}$$

where in (7.15) the supremum or maximum can be seen as a 'selection' from the information provided by $A'_{\text{ext}} \cap R$. Taking the maximum over all values in X, one may view B' described by $\mu_{B'}(y)$ as the *shadow* of fuzzy set $A'_{\text{ext}} \cap R$.

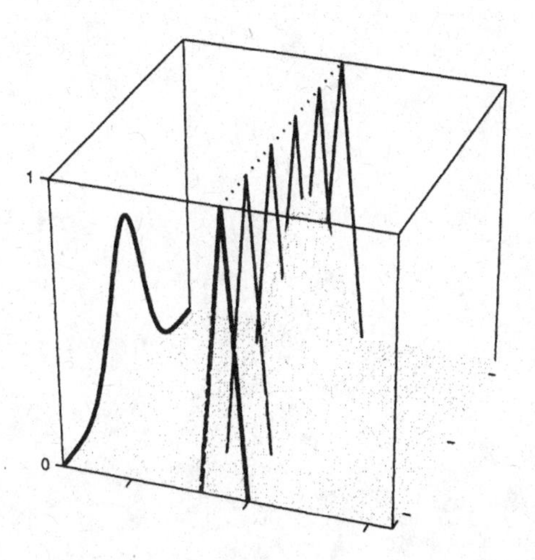

Fig. 7.2 Compositional rule of inference in approximate reasoning.

The link of (7.15) to the composition of fuzzy relations can be easily arrived at from the definition of a composite mapping (1.6) given in Section 1. Let g and h define two ordinary relations on $U \times \Omega$ and $\Omega \times Y$, respectively. The composition of g and h, denoted $h \circ g$, is defined as a relation in $U \times Y$ such that $(u, y) \in h \circ g$ if and only if there exists at least one $\omega \in \Omega$ such that $(u, \omega) \in g$ and $(\omega, y) \subset h$. Using characteristic function $\zeta_g : U \times \Omega \to \{0, 1\}$ and $\zeta_h : \Omega \times Y \to \{0, 1\}$, we have

$$\zeta_{h\circ g}(u, y) = \max_{\omega \in \Omega} T\big(\zeta_g(u, \omega), \zeta_h(\omega, y)\big) \tag{7.16}$$

for any $(u, y) \in U \times Y$ where T is any t-norm. Equation (7.16) is then generalized to fuzzy relations by simply replacing the characteristic function for crisp sets ζ by the fuzzy set-membership function μ:

$$\mu_{h\circ g}(u, y) = \max_{\omega \in \Omega} T\big(\mu_g(u, \omega), \mu_h(\omega, y)\big) . \tag{7.17}$$

Because the t-norm in (7.17) can take a variety of formulas, we obtain for each t-norm a particular composition. The two most commonly used compositions in the literature are the so-called *max-min composition* and *max-product composition* using T_{min} and T_{pro}, respectively.

8

Fuzzy Inference Engines

- ☐ *There are various ways to realize a fuzzy rule-based system, distinguished by the way rules are combined and the inference engine employed.*
- ☐ *Fuzzy systems are nonlinear mappings.*
- ☐ *Fuzzy systems are universal function approximators.*

All traditional logic habitually assumes that precise symbols are being employed. It is therefore not applicable to this terrestrial life but only to an imagined celestial existence.

—Bertrand Russell

The previous section introduced approximate reasoning as a methodology to encode rule-based knowledge and to process linguistic information using fuzzy sets and fuzzy logic. We have seen that for one rule the generalized modus ponens specifies a mapping. Because any realistic rule-base will consist of several rules, the question arises of how make an inference with a set of rules. Hereafter, two frequently used inference schemes, namely, the *composition-based* and the *individual-rule-based* inferences, are described. Depending on various assumptions on the type of input data (fuzzy or non-fuzzy) and fuzzy logics (implication,...) used, we obtain compact representations of fuzzy if-then rule-based systems as mappings. This section follows closely the comprehensive discussion in [Wan97].

8.1 COMPOSITION-BASED INFERENCE

In composition-based inference, all rules are combined into a single fuzzy relation R in $X \times Y$, which is then viewed as a single fuzzy if-then rule. Hereafter, we consider the 'linguistic model structure' (5.5), or its conjunctive form (5.6), with r input variables combined in vector $\mathbf{x}$. Then, A_i is a fuzzy set defined by a multi-variate membership function $\mu_{A_i}(\mathbf{x})\colon X_1 \times \cdots \times X_r \to [0,1]$.

The way rules are combined depends on the interpretation of what a set of rules should mean. If rules are viewed as *independent conditional statements*, then a reasonable mechanism for aggregating n_R individual rules R_i (fuzzy relations) is the union:

$$\begin{aligned} R &\doteq \bigcup_{i=1}^{n_R} R_i \\ &= S\big(\mu_{R^1}(\mathbf{x},y),\ldots,\mu_{R^{n_R}}(\mathbf{x},y)\big) . \end{aligned} \tag{8.1}$$

On the other hand, if rules are seen as *strongly coupled conditional statements*, their combination should employ an intersection operator:

$$\begin{aligned} R &\doteq \bigcap_{i=1}^{n_R} R_i \\ &= T\big(\mu_{R^1}(\mathbf{x},y),\ldots,\mu_{R^{n_R}}(\mathbf{x},y)\big) . \end{aligned} \tag{8.2}$$

Composition-based inference is summarized in Algorithm 8.1.

Let A' be any fuzzy set in X. We then obtain the output of composition-based fuzzy inference engine as follows. For the n_R fuzzy if-then rules of the conjunctive linguistic model structure

$$R_i: \quad \text{IF } x_1 \text{ is } A_{i1} \text{ AND } x_2 \text{ is } A_{i2} \ldots \text{AND } x_r \text{ is } A_{ir}, \text{ THEN } y \text{ is } B_i . \tag{5.6}$$

Step 1: Determine the fuzzy set membership functions

$$\mu_{A_{i1}\times\cdots\times A_{ir}}(x_1,\ldots,x_r) \doteq T\big(\mu_{A_{i1}}(x_1),\ldots,\mu_{A_{ir}}(x_r)\big) . \tag{8.3}$$

Step 2: Equation (8.3) is viewed as the fuzzy set μ_A in the fuzzy implications (7.9)-(7.14) and $\mu_{R_i}(\mathbf{x},y)$, $i = 1,\ldots,n_R$, is calculated according to any of the implications.

Step 3: $\mu_R(\mathbf{x},y)$ is determined according to (8.1) or (8.2).

Step 4: Finally, for an arbitrary input A', the output B' is obtained according to

$$\mu_{B'}(y) = \sup_{\mathbf{x}\in X} T\big(\mu_{A'}(\mathbf{x}),\mu_R(\mathbf{x},y)\big) . \tag{8.4}$$

Algorithm 8.1 Composition-based inference.

The first two steps are identical to the composition-based inference.

Step 1: Determine the fuzzy set membership functions

$$\mu_{A_{i1}\times\cdots\times A_{ir}}(x_1,\ldots,x_r) \doteq T(\mu_{A_{i1}}(x_1),\ldots,\mu_{A_{ir}}(x_r)) \ . \qquad (8.3)$$

Step 2: Equation (8.3) is viewed as the fuzzy set μ_A in the fuzzy implications (7.9)-(7.14) and $\mu_{R_i}(\mathbf{x},y)$, $i=1,\ldots,n_R$, is calculated according to any of the implications.

Step 3: For a given input fuzzy set A' in X, determine the output fuzzy set B'_i in Y for each rule R_i according to the generalized modus ponens (7.15), *i.e.*

$$\mu_{B'_i}(y) = \sup_{\mathbf{x}\in X} T(\mu_{A'}(\mathbf{x}),\mu_{R_i}(\mathbf{x},y)) \qquad (8.6)$$

for $i=1,\ldots,n_R$.

Step 4: The output of the fuzzy inference engine is obtained from either the union

$$\mu_{B'}(y) = S(\mu_{B'_1}(y),\ldots,\mu_{B'_r}(y)) \qquad (8.7)$$

or intersection

$$\mu_{B'}(y) = T(\mu_{B'_1}(y),\ldots,\mu_{B'_r}(y)) \qquad (8.8)$$

of the individual output fuzzy sets $B'_1,\ldots,B'_r$.

Algorithm 8.2 Individual-rule-based inference.

The fuzzy system represented by fuzzy relation R, defined on the Cartesian product space of the system variables $X_1\times X_2\times\cdots\times X_r\times Y$, describes a fuzzy graph $\widetilde{F}$ equivalent to (4.33), which explains why the compositional rule of inference can be regarded as a generalized function evaluation (cf. Section 1). Let the fuzzy graph $\widetilde{F}$ be defined by the fuzzy relation $R\subset X\times Y$, $R=\bigcup_{i=1}^{n_R} R_i$ such that

$$\widetilde{F} = \{(\mu_{B'}(y),A') \ : \ B' = A'\circ R\} \qquad (8.5)$$

in analogy to (3.9) (cf. (4.33)).

8.2 INDIVIDUAL-RULE-BASED INFERENCE

Instead of combining all rules into one, each rule may be evaluated individually to obtain for each rule an output fuzzy set which are then aggregated into one output fuzzy set by taking either their union or intersection.

That is, for a given input fuzzy set A' in X, determine the output fuzzy set B'_i in Y for each rule R_i according to the generalized modus ponens (7.15), that is,

$$\mu_{B'_i}(y) = \sup_{\mathbf{x}\in X} T(\mu_{A'}(\mathbf{x}),\mu_{R_i}(\mathbf{x},y)) \qquad (8.6)$$

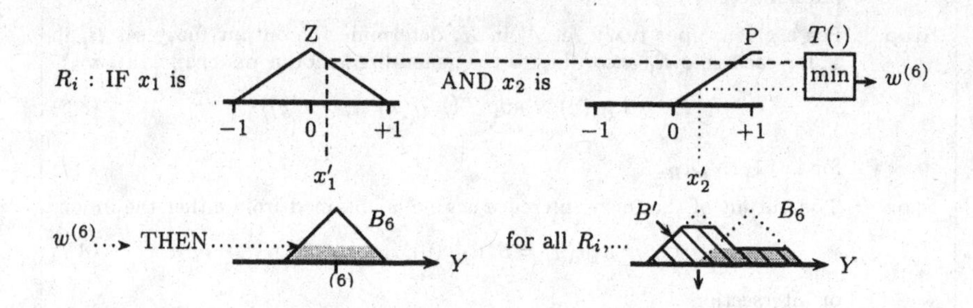

Fig. 8.1 Individual-rule-based inference.

for $i = 1, \ldots, n_R$. The output of the fuzzy inference engine is obtained from either the union

$$\mu_{B'}(y) = S\big(\mu_{B'_1}(y), \ldots, \mu_{B'_r}(y)\big) \tag{8.7}$$

or intersection

$$\mu_{B'}(y) = T\big(\mu_{B'_1}(y), \ldots, \mu_{B'_r}(y)\big) \tag{8.8}$$

of the individual output fuzzy sets $B'_1, \ldots, B'_r$. The basic idea of the individual-rule-based inference, for min-inference, singleton input, union intersection, is illustrated in Figure 8.1. A step-by-step summary is given by Algorithm 8.2.

Using individual-rule-based inference with the **union combination** in (8.7), the implications (7.13) and (7.14), and max for all the t-conorm operators, we obtain from (8.6) and (8.7) the following two inference engines:

▷ **Minimum Inference Engine:** Using (7.13) and the min for all t-norm operators,

$$\mu_{B'}(y) = \max_{i=1,\ldots,n_R} \left\{ \sup_{\mathbf{x}\in X} \min\big(\mu_{A'}(\mathbf{x}), \mu_{A_{i1}}(x_1), \ldots, \mu_{A_{ir}}(x_r), \mu_{B_i}(y)\big) \right\}. \tag{8.9}$$

▷ **Product Inference Engine:** Using (7.14) and the algebraic product for all t-norm operators,

$$\mu_{B'}(y) = \max_{i=1,\ldots,n_R} \left\{ \sup_{\mathbf{x}\in X} \left(\mu_{A'}(\mathbf{x}) \cdot \prod_{k=1}^{r} \mu_{A_{ik}}(x_k) \cdot \mu_{B_i}(y) \right) \right\}. \tag{8.10}$$

Let the fuzzy set A' be a singleton, that is, if we consider 'crisp' input data,

$$\mu_{A'}(\mathbf{x}) = \begin{cases} 1 & \text{if } \mathbf{x} = \mathbf{x}' \\ 0 & \text{otherwise,} \end{cases} \tag{8.11}$$

where $\mathbf{x}'$ is some point in X. Substituting (8.11) in (8.9) and (8.10), we find that the $\sup_{\mathbf{x}\in X}$ is achieved at $\mathbf{x} = \mathbf{x}'$. Hence, (8.9) reduces to, cf. (4.32),

$$\mu_{B'}(y) = \max_{i=1,\dots,n_R} \left\{ \min\left(\mu_{A_{i1}}(x_1'), \dots, \mu_{A_{ir}}(x_r'), \mu_{B_i}(y)\right) \right\} \tag{8.12}$$

and (8.10) reduces to

$$\mu_{B'}(y) = \max_{i=1,\dots,n_R} \left\{ \prod_{k=1}^{r} \mu_{A_{ik}}(x_k') \cdot \mu_{B_i}(y) \right\} . \tag{8.13}$$

A disadvantage of the minimum and product inference engines is that if for some $\mathbf{x} \in X$, $\mu_{A_ik}(x_k)$ is small, then $\mu_{B'}(y)$ obtained from (8.9) and (8.10) will be very small.

Using individual-rule-based inference, *intersection combination* (8.8), the implications (7.9), (7.10), (7.11), we obtain from (8.6) the following inference engines ($i = 1, \dots, n_R$):

▷ **Dienes-Rescher Inference Engine:** Using (7.9) and the min(·) t-norm in (8.8) and (8.3),

$$\mu_{B'}(y) = \min_i \left\{ \sup_{\mathbf{x}\in X} \min\left[\mu_{A'}(\mathbf{x}), \max\left(1 - \min_{k=1,\dots,r} (\mu_{A_{ik}}(x_k)), \mu_{B_i}(y)\right)\right] \right\}. \tag{8.14}$$

▷ **Zadeh Inference Engine:** Using (7.10), and t-norm min( ) in (8.8) and (8.3)

$$\mu_{B'}(y) = \min_i \left\{ \sup_{\mathbf{x}\in X} \min\left[\mu_{A'}(\mathbf{x}), \max\left(\min\left[\mu_{A_{i1}}(x_1), \dots, \mu_{A_{ir}}(x_r), \mu_{B_i}(y)\right], 1 - \min_{k=1,\dots,r} (\mu_{A_{ik}}(x_k))\right)\right] \right\}. \tag{8.15}$$

▷ **Lukasiewicz Inference Engine:** Using (7.11) and the min t-norm in (8.8) and (8.3),

$$\begin{aligned} \mu_{B'}(y) &= \min_i \left\{ \sup_{\mathbf{x}\in X} \min\left[\mu_{A'}(\mathbf{x}), \min\left(1, 1 - \min_k (\mu_{A_{ik}}(x_k)) + \mu_{B_i}(y)\right)\right] \right\} \\ &= \min_i \left\{ \sup_{\mathbf{x}\in X} \min\left[\mu_{A'}(\mathbf{x}), 1 - \min_{k=1,\dots,r} (\mu_{A_{ik}}(x_k)) + \mu_{B_i}(y)\right] \right\} . \end{aligned} \tag{8.16}$$

If the fuzzy set A' is a singleton, substituting (8.11) into the equations of the inference engines (8.14)-(8.16), the $\sup_{\mathbf{x}\in X}$ is obtained at $\mathbf{x} = \mathbf{x}'$, leading to the following singleton input inference engines. From (8.14) we obtain

$$\mu_{B'}(y) = \min_{i=1,\ldots,n_R} \left\{ \max[1 - \min_{k=1,\ldots,r} (\mu_{A_{ik}}(x'_k)), \mu_{B_i}(y)] \right\} .$$

From (8.14)

$$\mu_{B'}(y) = \min_{i=1,\ldots,n_R} \left\{ \max[1 - \min_{k=1,\ldots,r} (\mu_{A_{ik}}(x'_k)), \mu_{B_i}(y)] \right\} ,$$

and from (8.16),

$$\mu_{B'}(y) = \min_{i=1,\ldots,n_R} \left\{ 1, 1 - \min_{k=1,\ldots,r} (\mu_{A_{ik}}(x'_i)) + \mu_{B_i}(y) \right\} .$$

8.3 FUZZY SYSTEMS AS NONLINEAR MAPPINGS

This section highlights the dual role of fuzzy systems. Starting with logical or conditional statements to capture context-dependent expert knowledge, fuzzy systems are if-then rule-based systems constructed from a collection of linguistic rules. On the other hand, fuzzy systems are nonlinear mappings between two spaces, say X and Y. As such, they can be identified from sampled data by various methods including fuzzy clustering and we can also formally relate fuzzy systems to various other methodologies. One therefore has to consider the context in which a fuzzy system is employed and accordingly select the appropriate formal representation. Semantics can therefore play an important role; not only in separating various methodologies. For example, if a fuzzy system is identified from data and the rules are interpreted by the user in order to gain new knowledge or insight into the process that generated the data in the first place, then the choice of a framework to formalize implications, conditionals, or correlations is very important. If, on the other hand, the aim is to generalize from data regardless of the interpretation, we should select a convenient (simple, accurate,..) framework to identify such a relation from sample data.

A *defuzzifier* is a mapping from the fuzzy set B' in Y to a point y' in Y. To obtain a single-valued numerical output from the inference engines, one has to somehow capture the information given in $\mu_{B'}(y)$ by a single number. The *center of gravity defuzzifier* determines y' as the center of the area under the membership function $\mu_{B'}(y)$:

$$y' \doteq \frac{\int_Y \mu_{B'}(y) \cdot y \, \mathrm{d}y}{\int_Y \mu_{B'}(y) \, \mathrm{d}y} , \tag{8.17}$$

where $\int_Y$ is the conventional Riemann integral which, in case of a discrete space Y, is replaced by a finite sum. If $\mu_{B'}(y)$ is viewed as a density or distribution function, then the center of gravity gives the mean value. The main problem with this defuzzifier is the calculation of the integral for irregular shapes of $\mu_{B'}(y)$. Since the fuzzy set B' is the union or intersection of n_R fuzzy sets, the weighted average of the centers of the n_R fuzzy sets provides a reasonable approximation of (8.17). Let $y_0^{(i)}$ be the center of the ith fuzzy set and $w^{(i)}$ be its height. Then, the *center-average defuzzifier* calculates y' as

$$y' \doteq \frac{\sum_{i=1}^{n_R} y_0^{(i)} \cdot w^{(i)}}{\sum_{i=1}^{n_R} w^{(i)}} . \tag{8.18}$$

Consider a normalized fuzzy set B_i, that is, $\exists y\colon \mu_{B_i}(y) = 1$, in the 'Linguistic model'

$$R_i : \quad \text{IF } x_1 \text{ is } A_{i1} \text{ AND } x_2 \text{ is } A_{i2} \ldots \text{AND } x_r \text{ is } A_{ir}, \text{ THEN } y \text{ is } B_i . \tag{5.6}$$

Let the input data be crisp, that is, substituting (8.11) into the product inference engine (8.10), we have

$$\mu_{B'}(y) = \max_{i=1,\ldots,n_R} \left\{ \prod_{k=1}^{r} \mu_{A_{ik}}(x'_k) \cdot \mu_{B_i}(y) \right\} . \tag{8.19}$$

Suppose we use the center-average defuzzifier (8.18). Then, the center of the fuzzy set $\mu_{A_{ik}}(x'_k) \cdot \mu_{B_i}(y)$ determines the center of B_i, denoted by $y_0^{(i)}$ in (8.18). The height of the ith fuzzy set in (8.19) is $\prod_{k=1}^{r} \mu_{A_{ik}}(x'_k) \cdot \mu_{B_i}(y_0^{(i)}) = \prod_{k=1}^{r} \mu_{A_{ik}}(x'_k)$ and equals $w^{(i)}$ in (8.18). This reduces the fuzzy system (5.6) with product inference engine, singleton input data, and center-average defuzzier to the compact formulation

$$y' = \frac{\sum_{i=1}^{n_R} y_0^{(i)} \cdot \prod_{k=1}^{r} \mu_{A_{ik}}(x'_k)}{\sum_{i=1}^{n_R} \prod_{k=1}^{r} \mu_{A_{ik}}(x'_k)}$$

or, in general, we find that the fuzzy system is a nonlinear mapping (cf. Section 1)

$$\begin{aligned} f : X &\to Y \\ \mathbf{x} &\mapsto f(\mathbf{x}) , \end{aligned}$$

where the argument $\mathbf{x} \in X \subset \mathbb{R}^r$ maps to $f(\mathbf{x}) \in Y \subset \mathbb{R}$, a weighted average of the consequent fuzzy sets:

$$f(\mathbf{x}) = \frac{\sum_{i=1}^{n_R} y_0^{(i)} \cdot \prod_{k=1}^{r} \mu_{A_{ik}}(x_k)}{\sum_{i=1}^{n_R} \prod_{k=1}^{r} \mu_{A_{ik}}(x_k)} . \tag{8.20}$$

Similar to (8.20), we obtain for a fuzzy system (5.6), with minimum inference engine (8.9), singleton input (8.11) and center-average defuzzifier (8.18),

$$f(\mathbf{x}) = \frac{\sum_{i=1}^{n_R} y_0^{(i)} \cdot \min_{k=1}^{r} \mu_{A_{ik}}(x_k)}{\sum_{i=1}^{n_R} \min_{k=1}^{r} \mu_{A_{ik}}(x_k)} . \tag{8.21}$$

An important conclusion is that fuzzy rule-based systems enable us to encode 'logical' if-then relationships and to process fuzzy information. By describing such *knowledge processing* using fuzzy relations and representing a fuzzy system as a nonlinear mapping, we should be able to combine such *qualitative reasoning* system with *quantitative models* processing numerical data.

8.4 EXAMPLE: COMPARISON OF INFERENCE ENGINES

In this section, the input-output relationships of the fuzzy inference engines, introduced in the previous section, are visualized. Figure 8.2 illustrates the rule-base.

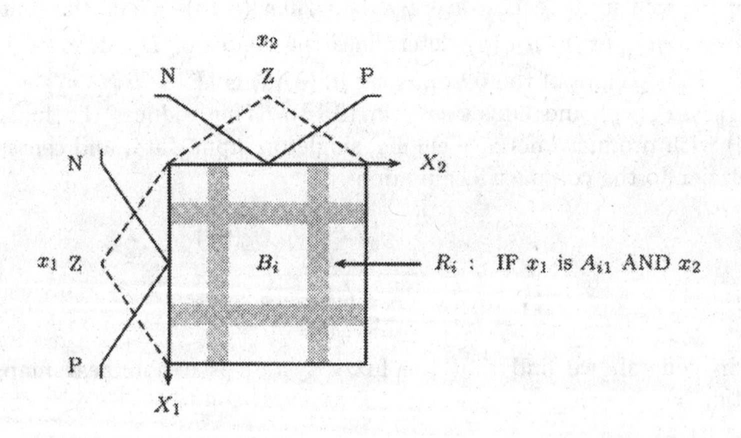

Fig. 8.2 Rule-based fuzzy reasoning.

For the Gaussian and triangular shaped input space fuzzy sets in Figure 8.3 and output sets B_i in Figure 8.4, Figures 8.7 and 8.6 describe the input-output behavior of the minimum inference engine (8.21) and the product inference engine (8.20), respectively. In both cases, we consider the conjunctive structure of the 'Linguistic' fuzzy model (5.6), with singleton input (8.11), and center-average defuzzifier (8.18). Figure 8.8 compares the contour plots of the minimum and product inference engines for triangular sets.

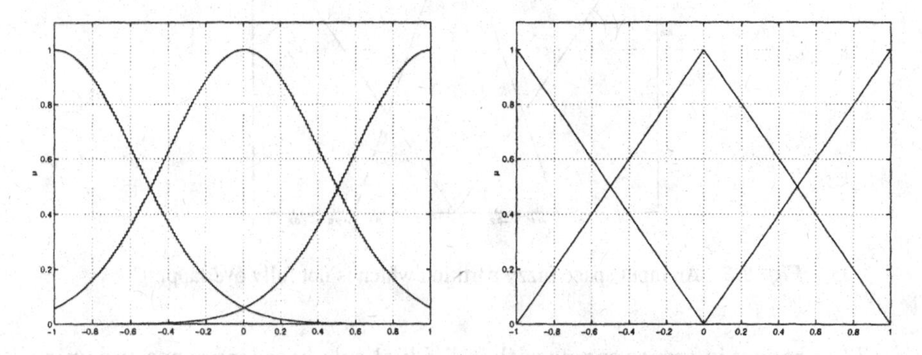

Fig. 8.3 Gaussian and trapezoidal input fuzzy sets.

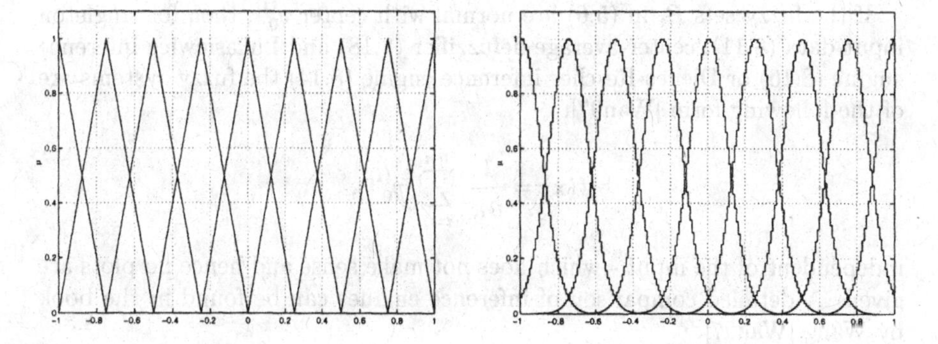

Fig. 8.4 Gaussian and trapezoidal outputs sets B_i.

From the figures, we find that the shape of the fuzzy sets has no major influence on the input-output behavior of the fuzzy system. For most engineering applications we may therefore chose fuzzy sets which are most convenient with respect to implementation or analysis. For triangular membership functions, the product inference engine has a linear overall behavior as seen in Figure 8.8. However, if we change the partition of the input space as shown in Figure 8.5, we obtain for the minimum and product inference the input-output behavior shown in Figures 8.9 and 8.10, respectively.

In general, we find that a disadvantage of the minimum and product inference engines is that for some $\mathbf{x} \in X$, $\mu_{A_ik}(x_k)$ is very small. Then, $\mu_{B'}(y)$ obtained from (8.9) and (8.10) will be very small. For individual-rule-based inference with intersection-based combination, this problem does not occur.

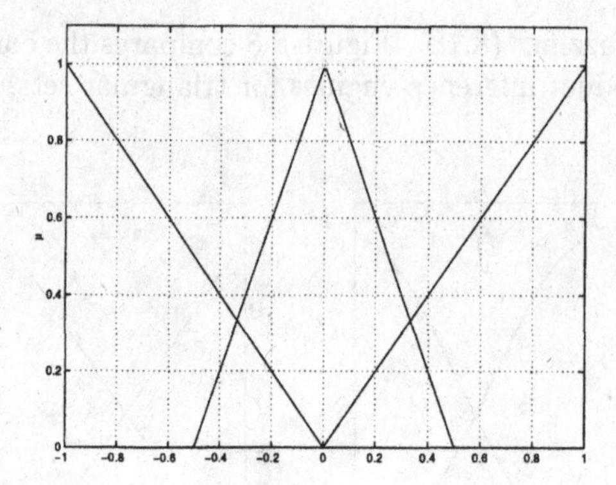

Fig. 8.5 An input space fuzzy partition which is not fully overlapping.

The product inference engine with individual-rule-based inference and union combination (8.7) is identical to the composition-based inference with union combination of rules (8.1).

If the fuzzy sets B_i in (5.6) are normal with center $y_0^{(i)}$, then for singleton input data (8.11), center-average defuzzifier (8.18) and Lukasiewicz inference engine (8.16) or Dienes-Rescher inference engine (8.14) the fuzzy systems are of the following form [Wan97]:

$$f(\mathbf{x}) = \frac{1}{n_R} \sum_{i=1}^{n_R} y_0^{(i)} ,$$

independent of the input - which does not make sense and hence no plots are given. A detailed comparison of inference engines can be found in the book by Wang [Wan97].

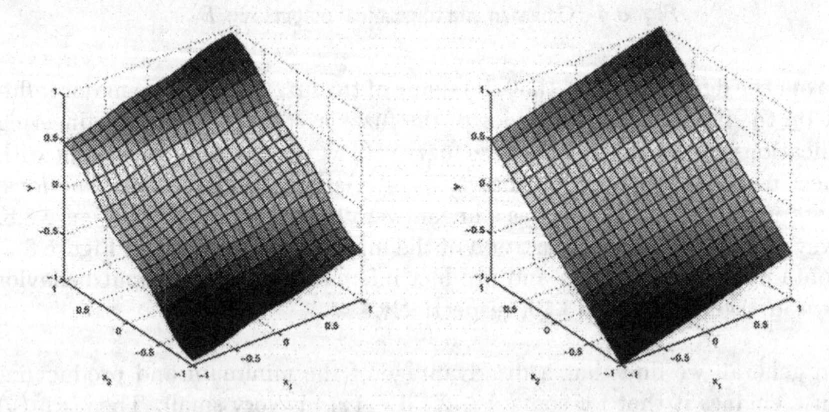

Fig. 8.6 Product inference with Gaussian and trapezoidal sets.

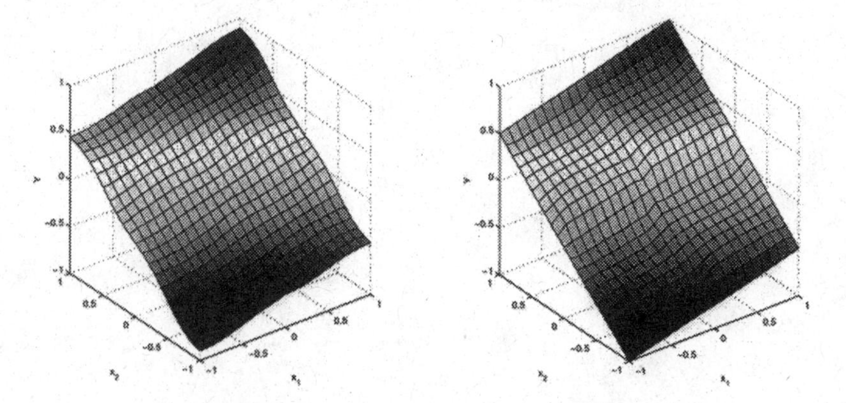

Fig. 8.7 Minimimum inference with Gaussian and trapezoidal sets.

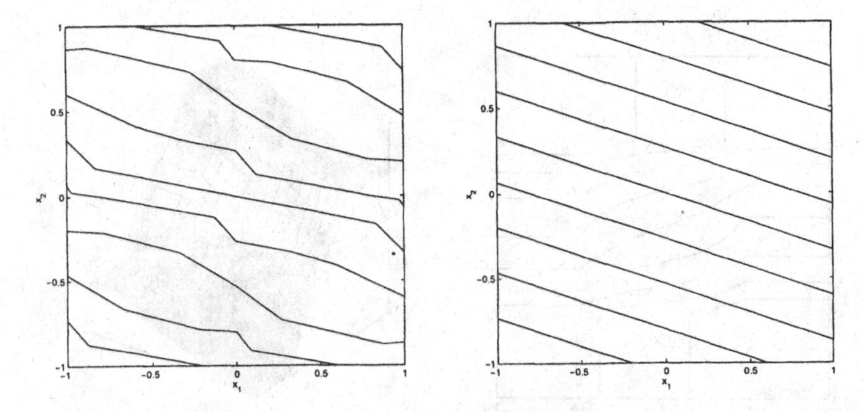

Fig. 8.8 Contour plots for minimum inference (left) vs. product inference (right).

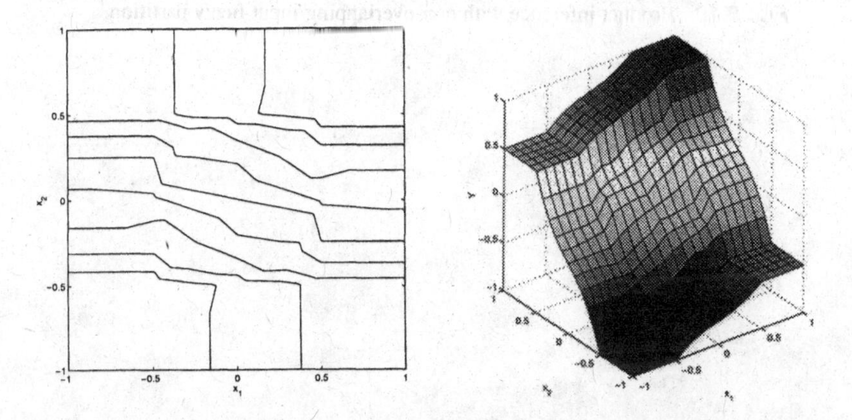

Fig. 8.9 Minimum inference with input fuzzy partition that does not fully overlap.

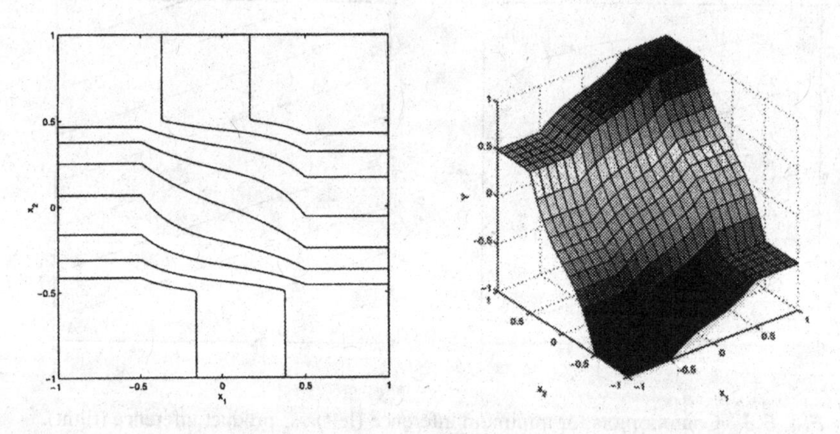

Fig. 8.10 Product inference with non-overlapping input fuzzy partition.

9

Fuzzy Classification

- ☐ *Fuzzy clustering groups unlabelled data into a fixed number of classes and hence can be used to design classifiers.*
- ☐ *Specific fuzzy classifiers can be shown to be formally equivalent to optimal statistical classifiers.*
- ☐ *If-then rule-based fuzzy classifiers provide an intuitive framework to interpret data.*

In Section 4.5, we devised a fuzzy classifier using clustering algorithms. Here we continue our discussion on *classification* from a fuzzy systems perspective. First, we show that a fuzzy classifier can be related to Parzen's kernel density estimator, introduced in Section 3.2. We then apply the concept of a fuzzy rule-base, introduced in the previous section, to design a classifier. We use the well-known Fisher-Anderson 'iris' data set to illustrate the concept of fuzzy classification.

9.1 EQUIVALENCE OF FUZZY AND STATISTICAL CLASSIFIERS

In Section 5.3, we demonstrated that a fuzzy system is in fact equivalent to a basis function expansion (Section 3.3). The present section will add another example of classification. We can show that the fuzzy system (8.20) is equivalent to a statistical classifier[1] using the density estimates from Section 3.2.

[1] For a more extensive discussion see [Kun96].

This example should be considered in conjunction with the example given in Section 4.5 in order to obtain a more complete understanding of classification.

The problems of classification and discrimination are closely related. In each instance, the data $\mathbf{m}_j \in \mathbb{R}^r$ are assumed to comprise c clusters. If the number of clusters c is known, and a training sample of data is available from each cluster, then the problem is to formulate rules for assigning new unclassified (unlabelled) observations to one of the clusters. In other words, we assign to an object (described as a point $\mathbf{x}$ in the feature space $X_1 \times \cdots \times X_r$) a class label C from the set $\mathcal{C} = \{C_1, \ldots, C_c\}$. Hereafter we assume that $X_1 \times \cdots \times X_r$ coincides with $\mathbb{R}^r$ and that have available a set of (labelled) training data $\mathbf{M} = \{\mathbf{m}_1, \ldots, \mathbf{m}_d\}$, $\mathbf{m}_j = [m_{1j}, \ldots, m_{rj}]^T \in \mathbb{R}^r$. We denote by $i \in \{1, 2, \ldots, c\}$ the index of the class label among $\{C_1, \ldots, C_c\}$, associated with $\mathbf{m}_j$. The problem is to design a classifier, that is, to specify a mapping ψ such that each object $\mathbf{x}$ is associated with one class C_i:

$$\psi : \mathbb{R}^r \to \mathcal{C} .$$

In a probabilistic framework, both $\mathbf{x}$ and C are random variables. Let $Pr(C_i)$ be the prior probability for class C_i, $i = 1, \ldots, c$ and denote by $p(\mathbf{x}|C_i)$ the class-conditional probability density function. In statistical (Bayesian) decision theory, the aim is to design an optimal classifier with a small error, that is, one that assigns to $\mathbf{x}$ a class label C^* corresponding to the highest posterior probability:

$$C^* = \arg\max_{C} \; Pr(C|\mathbf{x}) .$$

Here, the posterior probability is calculated by

$$Pr(C_i|\mathbf{x}) = \frac{Pr(C_i)\; p(\mathbf{x}|C_i)}{p(\mathbf{x})} , \tag{9.1}$$

$$p(\mathbf{x}) = \sum_k Pr(C_k)\; p(\mathbf{x}|C_k) .$$

In Section 3.2, Parzen's kernel estimator was introduced as a non-parametric approximation of a probability density function. Let $K(\mathbf{x})$ be a *kernel function* (also referred to as a *Parzen window*) which peaks at zero, is non-negative, and whose integral equals one over $\mathbb{R}^r$. The multi-dimensional kernel function centered around $\mathbf{m}_j \in \mathbb{R}^r$ can be expressed in the form $\frac{1}{h^r} K\left(\frac{\mathbf{x}-\mathbf{m}_j}{h}\right)$, where h determines the window with and hence is a *smoothing parameter*. Using the estimator (3.19), we can approximate the class-conditional probability density using the sample set $\mathbf{M}$ by

$$\hat{p}(\mathbf{x}|C_i) = \frac{1}{d_{C_i} h^r} \sum_{j:\; b_j = i} K\left(\frac{\mathbf{x} - \mathbf{m}_j}{h}\right) , \qquad \mathbf{m}_j \in \mathbf{M} ,$$

where d_{C_i} is the number of elements of $\mathbf{M}$ from class C_i. Finally, we estimate the prior probabilities in (9.1) by

$$\widehat{Pr}(C_i) = \frac{d_{C_i}}{d} .$$

Inserting both approximations into (9.1), we obtain the following estimate of the posterior probability:

$$\widehat{Pr}(C_i|\mathbf{x}) = \frac{1}{d \cdot h^r \cdot p(\mathbf{x})} \cdot \sum_{j:\, b_j=i} K\left(\frac{\mathbf{x}-\mathbf{m}_j}{h}\right) . \tag{9.2}$$

Introducing an indicator function $\zeta_{C_i}(\mathbf{m}_j)$,

$$\zeta_{C_i}(\mathbf{m}_j) = \begin{cases} 1 , & \text{if } b_j = i, \text{ i.e. } \mathbf{m}_j \text{ comes from class } C_i; \\ 0 , & \text{otherwise.} \end{cases}$$

We can rewrite (9.2) as

$$\widehat{Pr}(C_i|\mathbf{x}) = \frac{1}{d} \cdot a_1(\mathbf{x}) \cdot \sum_{j=1}^{d} \zeta_{C_i}(\mathbf{m}_j) K\left(\frac{\mathbf{x}-\mathbf{m}_j}{h}\right) , \tag{9.3}$$

where factor $a_1(\mathbf{x})$ depends on $\mathbf{x}$ but not on the class label. Using the multi-dimensional Gaussian kernel

$$\frac{1}{h^r} K_G\left(\frac{\mathbf{x}-\mathbf{m}_j}{h}\right) = \frac{1}{h^r \sqrt{(2\pi)^r} \sqrt{|\boldsymbol{\Sigma}|}} \exp\left(-\frac{1}{2h^2}(\mathbf{x}-\mathbf{m}_j)^T \boldsymbol{\Sigma}^{-1} (\mathbf{x}-\mathbf{m}_j)\right) , \tag{9.4}$$

where $\boldsymbol{\Sigma}$ is the covariance matrix. Using the Gaussian kernel we have for the posterior probabilities (9.3)

$$\widehat{Pr}(C_i|\mathbf{x}) = \frac{1}{d} \cdot a_1(\mathbf{x}) \cdot \sum_{j=1}^{d} \zeta_{C_i}(\mathbf{m}_j) K_G\left(\frac{\mathbf{x}-\mathbf{m}_j}{h}\right) . \tag{9.5}$$

Let us now consider a fuzzy classifier where the rule-base takes the form of (8.20). For any class C_i, we consider d rules of the form

$$\begin{aligned} R_j \;:\; & \text{IF } x_1 \text{ is } A_{j1} \text{ AND } \ldots \text{ AND } x_r \text{ is } A_{jr}, \\ & \text{THEN } y^j_{i'} = 1 \text{ and } y^j_i = 0,\ \forall i \neq i',\ i = 1,\ldots,c,\ j = 1,\ldots,d , \end{aligned} \tag{9.6}$$

where y^j_i denotes the ith component of the output vector $\mathbf{y}_j$ associated with the jth rule. Each A_{jk} is a fuzzy set with membership function

$$\mu_{A_{jk}} : \mathbb{R} \to [0, 1] .$$

Specifically, we define

$$\mu_{A_{jk}}(\mathbf{x}) = \exp\left(-\frac{(x_k - m_{kj})^2}{2h^2}\right),$$

where h is a parameter and the membership functions evaluate the similarity of any given $\mathbf{x}$ with $\mathbf{m}_j$. Let the activation strength ('firing level') of the jth rule be

$$\begin{aligned}\beta_j(\mathbf{x}) &= \prod_{k=1}^{r} \mu_{A_{jk}}(x_k) \\ &= \exp\left(-\frac{1}{2h^2}\sum_{k=1}^{r}(x_k - m_{kj})^2\right) \\ &= \exp\left((\mathbf{x}-\mathbf{m}_j)^T \mathbf{A}^{-1}(\mathbf{x}-\mathbf{m}_j)\right).\end{aligned}$$

For $\mathbf{\Sigma}$ being an identity matrix, we notice that $\beta_j(\mathbf{x})$ differs from the Gaussian kernel (9.4) only by a constant. We therefore write

$$\beta_j(\mathbf{x}) = a_2 \cdot K\left(\frac{\mathbf{x}-\mathbf{m}_j}{h}\right).$$

The output of the fuzzy classifier w.r.t. class C_i is obtained as

$$\begin{aligned} y^i &= \frac{\sum_{j=1}^{d} y_i^j \cdot \beta_j(\mathbf{x})}{\sum_{j=1}^{d} \beta_j(\mathbf{x})} \qquad \text{...equivalent to (8.20) and (5.19)!} \\ &= a_3(\mathbf{x}) \cdot \sum_{j=1}^{d} y_i^j \cdot K_G\left(\frac{\mathbf{x}-\mathbf{m}_j}{h}\right). \end{aligned} \tag{9.7}$$

Since y_i^j functions as an indicator function for $\mathbf{m}_j$ with respect to C_i, we find that equations (9.7) and the posterior probability of the statistical classifier (9.5) differ only by a factor which does not depend on the class i. In both cases, for the fuzzy classifier and the statistical classifier a decision is obtained by choosing the class label for which (9.7) and (9.5) are largest. We conclude that a fuzzy system can be shown to be equivalent to a probabilistic classifier (which is known to be asymptotically optimal in the Bayesian sense).

9.2 FUZZY RULE-BASED CLASSIFIER DESIGN

In Section 4, we used clustering algorithms to identify groups of data in the data space. Clusters may also be considered as *classes* and a clustering algorithm can therefore be used to design a classifier. The principal idea of fuzzy classification is to use fuzzy sets and fuzzy if-then rules to characterize classes. One motivation is that the fuzzy classifier provides some degree of linguistic interpretability of the data in terms of fuzzy labels (like "small" or "low"). We shall develop the concept of a rule-based fuzzy classifier using the 'iris' data set.

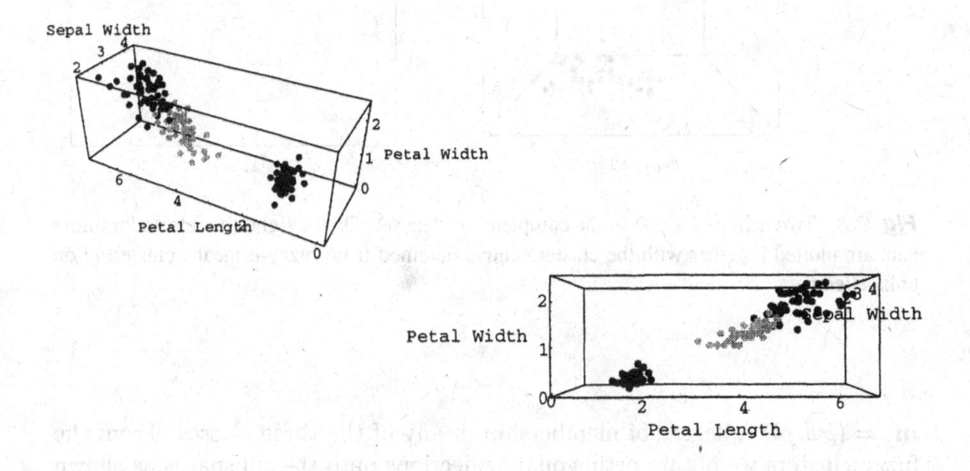

Fig. 9.1 Two different views of the complete Iris data set.

In his pioneering work on discriminant analysis, R.A. Fisher analyzed data, originally collected by E. Anderson, on three species of iris flowers[2]. Let the classes be defined as:

C_1: Iris sestosa; C_2: Iris versicolor; C_3: Iris virginica.

The following four variables were measured: sepal length (sl), sepal width (sw), petal length (pl), and petal width (pw). Since we cannot visualize all four variables at the same time, we choose to plot all 150 data sets for the variables pl, pw, and sw in Figure 9.1. In Figure 9.2, we have selected two particular angles which demonstrate that the two variables, petal width and petal length alone may well be sufficient to build a classifier. We therefore chose 75 sets of training (pw, pl)-data (25 from each class).

In the original studies concerning the iris data, a question was whether versicolor is not a sestosa-virginica hybrid. This was confirmed by Fisher. Applying the fuzzy-c-means algorithm with $w = 1.5$, $c = 3$ and $\delta = 0.01$ to the data, we obtained after 11 iterations the cluster centers depicted on the right in Figure 9.2. The partition matrix $\mathbf{U}$ contains for each training vector

[2]See [JW98] for the data set and classification based on discriminant analysis.

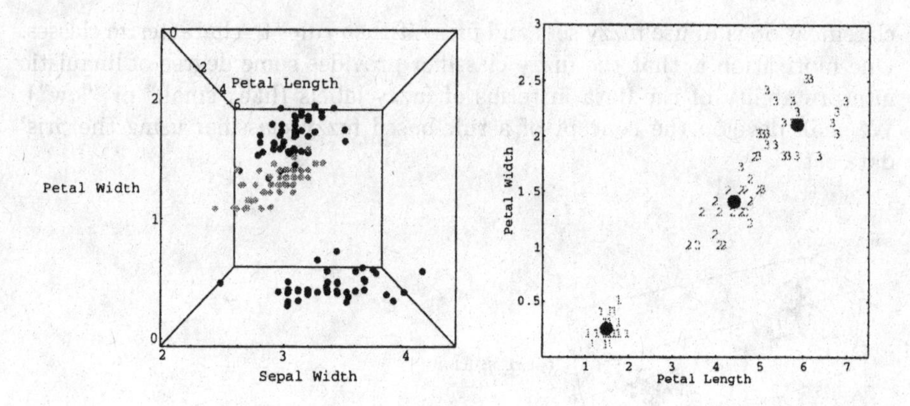

Fig. 9.2 Two different views of the complete iris data set. On the right, the selected training data are plotted together with the cluster centers obtained from fuzzy-c-means clustering on unlabelled data.

$\mathbf{m}_j = (pw, pl)$ a degree of membership in any of the three classes. From the fuzzy clusters we obtain orthogonal projections onto the subspaces as shown in Figure 9.3. The projected membership degrees are used to fit piecewise-linear fuzzy set-membership functions as described in Section 11. We would refer to the linguistic terms, from left to right, as "small", "medium", "large", describing the extent of the petal width and length.

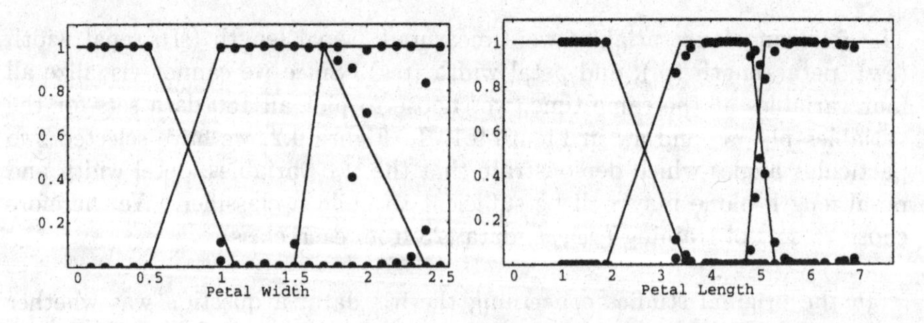

Fig. 9.3 Projections of the fuzzy clusters in $\mathbf{U}$ and fitted membership functions.

The next step is to design a fuzzy classifier using the fuzzy sets. From the plot on the right in Figure 9.2, we can intuitively see that the following rule-base describes an appropriate classification:

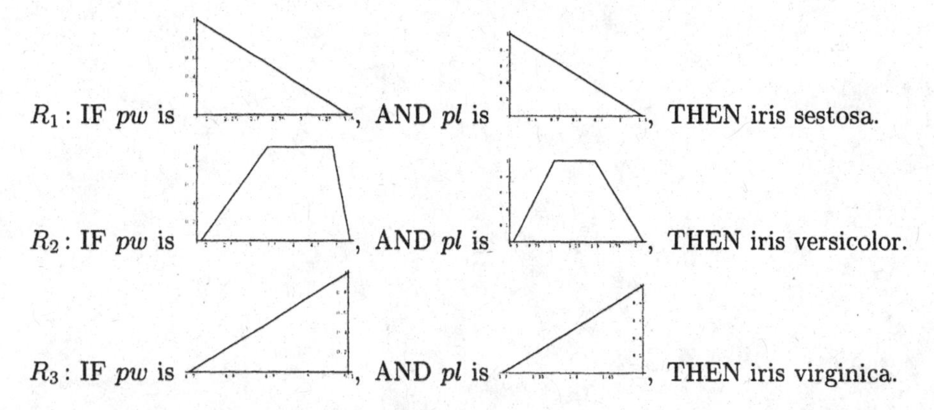

R_1 : IF pw is , AND pl is , THEN iris sestosa.
R_2 : IF pw is , AND pl is , THEN iris versicolor.
R_3 : IF pw is , AND pl is , THEN iris virginica.

Each rule describes one class and the fuzzy sets in the if-part of the rules correspond to those in Figure 9.3. We used the t-norm min for the conjunction such that the aggregate agreement of any given data vector with a rule is calculated by

$$\beta_i(\mathbf{x}) \doteq \mu_{A_{i1}}(x_1) \wedge \mu_{A_{i2}}(x_2) \wedge \cdots \wedge \mu_{A_{ir}}(x_r) , \tag{5.7}$$

where $\wedge$ is a suitable t-norm. The degree of fulfillment $\beta_i(\mathbf{x})$ of the if-part is therefore the degree of confidence in that a given data vector $\mathbf{x}$ belongs to class C_i. The final decision assigns the $\mathbf{x}$ to the class which maximizes the degree of confidence $\beta(\mathbf{x})$:

$$C^* = \arg\max_i \beta_i(\mathbf{x}) .$$

With this classification we can validate the rule-based classifier against all 150 (labelled) vectors. For the given parameters and fuzzy sets we have five misclassifications. If there are two or more classes that are assigned the maximal degree of confidence by the rules, we may either refrain from a classification and label it as 'unknown', or we break the tie by selecting one these classes randomly.

10

Fuzzy Control

- ☐ *Fuzzy rule-based systems can also be used to devise control laws.*
- ☐ *Fuzzy control can be particularly useful if no linear parametric model of the process under control is available.*
- ☐ *Fuzzy control is not 'model-free' as a good understanding of the process dynamics may be required.*
- ☐ *Fuzzy control lacks of design methodologies.*
- ☐ *Fuzzy controllers are easy to understand and simple to implement.*

Feedback control theory arose from *cybernetics*[1], the interdisciplinary science dealing with communication and control systems in living organisms, machines, and organizations founded by Norbert Wiener in the 1940s. Cybernetics developed as the investigation of the techniques by which information is transformed into desired performance. Systems of communication and control in living organisms and those in machines are considered analogous in cybernetics. To achieve desired performance of a system, the actual results of control actions must be made available as *feedback*[2] for future action. A standard feedback control system is illustrated in Figure 10.1.

[1]The term cybernetics is derived from the Greek word kybernetes, meaning "steersman" or "governor".

[2]Feedback: When we desire a system to follow a given pattern the difference between this pattern and the actual behavior is used as a new input to cause the part regulated to change in such a way as to bring its behavior closer to that given by the pattern.

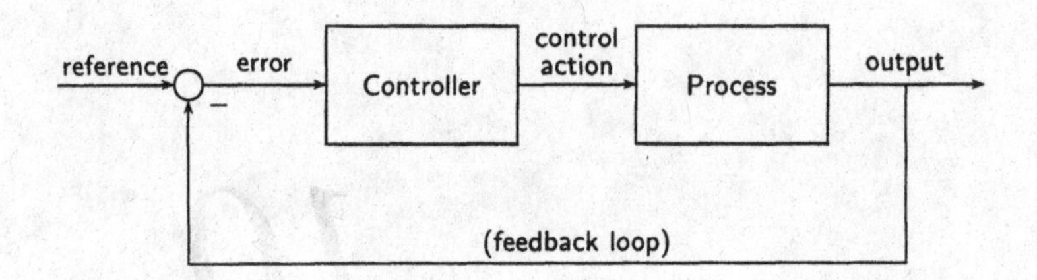

Fig. 10.1 Standard configuration of a feedback control system.

Conventional approaches such as PID, adaptive, optimal and robust control are often referred to as model-based since for the design of the control law and its parameters a linear transfer function model, based on differential equations, is required. Assuming, or knowing, that the model is sufficiently accurate, a powerful theory is available not only to guarantee stability, robustness, optimality and so forth, but also to guide the non-expert through the design in an algorithmic manner. Fuzzy control starts from the assumption that a linear model is not available - either because it is not accurate, not feasible or too cumbersome to determine. It is then assumed that on the scale of input variables alone it is possible to decide upon the control action. Such an approach does not require a parametric model but is not model-free either. We may be required to have a good understanding of the behavior of the process or at least have to accept heuristics. In this section we apply the previously introduced fuzzy rule-based inference schemes to control a dynamic process in a closed-loop configuration. The concepts presented here are only a basic example of fuzzy control, chosen to make comparisons and to introduce the main ideas. Many more sophisticated and advanced nonlinear, adaptive fuzzy control schemes have been developed and described in numerous textbooks [Wan97, Ped93, KGK94, Bab98, PY98].

10.1 PI-CONTROL VS. FUZZY PI-CONTROL

This section is based on the approach presented in [SY89]. A conventional linear proportional-integral (PI) controller with two inputs (error and rate of change in error) and one output is compared to a fuzzy PI-controller with piecewise-linear fuzzy set membership functions and different combinations of logical fuzzy connectives. It is shown that, in general, a fuzzy controller is nonlinear, but with a specific combination of t-norms and t-conorms the fuzzy PI-controller is equivalent to its linear counterpart. The configuration of the closed-loop system is shown in Figure 10.2.

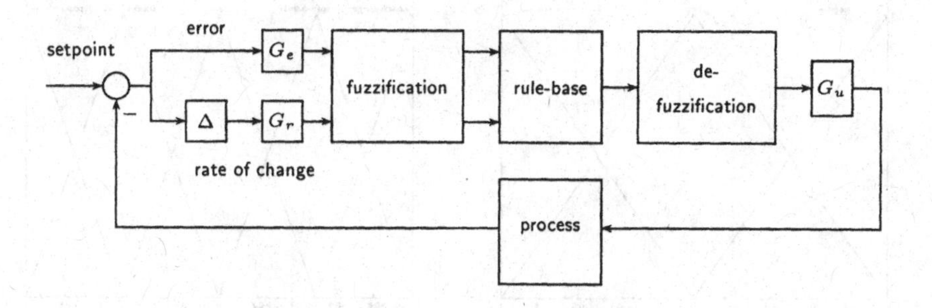

Fig. 10.2 Block diagram of a closed-loop fuzzy control system.

A non-fuzzy PI-controller changes its control action $u(t)$ depending on the error $e(t)$, that is, the difference $(s - y)$ between the setpoint $s(t)$ and current output, and the integral of the error

$$\frac{\mathrm{d}u(t)}{\mathrm{d}t} = K_p \frac{\mathrm{d}e(t)}{\mathrm{d}t} + K_i \, e(t) , \tag{10.1}$$

where $e(t) = s(t) - y(t)$. To obtain a control action the term $\mathrm{d}u(t)/\mathrm{d}t$ is integrated. A fuzzy PI-controller is developed analogously:

$$deriv'(k) = K_p \cdot rate'(k) + K_i \cdot error'(k) , \tag{10.2}$$

where *error*, *rate*, *deriv* are ***fuzzy (or linguistic) variables*** partitioning the underlying spaces by piecewise-linear (triangular) fuzzy sets as shown in Figure 10.3. Each of these fuzzy sets is called a ***linguistic term***. The (ordered) sets of linguistic terms, partitioning the underlying domain (universe of discourse), are denoted RATE and ERROR = ⟨negative, zero, positive⟩, whereas for DERIV, negative and positive are further refined into 'negative small' and 'negative large'. We write DERIV_i to refer to the ith term.

Equation (10.2) can be redefined to scale the fuzzy variables into a range from −1 to +1. This requires the introduction of the scaling gains G_u, G_r and G_e, where $deriv' = G_u \cdot deriv$; $G_r \cdot rate' = rate$ and $G_e \cdot error' = error$. Substituting these into equation (10.2), we obtain the following equation for the incremental control output of the fuzzy PI-controller:

$$deriv(k) = \frac{K_p}{G_u G_r} \, rate(k) + \frac{K_i}{G_u G_e} \, error(k) . \tag{10.3}$$

The constants $K_p\,(G_u G_r)$ and $K_i/(G_u G_e)$ are assumed equal to 0.5 to make *deriv* fall into the interval $[-1, 1]$. The fuzzy controller is then equivalent to a conventional PI-controller with proportional gain $K_p = 0.5 \cdot G_u \cdot G_r$ and integral gain $K_i = 0.5 \cdot G_u \cdot G_e$. We note that there are infinitely many

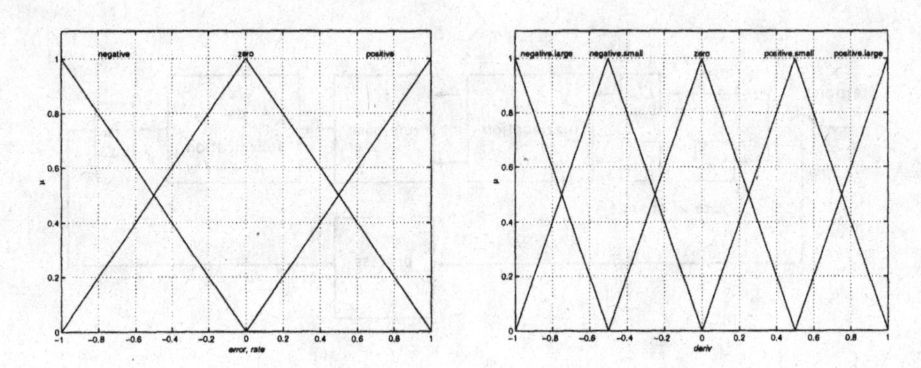

Fig. 10.3 Fuzzy sets for the variables error, rate and the output, respectively.

combinations of G_e, G_r, and G_u to hold true for these expressions. The complete rule-base, that is the set of control rule takes the form:

R_1 : IF *error* is 'negative' AND *rate* is 'negative',
THEN *deriv* is 'negative large'

R_2 : IF *error* is 'negative' AND *rate* is 'zero', OR *error* is 'zero'
AND *rate* is 'negative', THEN *deriv* is 'negative small'

R_3 : IF *error* is 'negative' AND *rate* is 'positive', OR *error* is 'zero'
AND *rate* is 'zero', OR *error* is 'positive'
AND *rate* is 'negative', THEN *deriv* is 'zero'

R_4 : IF *error* is 'zero' AND *rate* is 'positive', OR *error* is 'positive'
AND *rate* is 'zero', THEN *deriv* is 'positive small'

R_5 : IF *error* is 'positive' AND *rate* is 'positive',
THEN *deriv* is 'positive large'

where 'negative', 'zero', 'positive', and so on are fuzzy terms represented by fuzzy sets. The logical connectives 'AND' and 'OR' are subsequently replaced by some *t*- and *t*-conorm respectively to determine the *firing level* of the ith rule, denoted $\mu_{\text{DERIV}_i}(deriv)$. The rule-base can conveniently be illustrated in matrix form:

rate \ *error*	N	Z	P
P	Z	PS	PL
Z	NS	Z	PS
N	NL	NS	Z

Assuming n_s fuzzy sets for the error and rate of change in error, we require $2(n_s - 1)$ fuzzy sets (and rules) for the output *deriv*. The principal values for fuzzy sets in the consequence part of the rules (at which $\mu_{\text{DERIV}_i}(deriv) = 1$) are equally spaced, but at half the interval of the members of the antecedent fuzzy sets. There are three fuzzy sets defined on each input space for the error and rate of error. The principal values of the ith member of the fuzzy partition DERIV_i are given by $-1 + (i-1)/(n_s - 1)$. Following the paper by W. Siler we employ a linear defuzzification strategy:

$$deriv(k) = \sum_{i=1}^{2n_s - 1} \mu_{\text{DERIV}_i}(deriv) \cdot \left(-1 + \frac{(i-1)}{n_s - 1} \right) . \tag{10.4}$$

This value is integrated and scaled to obtain the control action required to drive the plant. We first evaluate the control rules using the *Zadeh logic*:

$$T(\mu_A(\cdot), \mu_B(\cdot)) = \min(\mu_A(\cdot), \mu_B(\cdot)), \quad S(\mu_A(\cdot), \mu_B(\cdot)) = \max(\mu_A(\cdot), \mu_B(\cdot)) .$$

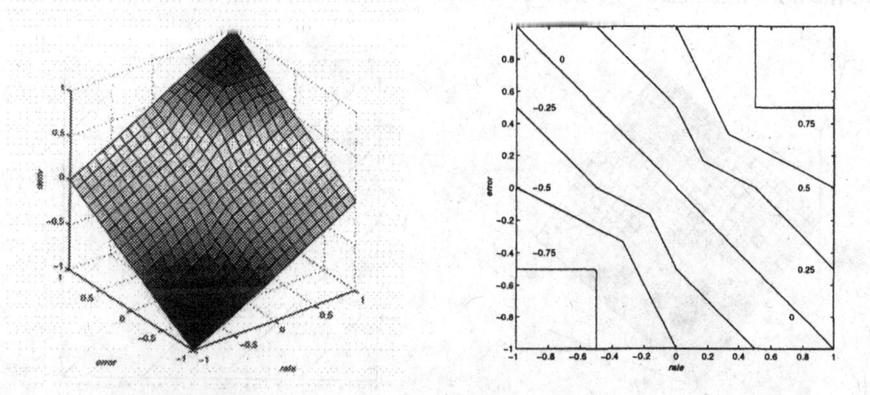

Fig. 10.4 Characteristics of the fuzzy PI-controller using Zadeh logic.

As can already be seen, the fuzzy PI-controller is structurally relatively simple and is, in fact, static. All dynamic elements, such as the rate of change or the integration of the output signal, are outside the inference engine. This fact allows us to study the characteristics of the controller by means of the *control surface* and *phase-plane* plot (isocontours of constant values of *deriv*) shown in Figure 10.4. The control is linear in the second and fourth quadrants but hyperlinear in the first and third quadrants.

Using a mixed logic, that is, Zadeh logic for R_1 and R_5, and Lukasiewicz logic for R_2 and R_4

$$T(\mu_A(\cdot), \mu_B(\cdot)) = \max(0, (\mu_A(\cdot) + \mu_B(\cdot)) - 1),$$
$$S(\mu_A(\cdot), \mu_B(\cdot)) = \min(1, \mu_A(\cdot) + \mu_B(\cdot))$$

and ignoring rule R_3 since its consequent is multiplied by zero, we obtain the controller characteristic shown in Figure 10.5. Similarly, using the ***probability logic***

$$T(\mu_A(\cdot), \mu_B(\cdot)) = \mu_A(\cdot) \cdot \mu_B(\cdot),$$
$$S(\mu_A(\cdot), \mu_B(\cdot)) = \mu_A(\cdot) + \mu_B(\cdot) - \mu_A(\cdot) \cdot \mu_B(\cdot)$$

for R_1 and R_5, and the probability 'AND' and the Lukasiewicz 'OR' for rules R_2 and R_4, the fuzzy controller is linear. We can conclude that using some appropriate mixed logic, the fuzzy PI-controller is theoretically and numerically identical to a linear PI-controller. Hence, choosing such mixed logic we would not have gained any possible advantage with the fuzzy controller in comparison to the conventional PI-controller. Even worse, while for the linear PI-controller a design methodology is available, with the fuzzy PI-controller we have no fixed strategy how to decide upon its parameters.

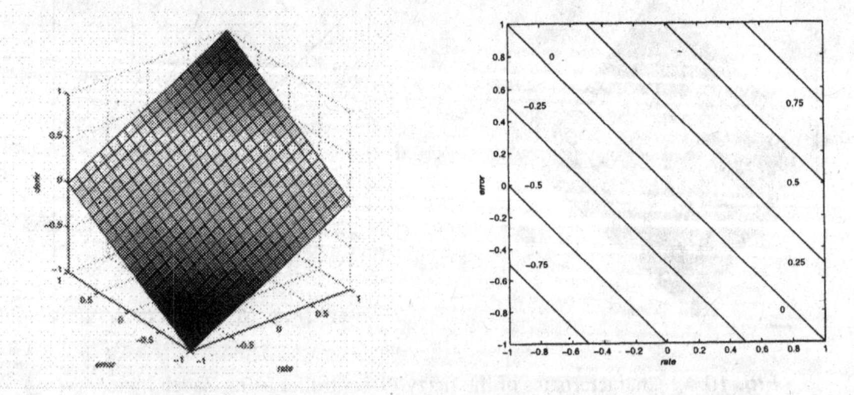

Fig. 10.5 Characteristics of the fuzzy PI-controller using a mixed logic.

10.2 EXAMPLE 1: FIRST-ORDER SYSTEM WITH DEAD-TIME

Based on the controller introduced in the previous section and the paper [YSB90], this section will first replace the linear defuzzification by the center-average defuzzification by a nonlinear strategy and then compare the nonlinear fuzzy PI-controller with its linear counterpart. The following notation will be employed:

$$\begin{aligned}
error'(k) &= s(k) - y(k) \\
error(k) &= G_e \cdot error'(k) \\
rate'(k) &= error'(k) - error'(k-1) \\
rate(k) &= G_r \cdot rate'(k) \\
deriv'(k) &= G_u \cdot deriv(k) \\
u(k) &= u(k-1) + deriv'(k) \, .
\end{aligned}$$

As in the previous section, we assume the sampling period to be equal to one. Hence, the rate of change can be calculated by simply taking the difference between two error measurements. Considering the last equation for $u(k)$ and the block diagram in Figure 10.2, one should also bear in mind that the output of the fuzzy inference engine is an incremental control action not the control action itself. The two input variables *error* and *rate* are characterized by two fuzzy sets describing a 'negative' and 'positive' error and rate, respectively. With only two fuzzy sets describing the inputs, the output space is partitioned into three fuzzy sets ('negative','zero','positive') as shown in Figure 10.6.

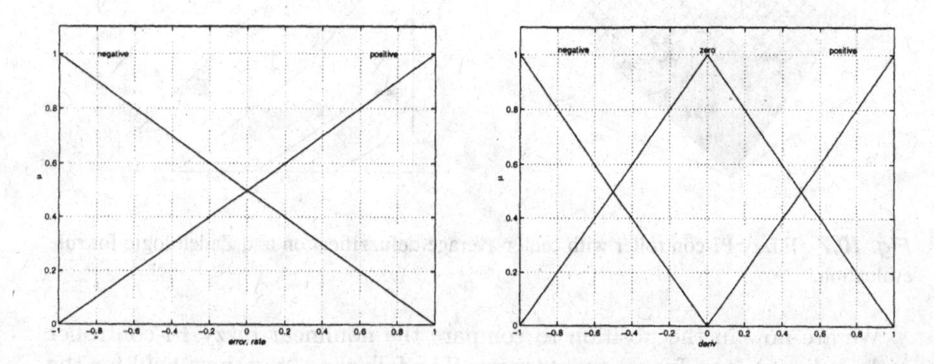

Fig. 10.6 Fuzzy sets for the variables *error* and *rate* (left) and the output (right).

There are three unique fuzzy control rules composed out of four if-then rules in total:

R_1 : IF *error* is 'negative' AND *rate* is 'negative', THEN *deriv* is 'negative'
R_2 : IF *error* is 'negative' AND *rate* is 'positive', THEN *deriv* is 'zero'
R_3 : IF *error* is 'positive' AND *rate* is 'negative', THEN *deriv* is 'zero'
R_4 : IF *error* is 'positive' AND *rate* is 'positive', THEN *deriv* is 'positive'

In contrast to the linear defuzzification strategy (10.4), employed in the previous section, here we use center-average defuzzification (8.18). The denominator, normalizing the membership degrees to one, introduces some nonlinearity:

$$deriv(k) = \frac{\sum_{i=1}^{2n_s-1} \mu_{DERIV_i}(deriv) \cdot \left(-1 + \frac{i-1}{n_s-1}\right)}{\sum_{i=1}^{2n_s-1} \mu_{DERIV_i}(deriv)} . \tag{10.5}$$

Employing the Zadeh logic the control characteristics are shown in Figure 10.7. The fuzzy controller with the nonlinear defuzzification shows an opposite behavior to the fuzzy PI-controller with linear defuzzification (Figure 10.4). From the step-responses of the plant under control one would consider whether an increased or decreased gain near (away) the steady-state (center of the phase-plane) is desirable and would choose the defuzzification strategy accordingly.

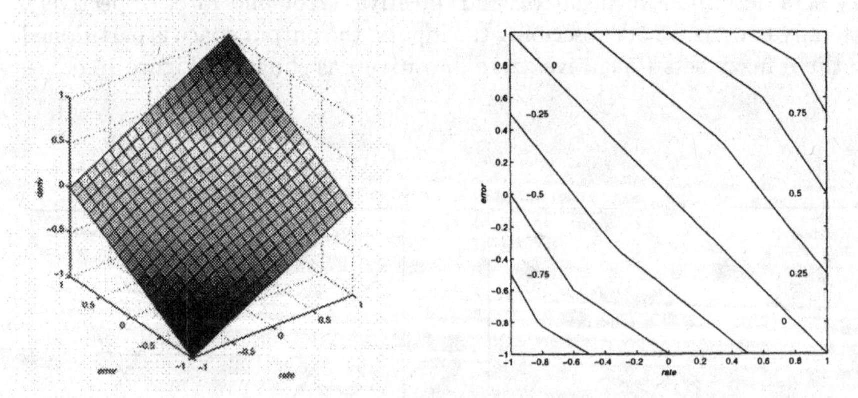

Fig. 10.7 Fuzzy PI-controller with center-average defuzzification and Zadeh logic for rule evaluation.

We are now in the position to compare the nonlinear fuzzy PI-controller with its linear (non-fuzzy) counterpart. The following equations hold for the membership functions associated with the error and rate of change of the error:

$$\mu_{error\ is\ pos}\big(e(k)\big) = \frac{error(k)+1}{2} = \frac{G_e \cdot error'(k)+1}{2} \tag{10.6}$$

$$\mu_{error\ is\ neg}\big(e(k)\big) = \frac{-error(k)+1}{2} = \frac{-G_e \cdot error'(k)+1}{2} \tag{10.7}$$

$$\mu_{rate\ is\ pos}\big(r(k)\big) = \frac{rate(k)+1}{2} = \frac{G_r \cdot rate'(k)+1}{2} \tag{10.8}$$

$$\mu_{rate\ is\ neg}\big(r(k)\big) = \frac{-rate(k)+1}{2} = \frac{-G_r \cdot rate'(k)+1}{2}. \tag{10.9}$$

These equations suggest a partition of the phase-plane into sectors for which, depending on the rule, specific fuzzy sets are relevant (see Figure 10.8):

Sector	R_1	R_2	R_3	R_4
1	*error* is 'neg.'	*rate* is 'neg.'	*error* is 'neg.'	*rate* is 'pos.'
2	*error* is 'neg.'	*rate* is 'neg.'	*error* is 'neg.'	*rate* is 'pos.'
3	*rate* is 'neg.'	*rate* is 'neg.'	*error* is 'neg.'	*error* is 'pos.'
4	*rate* is 'neg.'	*rate* is 'neg.'	*error* is 'neg.'	*error* is 'pos.'
5	*rate* is 'neg.'	*error* is 'pos.'	*rate* is 'pos.'	*error* is 'pos.'
6	*rate* is 'neg.'	*error* is 'pos.'	*rate* is 'pos.'	*error* is 'pos.'
7	*error* is 'neg.'	*error* is 'pos.'	*rate* is 'pos.'	*rate* is 'pos.'
8	*error* is 'neg.'	*error* is 'pos.'	*rate* is 'pos.'	*rate* is 'pos.'

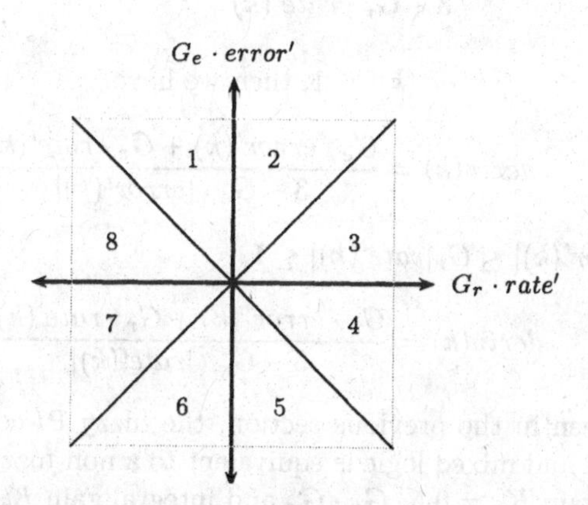

Fig. 10.8 Partioning of the phase-plane for a fuzzy PI-controller.

From equations (10.6)-(10.9) and (10.5), we obtain the following equations for the output w.r.t. the sectors defined in Figure 10.8:

Sectors 1 and 2:

$$deriv(k) = \frac{-\mu_{error\ \mathrm{is}\ neg}\big(e(k)\big) + \mu_{rate\ \mathrm{is}\ pos}\big(r(k)\big)}{\mu_{error\ \mathrm{is}\ neg}\big(e(k)\big) + \mu_{rate\ \mathrm{is}\ neg}\big(r(k)\big) + \mu_{rate\ \mathrm{is}\ pos}\big(r(k)\big)}$$
$$= \frac{G_e \cdot error'(k) + G_r \cdot rate'(k)}{3 - G_e \cdot error'(k)} \ . \qquad (10.10)$$

Sectors 3 and 4:

$$deriv(k) = \frac{-\mu_{rate\ \mathrm{is}\ neg}\big(r(k)\big) + \mu_{error\ \mathrm{is}\ pos}\big(e(k)\big)}{\mu_{rate\ \mathrm{is}\ neg}\big(r(k)\big) + \mu_{error\ \mathrm{is}\ neg}\big(e(k)\big) + \mu_{error\ \mathrm{is}\ pos}\big(e(k)\big)}$$
$$= \frac{G_r \cdot rate'(k) + G_e \cdot error'(k)}{3 - G_r \cdot rate'(k)} \ . \qquad (10.11)$$

Sectors 5 and 6:

$$deriv(k) = \frac{-\mu_{rate\ \mathrm{is}\ neg}\big(r(k)\big) + \mu_{error\ \mathrm{is}\ pos}\big(e(k)\big)}{\mu_{rate\ \mathrm{is}\ neg}\big(r(k)\big) + \mu_{rate\ \mathrm{is}\ pos}\big(r(k)\big) + \mu_{error\ \mathrm{is}\ pos}\big(e(k)\big)}$$
$$= \frac{G_r \cdot rate'(k) + G_e \cdot error'(k)}{3 + G_r \cdot error'(k)} \ . \qquad (10.12)$$

Sectors 7 and 8:

$$deriv(k) = \frac{-\mu_{error\ \mathrm{is}\ neg}\big(e(k)\big) + \mu_{rate\ \mathrm{is}\ pos}\big(r(k)\big)}{\mu_{error\ \mathrm{is}\ neg}\big(e(k)\big) + \mu_{error\ \mathrm{is}\ pos}\big(e(k)\big) + \mu_{rate\ \mathrm{is}\ pos}\big(r(k)\big)}$$
$$= \frac{G_e \cdot error'(k) + G_r \cdot rate'(k)}{3 + G_r \cdot rate'(k)} \ . \qquad (10.13)$$

If $G_r|rate'(k)| \leq G_e|error'(k)| \leq 1$, then we have

$$deriv(k) = \frac{G_e \cdot error'(k) + G_r \cdot rate'(k)}{3 - G_e \cdot |error'(k)|} \qquad (10.14)$$

and if $G_e|error'(k)| \leq G_r|rate'(k)| \leq 1$,

$$deriv(k) = \frac{G_e \cdot error'(k) + G_r \cdot rate'(k)}{3 - G_r \cdot |rate'(k)|} \ . \qquad (10.15)$$

As we have seen in the previous section, the fuzzy PI-controller with linear defuzzification and mixed logic is equivalent to a non-fuzzy PI-controller with proportional gain $K_p = 0.5 \cdot G_u \cdot G_r$ and integral gain $K_i = 0.5 \cdot G_u \cdot G_e$:

$$deriv'(k) = K_p \cdot rate'(k) + K_i \cdot error'(k) \ . \qquad (10.16)$$

Comparing (10.16) with equations (10.14) and (10.15), we notice that the fuzzy PI-controller with nonlinear defuzzification and Zadeh logic for rule evaluation is equivalent to a linear PI-controller with changing gains K_p and K_i given by

$$K_p = \frac{G_r \cdot G_u}{3 - G_e|error'(k)|} \tag{10.17}$$

$$K_i = \frac{G_e \cdot G_u}{3 - G_e|error'(k)|} \tag{10.18}$$

when $G_r|rate'(k)| \leq G_e|error'(k)| \leq 1$, and

$$K_p = \frac{G_r \cdot G_u}{3 - G_r|rate'(k)|} \tag{10.19}$$

$$K_i = \frac{G_e \cdot G_u}{3 - G_r|rate'(k)|} \tag{10.20}$$

when $G_e|error'(k)| \leq G_r|rate'(k)| \leq 1$. If we define the static gains K_{p_s} and K_{i_s} as the proportional and integral gains when both *error'* and *rate'* are equal to zero, we have

$$K_{p_s} = \frac{G_r \cdot G_u}{3} \tag{10.21}$$

$$K_{i_s} = \frac{G_e \cdot G_u}{3} \tag{10.22}$$

and find for the conventional PI-controller

$$\begin{aligned} deriv(k) &= \frac{K_{p_s}}{G_u} \cdot rate'(k) + \frac{K_{i_s}}{G_u} \cdot error'(k) \\ &= \frac{G_r \cdot rate'(k) + G_e \cdot error'(k)}{3} . \end{aligned} \tag{10.23}$$

Comparing equality (10.23) with equations (10.14) and (10.15), the following inequalities are obtained:

$$\frac{1}{3 - G_e \cdot error'(k)} \geq \frac{1}{3}$$

when $G_r|rate'(k)| \leq G_e|error'(k)| \leq 1$, and

$$\frac{1}{3 - G_r \cdot rate'(k)} \geq \frac{1}{3}$$

when $G_r|error'(k)| \leq G_e|rate'(k)| \leq 1$. In other words, the (absolute value of the) incremental control action of the fuzzy PI-controller is equal to, or greater than, the (absolute value of the) incremental control action of the non-fuzzy PI-controller when $G_e|error'(k)| \leq 1$ and $G_r|rate'(k)| \leq 1$. We can conclude that the larger the (absolute values of) error (rate) values, the larger the difference between the outputs of the two controllers. The nonlinearity of the

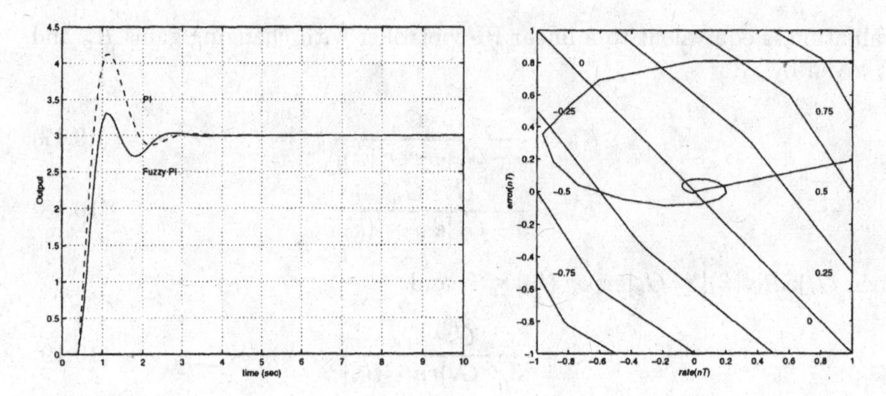

Fig. 10.9 Step-responses, phase-plane and trajectory for (fuzzy) PI-controller.

fuzzy PI-controller can therefore be used to improve the control performance in comparison to a non-fuzzy and linear PI-controller.

For a comparison of the step-responses of two PI-controllers, fuzzy (with Zadeh logic and nonlinear defuzzification) and non-fuzzy, the static proportional gain K_{p_s} and the static integral gain K_{i_s} of the fuzzy controller were set equal the proportional and integral gains $K_p = 2.38$ and $K_i = 4.43$ of the conventional PI-controller. The process plant is taken to be the following first-order system with time delay and transfer function:

$$\frac{Y(s)}{U(s)} = \frac{1}{s+1} \cdot e^{-0.2s} .$$

The step-response, phase-plane and trajectory are shown in Figure 10.9. We note that there are infinitely many combinations of G_e, G_r and G_u for expressions (10.21) and (10.22) to hold true. The fact that there is no *design methodology* providing step-by-step guidance how to select the parameters of the fuzzy PI-controller, is a major disadvantage.

10.3 EXAMPLE 2: COUPLED TANKS

Using the coupled tanks model from Section 1, we first describe two linear control designs and then develop a fuzzy PI-controller with equivalent performance.

Design of a Fluid Level Proportional Controller

The design procedure is to plot the root locus of the open-loop transfer function $G_v(z)$, and select the gain K_p that gives a closed-loop damping factor of $\vartheta = 0.7$. The closed-loop natural frequency ω_n can be read off the root locus diagram. The closed-loop transfer function is

$$H_v(z) = \frac{v_{d_2}(z)}{v_r(z)} = \frac{K_p\, G_v(z)}{1 + K_p\, G_v(z)}\,, \tag{10.24}$$

where

$$G_v(z) = \frac{z-1}{z}\, Z\left(\frac{G_v(s)}{s}\right)\,. \tag{10.25}$$

The steady-state error can be calculated using

$$e_{ss} = \left[v_r(z) - v_{d_2}(z)\right]_{z\to 1} = v_r(1)\left[1 - H_v(z)\right]_{z\to 1}\,, \tag{10.26}$$

where $v_r(1)$ is the steady-state reference input.

Design of a Proportional Plus Integral Controller

The design procedure is to set the integral action time constant to a reasonable value (in this case $T_i = 50\text{s}$), plot the root locus of the open-loop system in cascade with the compensator $1 + \frac{1}{T_i}\frac{z}{z-1}$, and select the gain k that gives a closed-loop damping factor of $\vartheta = 0.7$. The purpose of the integral action is to make the steady-state error zero. The proportional and integral gains K_p and K_i can be computed by comparing the coefficients of the compensator transfer functions

$$K\left(1 + \frac{z}{(z-1)\,T_i}\right) = K_p + \frac{K_i z}{(z-1)}\,. \tag{10.27}$$

Writing the open-loop system in series with a proportional plus integral action compensator as

$$G_c(s) = \frac{K\,(sT_i + 1)}{sT_i} \cdot \frac{\frac{g_p g_{d_2}}{K_2}}{T_1 T_2 s^2 + (T_1 + T_2)\, s + 1} \tag{10.28}$$

the closed-loop transfer function becomes

$$H_v(s) = \frac{K g_p g_{d_2}}{k_2} \cdot \frac{sT_i + 1}{T_i T_1 T_2 s^3 + T_i\,(T_1 + T_2)\, s^2 + T_i\left(1 + \frac{K g_p g_{d_2}}{K_2}\right) s + \frac{K g_p g_{d_2}}{K_2}}\,. \tag{10.29}$$

Assuming

$$T_1 > 0;\quad T_2 > 0;\quad g_p > 0;\quad g_{d_2} > 0;\quad k_2 > 0;\quad T_i > 0;\quad K > 0 \tag{10.30}$$

the closed-loop system is stable for

$$T_i\,(T_1 + T_2)\left(1 + \frac{K g_p g_{d_2}}{K_2}\right) > T_1 T_2 \frac{K g_p g_{d_2}}{K_2}\,. \tag{10.31}$$

Hence, the closed-loop system can become unstable for sufficiently large gain if

$$T_i < \frac{T_1 T_2}{T_1 + T_2} \tag{10.32}$$

and the gain required to make the closed-loop system unstable is

$$K = \frac{K_2}{g_p g_{d_2}} \cdot \frac{T_i (T_1 + T_2)}{T_1 T_2 - T_i (T_1 + T_2)} . \tag{10.33}$$

The closed-loop system under proportional plus integral control has three poles, and one zero. Using the design procedure outlined above, two of the poles will be complex conjugate (at $\vartheta = 0.7$); the remaining pole and zero are negative real. •

Design of a Fuzzy PI-Controller

In contrast to conventional control theory which, based on a linear parametric model of the process, provides a step-by-step methodology to select parameters, in fuzzy control we lack such a desirable methodology and usually rely on (often unsatisfying) trial-and-error procedures. In the following simulations, using the notation introduced in Section 10.1, the constants $\frac{K_p G_r}{G_u}$ and $\frac{K_i G_e}{G_u}$ are set to make the output to fall within the interval $[-1, +1]$. The fuzzy sets for the input and output are identical to the sets in Figure 10.3. Two step-responses of a conventional and the equivalent fuzzy controller are shown in Figure 10.10. In the simulation, $G_e = 0.02$, $G_r = 1$, and $G_u = 15$.

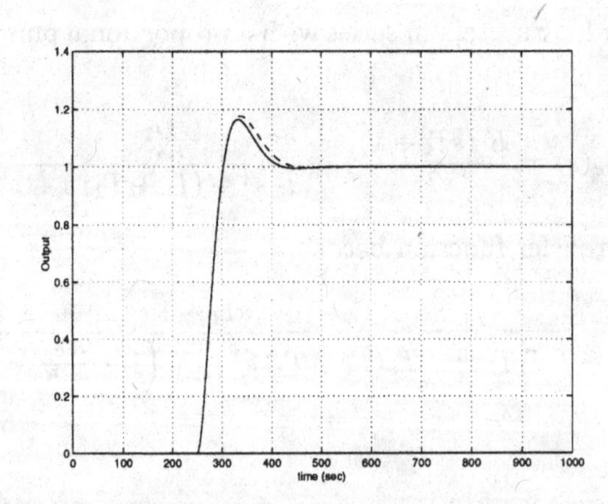

Fig. 10.10 Step-responses: non-fuzzy PI-controller with $K_p = 7.5$ and $K_i = 0.15$ (dashed line), and its fuzzy equivalent.

On the basis of the proportional-integral control schemes evaluated here we can summarize fuzzy control design as follows. Assuming the process under consideration is nonlinear or a linear parameterized model is not available, a fuzzy controller should be designed in the phase-plane according to the procedure below.

1. Fix the number of (triangular, fully overlapping) fuzzy sets partitioning the input spaces.

2. On the basis of the phase-plane characteristic and/or trajectory, decide on the fuzzy logic to be employed. This will determine the overall gain structure of the phase-plane (cf. Figures 10.4 and 10.7).

3. Adjust input-output gains to have trajectories of the system to fall within the $[-1, 1]$ range.

4. Change positions of principal values for input fuzzy sets to fine-tune gain structure in the quadrants of the phase-plane.

From the control surface or phase-plane characteristics of the fuzzy controllers, we have seen that a fuzzy controller, though overall nonlinear, is, in fact, a piecewise-linear system. It is therefore possible to use textbook knowledge on nonlinear control design and analysis in the phase-plane.

11

Fuzzy Mathematics

- ☐ *Fuzzy concepts are 'natural' generalizations of conventional mathematical concepts.*
- ☐ *Probability and possibility are complementary.*
- ☐ *Possibility theory is the attempt to be precise about uncertainty, to related statistical objects with rule-based and fuzzy concepts.*

Everything is a matter of degree.

—The Fuzzy Principle

Since almost all objects of mathematics can be described by sets (e.g., a function as a set of ordered pairs), one can establish fuzzy generalizations for nearly all of those. In the previous sections, we have seen that fuzzy mathematical concepts occur 'naturally', that is, considering the uncertainty involved in the application, the need to generalize creates fuzzy mathematical objects. We considered

- fuzzy sets
- fuzzy mappings
- fuzzy relations and their composition
- fuzzy graphs
- fuzzy clustering
- possibility measures and distributions

To apply formal mathematical objects to an engineering problem, mathematical statements are formulated using *propositional calculus* for which the standard logical connectives and their set-theoretic equivalents for propositions p and q are:

Conjunction	$p \wedge q$	"p and q"	$A_p \cap B_q$	Intersection
Disjunction	$p \vee q$	"p or q"	$A_p \cup B_q$	Union
Negation	$\neg p$	"not p"	A_p^c	Complement
Implication	$p \Longrightarrow q$	"p implies q"	$A_p^c \cup B_q$	Entailment

Classically, the subsets A of a set X form a Boolean algebra under the operations union, intersection, and complement. In particular, the double complement of a subset A is A itself, $(A^c)^c = A$, in parallel to the tautology $\neg\neg p = p$ of the propositional calculus. Fuzzy versions of these operations above will abandon the *law of the excluded middle*, $A \cap A^c = \emptyset$, and consequently allow us to model vague, fuzzy, and ambiguous concepts, avoiding paradoxes. From these extensions, we find that there are at least five ways to develop rule-based systems:

1. Composite fuzzy relations (approximate reasoning),
2. Functional approximation (e.g., Takagi-Sugeno models),
3. Similarity-based reasoning,
4. Multi-valued logic,
5. Possibilistic reasoning,

of which the first three were addressed in this book. An overview of different fuzzy mathematical extensions to classical concepts is given in Figure 11.1. In this book, fuzzy sets have been described in different ways. Most commonly a fuzzy set is considered to be a family of pairs, $R = \{(x, \mu_R(x))\}$ and membership function $\mu(\cdot)$. $\mu_R(x)$ describes the degree of membership of x in fuzzy set R. Viewing degrees of membership as some kind of weighting on elements of the underlying reference space X, a *fuzzy restriction* is the mapping μ

$$\begin{aligned} \mu : X &\rightarrow [0,1] \\ x &\mapsto \mu(x) \ . \end{aligned} \tag{11.1}$$

Especially in engineering, triangular and trapezoidal fuzzy set-membership functions are sufficiently accurate and have the advantage of a simple implementation. The following equations are useful in this context. Let $a \leq b \leq c \leq d$ denote characteristic points:

"Left-open" set: $\mu(x; a, b) = \max\left(\min\left(\frac{b-x}{b-a}, 1\right), 0\right)$

"Right-open": $\mu(x; a, b) = \max\left(\min\left(\frac{x-a}{b-a}, 1\right), 0\right)$

"Triangular": $\mu(x; a, b, c) = \max\left(\min\left(\frac{x-a}{b-a}, \frac{c-x}{c-b}\right), 0\right)$

"Trapezoidal": $\mu(x; a, b, c, d) = \max\left(\min\left(\frac{x-a}{b-a}, 1, \frac{d-x}{d-c}\right), 0\right)$

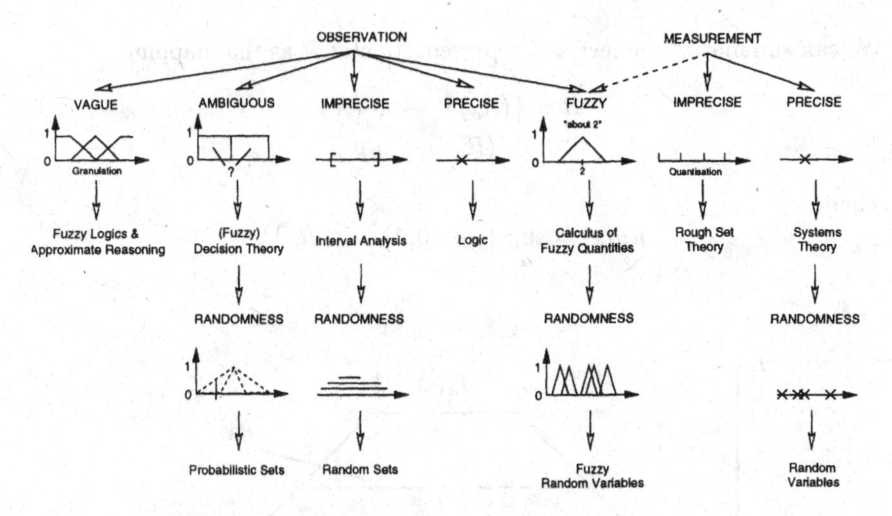

Fig. 11.1 Fuzzy mathematical extensions in context [Wol98].

Instead of taking elements of the universe of discourse as arguments, we may consider the co-domain of μ to describe subsets of X in terms of α-cuts, R^α where $R^\alpha = \{x\colon \mu_R(x) \geq \alpha\}$ is also called *level-set*. The fact that a family of level-sets can describe a fuzzy set is manifested in the decomposition or *representation theorem*. Instead of taking a set-membership perspective we may view $\mu(\cdot)$ as a mapping, or *fuzzy restriction* R:

$$\begin{aligned} \mu_R : X &\to L \\ x &\mapsto \alpha , \end{aligned}$$

where here we assume $L = [0, 1]$. While in the set-membership setting we first identify a value x and then determine its degree of membership, we may also start with a *level* $\alpha \in L$ to find out which elements in X satisfy this condition. This leads to the definition of a *level-set* or α-cut R_α (see Figure 11.2):

$$R_\alpha = \{x \in X \ : \ \mu(x) \geq \alpha\} . \tag{11.2}$$

The representation theorem [NW97] shows that a family of sets $\{R^\alpha\}$ with the assertion "**x** is in R^α" has the "degree of truth" α, and composes a fuzzy set, or equivalently, a fuzzy restriction

$$R(x) = \sup_{\alpha \in (0,1]} \min\bigl(\alpha, \zeta_{R^\alpha}(\mathbf{x})\bigr) , \tag{11.3}$$

where

$$\alpha = \min\{\alpha, \zeta_{R^\alpha}(\mathbf{x})\}$$

and hence

$$R = \{(x, \alpha) \ : \ x \in X, \ \mu_R(x) = R(x) = \alpha\} . \tag{11.4}$$

We can summarize the level-set representation of R as the mapping

$$\begin{aligned} R: \ \{\langle R_\alpha \rangle\} &\to \mathcal{F}(X) \\ \langle R_\alpha \rangle &\mapsto \mu_R \ , \end{aligned}$$

where

$$\mu_R(x) = \sup_\alpha \{\alpha \in [0,1], \ x \in R_\alpha\} \ .$$

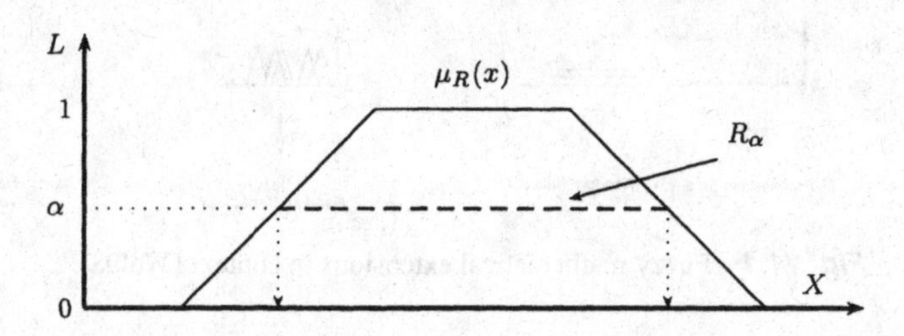

Fig. 11.2 Level-set representation of fuzzy restriction (set) R.

Figure 11.2 suggests yet another perspective on fuzzy restrictions as multi-valued maps. Let Γ be a map from L to X such that for any $\alpha \in L$, the image is made up of those $x \in X$ which are compatible with α:

$$\begin{aligned} \Gamma: \ L &\to \mathcal{P}(X) \\ \alpha &\mapsto \Gamma(\alpha) \ . \end{aligned} \tag{11.5}$$

Expressed in terms of ordered pairs $(\alpha, x) \in L \times X$,

$$R = \{(\alpha, x) : \ x \in \Gamma(\alpha)\} \ , \tag{11.6}$$

which is called *compatibility relation* and is related to $\Gamma(\alpha)$ by

$$\Gamma(\alpha) = \{x : \ (\alpha, x) \in R\} \ . \tag{11.7}$$

We also considered generalizations of ordinary relations to fuzzy equivalents. For example, considering the binary relation

$$\begin{aligned} R: \ X \times Y &\to \{0,1\} \\ (x,y) &\mapsto \zeta_R(x,y) \ , \end{aligned}$$

a fuzzy relation R is a fuzzy set defined in $X \times Y$ as

$$\begin{aligned} R: \ X \times Y &\to [0,1] \\ (x,y) &\mapsto \mu_R(x,y) \ . \end{aligned}$$

11.1 THE ALGEBRA OF FUZZY SETS

In this section, we first summarize the usual use and notation of (fuzzy) sets and then look at more formal aspects of how to describe sets.

For 'crisp' sets, if X is any set and $x \in X$, the algebra of the power set $\mathcal{P}(X)$ of X, that is, of the set of (crisp) subsets of X, is usually formulated in terms of $A \in \mathcal{P}(X)$ and $B \in \mathcal{P}(X)$ as follows:

$$\begin{aligned}
\text{Containment:} &\quad A \subset B \;\Leftrightarrow\; x \in A \;\Rightarrow\; x \in B \\
\text{Equality:} &\quad A = B \;\Leftrightarrow\; A \subset B \text{ and } B \subset A \\
\text{Complement:} &\quad A^c = \{x \in X :\; x \notin A\} \\
\text{Intersection:} &\quad A \cap B = \{x \in X :\; x \in A \text{ and } x \in B\} \\
\text{Union:} &\quad A \cup B = \{x \in X :\; x \in A \text{ or } x \in B \text{ or both}\} .
\end{aligned}$$

For fuzzy sets realized via functions, the set-theoretic operations and relations above have their equivalents in $\mathcal{F}(X)$, namely, the set of all fuzzy sets (the set of all membership functions). Let μ_A and μ_B be in $\mathcal{F}(X)$:

$$\begin{aligned}
\text{Containment:} &\quad A \subset B \;\Leftrightarrow\; \mu_A(x) \leq \mu_B(x) \\
\text{Equality:} &\quad A = B \;\Leftrightarrow\; \mu_A(x) = \mu_B(x) \\
\text{Complement:} &\quad \mu_{A^c}(x) = 1 - \mu_A(x) \\
\text{Intersection:} &\quad \mu_{A\cap B}(x) = T\big(\mu_A(x), \mu_B(x)\big) \\
\text{Union:} &\quad \mu_{A\cup B}(x) = S\big(\mu_A(x), \mu_B(x)\big) .
\end{aligned}$$

These are by no means the only definitions but those most commonly used. Similarly, the intersection and union operators are most commonly defined by

$$\mu_{A\cap B}(x) = \min\big(\mu_A(x), \mu_B(x)\big) \quad \text{and} \quad \mu_{A\cap B}(x) = \max\big(\mu_A(x), \mu_B(x)\big) .$$

With this definition, the fuzzy union of A and B is the smallest fuzzy set containing both A and B, and the intersection is the largest fuzzy set contained by both A and B. This pair of operators is the only pair which preserves the equalities $\mu \cap \mu = \mu$ and $\mu \cup \mu = \mu$. In other words, it is the only pair of distributive and thus absorbing and idempotent pair of t-norm and t-conorm [FR94], that is, $(\mathcal{F}(X), \cap, \cup)$ is a distributive lattice. However, $(\mathcal{F}(X), \cap, \cup)$ is not a Boolean algebra, since not all elements of the lattice have complements in it. Considering possibility distributions on product spaces, using max, min, joint distributions are separable, that is, it is possible to project marginal distributions onto subspaces. Subsequently the concept of *non-interactivity* can be defined in analogy to *independence* in probability theory.

We also made frequent use of the fact that there are two distinct ways of formalizing the notion of a set. A subset A of X may be formulated as a list describing its members

$$A = \{x \in X : x \text{ satisfies some condition}\} .$$

On the other hand, the subset A of X can be considered as an inclusion $A \subset X$ and hence as a mapping. Equivalent subobjects of X are represented by the same "arrow" from X, by taking the arrow to be the *characteristic function*

$$\zeta_A : X \to \{0,1\}$$

$$x \mapsto \zeta_A(x) \quad \text{where} \quad \zeta_A(x) = \begin{cases} 1 & \text{if } x \in A, \\ 0 & \text{if } x \notin A . \end{cases}$$

Moreover, from the characteristic function ζ, one can reconstruct the subset A: it consists of exactly those elements $x \in X$ which land on 1 under ζ. In other words, A is the *pullback*[1] of the inclusion $\{1\} \subset \{0,1\}$,

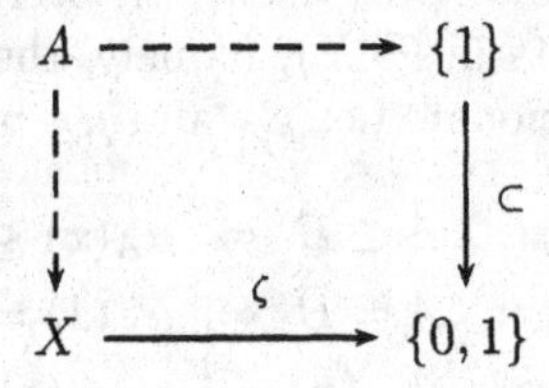

This can be generalized by replacing the set $\{0,1\}$ of 'truth values' by a non-negative interval L leading to 'categorical' formulations of fuzzy sets [Höh88]. This includes the case of $L = [0,1]$ being the unit-interval or some lattice defined on it. Then, the three operations $\wedge$, $\vee$, and $\neg$, defined by truth tables, are functions as $L \times L \to L$.

11.2 THE EXTENSION PRINCIPLE

In Section 7.2, instead of viewing the fuzzy systems as an algorithm based on formal multi-valued logic, the rule-base and inference mechanism was described as a mapping from a fuzzy set A' in X to a fuzzy set B' in Y. The compositional rule of inference generalized the 'crisp' rule

$$\text{IF } \mathbf{x} = a \text{ AND } \mathbf{y} = f(\mathbf{x}), \text{ THEN } y = f(a)$$

[1]In the commutative diagram, A completes the solid line 'corner' by means of the dashed lines and is thus called pullback. The terminology originates from Category Theory [LS97] in which algebra is organized to look not just at the objects but also at the mappings (arrows) between them. Examples are functions between sets or continuous maps between spaces. A category then consists of *objects* and *arrows* between them. Category Theory subsequently starts with functions between sets rather than with sets and their elements.

to be valid for fuzzy sets

$$\mu_{B'}(y) = \sup_{x \in X} T\big(\mu_{A'}(x), \mu_R(x,y)\big) . \tag{7.15}$$

The fuzzy system, defined by the compositional rule of inference, maps fuzzy sets in X to fuzzy sets in Y. In other words, the fuzzy model describes a *fuzzy mapping*

$$\begin{aligned} \tilde{f} : \mathcal{F}(X) &\rightarrow \mathcal{F}(Y) \\ \mu_A(\mathbf{x}) &\mapsto \tilde{f}(A) , \end{aligned} \tag{11.8}$$

where we obtain $\mu_{\tilde{f}(A)}(y)$ as a special case of the composition of two fuzzy relations

$$\mu_{\tilde{f}(A)}(y) = \sup_{\mathbf{x} \in X} T\big(\mu_{A_{\text{ext}}}(\mathbf{x}, y), \mu_R(\mathbf{x}, y)\big) \tag{11.9}$$

with extension $\mu_{A_{\text{ext}}}(\mathbf{x}, y) = \mu_A(\mathbf{x})$, equivalent to (7.15) or the individual-rule-based inference (8.6).

We can take the extension of the mapping to the fuzzy mapping as a blueprint for a general *extension principle*. Let f be a mapping from X to Y, $y = f(x)$. Consider the situation where we are given a fuzzy number A ("approximately x_0") instead of a real number. We wish to find the fuzzy image B by a generalization of f; how do we construct $B = f(A)$? We would require that the membership values of B should be determined by the membership values of A. Also $\sup B$ should be the image of $\sup A$ as defined by f. If the function f is surjective (onto), that is not injective (not a one-to-one mapping), we need to choose which of the values $\mu_A(x)$ to take for $\mu_B(y)$. Zadeh proposed the sup-union of all values x with $y = f(x)$ that have the membership degree $\mu_A(x)$. In other words,

$$\mu_B(y) = \sup_{x:\, y=f(x)} \mu_A(x) . \tag{11.10}$$

In general, we have the mapping

$$\begin{aligned} f : X_1 \times \cdots \times X_r &\rightarrow Y \\ (x_1, \ldots, x_r) &\mapsto y = f(x_1, \ldots, x_r) , \end{aligned}$$

which we aim to generalize to a function $\tilde{f}(\cdot)$ of fuzzy sets. The extension principle is defined as

$$\begin{aligned} \tilde{f} : \mathcal{F}(X) &\rightarrow \mathcal{F}(Y) \\ \big(\mu_{A_1}(x_1), \ldots, \mu_{A_r}(x_r)\big) &\mapsto \tilde{f}\big(\mu_{A_1}(x_1), \ldots, \mu_{A_r}(x_r)\big) , \end{aligned}$$

where

$$\mu_{\tilde{f}(A_1,\ldots,A_r)}(y) = \sup_{(x_1,\ldots,x_r) \in f^{-1}(y)} \{\mu_{A_1}(x_1) \wedge \cdots \wedge \mu_{A_r}(x_r)\} . \tag{11.11}$$

In Section 1, we introduced the functional representation (model $\mathfrak{M}$) of a system $\mathfrak{S}$, $\mathbf{y} = f(\mathbf{x})$ and its associated *graph*, $F = \{(\mathbf{x}, f(\mathbf{x}))\}$. We then introduced various (equivalent) generalizations into fuzzy models described by the fuzzy graph $\widetilde{F}$: In Section 4.6, equation (4.33), in Section 5, equation (5.5), in Section 6, by (6.19), and in Section 8 by equation (8.5).

11.3 FUZZY RULES AND FUZZY GRAPHS

There is a close relation between the concept of *approximate reasoning* (Section 7.2), the fuzzy inference engines (Section 8), the fuzzy mapping introduced in the previous section and a fuzzy graph $\widetilde{F}$. Fuzzy rules and a fuzzy graph may both be interpreted as granular representations of functional dependencies and relations.

A fuzzy graph $\widetilde{F}$, serves as an approximate or compressed representation of a functional dependence $f\colon X \to Y$, in the form

$$\widetilde{F} = A_1 \times B_1 \vee A_2 \times B_2 \vee \cdots \vee A_{n_R} \times B_{n_R} \tag{11.12}$$

or more compactly

$$\widetilde{F} = \bigvee_{i=1}^{n_R} A_i \times B_i ,$$

where the A_i and B_i are fuzzy subsets of X and Y, respectively, $A_i \times B_i$ is the Cartesian product of A_i and B_i, and $\vee$ is the operation of disjunction which is usually taken to be the union. In terms of membership functions, we may write

$$\mu_{\widetilde{F}}(x, y) = \bigvee_i \left(\mu_{A_i}(x) \wedge \mu_{B_i}(y)\right) ,$$

where $x \in X$, $y \in Y$, $\vee$ and $\wedge$ are any triangular t- and t-conorm, respectively. Usually $\vee = \max$ and $\wedge = \min$ establishing the relationship to the extension principle, approximate reasoning and so forth. A fuzzy graph may therefore be represented as a fuzzy relation or a collection of fuzzy if-then rules

$$\text{IF } \mathbf{x} \text{ is } A_i, \text{ THEN } \mathbf{y} \text{ is } B_i \quad i = 1, 2, \ldots, n_R .$$

Each fuzzy if-then rule is interpreted as the joint constraint on $\mathbf{x}$ and $\mathbf{y}$ defined by

$$(\mathbf{x}, \mathbf{y}) \quad \text{is} \quad A_i \times B_i .$$

In Section 4.6 we referred to such constraint as a *fuzzy point* in the data space $\Xi = X \times Y$. The concept of a fuzzy points and a fuzzy graph are illustrated in Figure 11.3.

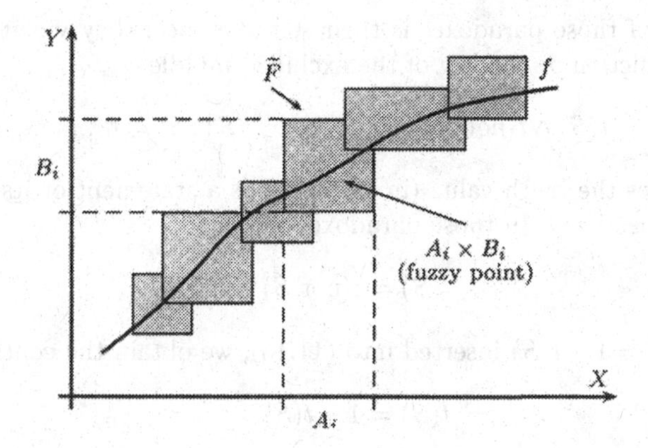

Fig. 11.3 Representation of a function and its fuzzy graph.

11.4 FUZZY LOGIC

> Any useful logic must concern itself with Ideas with a fringe of vagueness and a Truth that is a matter of degree.
>
> —Norbert Wiener

The basic claim of fuzzy theorists is that *everything is a matter of degree*. This view is based on the analysis of 'real-world' problems and the resulting paradox when applying conventional mathematics rooted in set theory. The following examples were first discussed by Bertrand Russell:

Liar paradox: Does the liar from Crete lie when he says that all Cretans are liars? If he lies, he tells the truth. But if he tells the truth, he lies.

Barber paradox: A barber advertises: "I shave all, and only, those men who don't shave themselves". Who shaves the barber? If he shaves himself, then according to the ad he doesn't. If he does not, then according to him he does.

Set as a collection of objects:

i. Consider the collection of books. This collection itself is not a book, thus it is not a member of another collection of books.

ii. The collection of all things that are not books is itself not a book and therefore a member of itself.

iii. Now consider the set of all sets that are not members of themselves. Is this a member of itself or not?

The problem of those paradoxes is their self-reference; they violate the laws of non-contradiction or the law of the excluded middle

$$t(S) \wedge t(\text{not } S) = 0 \qquad \text{or} \qquad S \cap S^c = \emptyset ,$$

where t denotes the truth value (in $\{0,1\}$), S is a statement or its set representation, respectively. In those paradoxes we find

$$t(S) = t(\text{not } S) . \tag{11.13}$$

With $t(\text{not } S) = 1 - t(S)$ inserted into (11.13), we obtain the contradiction

$$t(S) = 1 - t(S) . \tag{11.14}$$

However, in fuzzy logic we simply solve (11.14) for $t(S)$:

$$2 \cdot t(S) = 1 \qquad \text{or} \qquad t(S) = \frac{1}{2} .$$

...the truth lies somewhere in between!

Throughout the book, we have suggested that fuzzy sets and fuzzy rule-based systems are a suitable means to capture and process vague and fuzzy information. In Section 5, the fuzzy partition $A_{i1}, \ldots, A_{ir}$ in the if-antecedent part of a rule was identified by projecting fuzzy clusters onto subspaces. In Section 6, the identified model provided forecasts as fuzzy restrictions and we suggested that such information can be readily processed in a fuzzy rule-based system as detailed in Section 7.2. It should, however, be noted that we have not considered fuzzy logic as a multi-valued generalization of a pure (formal) logic. Instead, we motivated our rule-based systems in terms of approximate reasoning (see Section 7.2). Next, we provide more examples of how a fuzzy partition and fuzzy data can be generated from numerical data sets.

11.5 A BIJECTIVE PROBABILITY - POSSIBILITY TRANSFORMATION

Possibility theory is an information theory which is related to but independent of both fuzzy sets and probability theory. Technically, a possibility distribution is a fuzzy set. For example, all fuzzy numbers are possibility distributions. However, possibility theory can also be derived without reference to fuzzy sets. Many of the rules of possibility theory are similar to probability theory, but use either max/min or max/times calculus, rather than the plus/times calculus of probability theory. Research into possibility theory has been largely concerned with being precise about uncertainty and reasoning in the presence of uncertainty. Although not as well defined as probability theory, possibility theory has made important contributions to establish formal relationships between a number of paradigms including belief functions, rough sets, random sets, fuzzy sets and probabilistic sets. The publications of Didier Dubois provide a rich and comprehensive source for most aspects of possibility theory. Here we note only one particular aspect, a formal link between probability and possibility distribution.

A possibility distribution $\pi(\cdot)$ on X is a mapping from the reference set or universe X into the unit-interval,

$$\pi : X \rightarrow [0,1] .$$

A usual convention is to assume that there exists at least one $x \in X$ for which $\pi(x) = 1$. This is called the normalization condition. Like a probability distribution is related to its associated probability measure, the possibility distribution is described in terms of a ***possibility measure*** by

$$\pi(x) = \Pi(\{x\}) ,$$

where the possibility of some event A is defined by

$$\Pi(A) = \sup_{x \in X} \pi(x) \qquad \text{for an ordinary set } A, \tag{2.14}$$

$$= \sup_{x \in X} \min\{\mu_A(x), \pi(x)\} \qquad \text{for a fuzzy set } A . \tag{11.15}$$

A dual necessity (or certainty) measure is defined by

$$Ne(A) = 1 - \Pi(A^c),$$

$$= \inf_{x \notin A} \{1 - \pi(x)\} \qquad \text{for an ordinary set } A, \tag{11.16}$$

$$= \inf_{x \in X} \min\{\mu_A(x), (1 - \pi(x))\} \qquad \text{for a fuzzy set } A . \tag{11.17}$$

In analogy to axioms in probability theory, $Pr(X) = \Pi(X) = 1$, $Pr(\emptyset) = \Pi(\emptyset) = 0$. The probability of the statement "A or B" is given by $Pr(A \cup B) = Pr(A) + Pr(B)$. The events A and B are assumed to be mutually exclusive, that is, $A \cap B = \emptyset$. If A^c denotes the event "not A", in probability theory it is consequently required that probabilities are additive on disjoint sets, $Pr(A) + Pr(A^c) = 1$. In contrast, for a subjective judgement one may view both propositions "A occurs" and "A does not occur" as possible and we therefore only require

$$\Pi(A) + \Pi(A^c) \geq 1 . \tag{11.18}$$

To be able to operate with such weak evidence we require the measure of necessity to act as a lower bound:

$$Ne(A) = 1 - \Pi(A^c) , \quad \text{and} \quad Ne(A) + Ne(A^c) \leq 1 , \tag{11.19}$$

as the impossibility of the opposite event. "A is necessary" is interpreted as A is bound to occur. We can summarize these intuitive requirements by the following inequality:

$$Ne(A) = 1 - \Pi(A^c) \leq Pr(A) \leq \Pi(A) . \tag{11.20}$$

And for the union of two events,

$$\begin{aligned}\Pi(A \cup B) &= \sup\ \{\pi(x)\ :\ x \in A \cup B\}\\ &= \sup\ \{\pi(x)\ :\ x \in A \vee x \in B\}\\ &= \max(\sup\{\pi(x)\ :\ x \in A\}, \sup\{\pi(x)\ :\ x \in B\})\\ \therefore\quad \Pi(A \cup B) &= \max(\Pi(A), \Pi(B))\ . \qquad (11.21)\end{aligned}$$

From these definitions, D. Dubois and H. Prade [DP83] derived a necessity measure as a bias in probabilities: the necessity of an event is the extra amount of probability of elementary events in a set over the amount of probability assigned to the most frequent outcome outside the event of concern. For this view, the x_j's are (without loss of generality) ranked so that for $p_j = Pr(\{x_j\})$, $p_1 \geq p_2 \geq \cdots \geq p_{n_b}$ and A_j denotes the set $\{x_1, x_2, \ldots, x_j\}$. Thus,

$$Ne(A) = \sum_{x_i \in A} \max(p_i - p', 0)\ , \quad p' = \max_{x_k \notin A} p_k\ . \qquad (11.22)$$

From (11.22), $\forall\ j = 1, \ldots, n_b$,

$$\begin{aligned}\pi_j &= 1 - Ne(X - \{x_j\})\\ &= 1 - \sum_{x_i \in X - \{x_j\}} \max\{p_i - p_j, 0\},\\ &= p_j + \sum_{i \neq j} (p_i - \max\{p_i - p_j, 0\})\\ &= p_j + \sum_{i \neq j} \min(p_i, p_j)\\ &= \sum_{i=1}^{n_b} \min(p_i, p_j)\ . \qquad (11.23)\end{aligned}$$

11.6 EXAMPLE: MAINTENANCE DECISION MAKING

One of the major problems in maintenance practice is the lack of a systematic, adaptable and data-driven approach in setting preventive maintenance instructions. A computerized maintenance management system (CMMS) is a decision support system prioritizing machines based on the two most important criteria - the downtime of machines and the frequency of faults. The knowledge-base consisting of nine fuzzy rules is shown in Figure 11.4. For example, if a machine fails regularly, but the downtime is relatively small, that is, it does not take much time to solve the problem, the operator should obtain a skill upgrade to deal with the problem himself rather than calling out a maintenance engineer. On the other hand, a machine which, when it fails, does take long to repair, should be monitored for such faults if they do

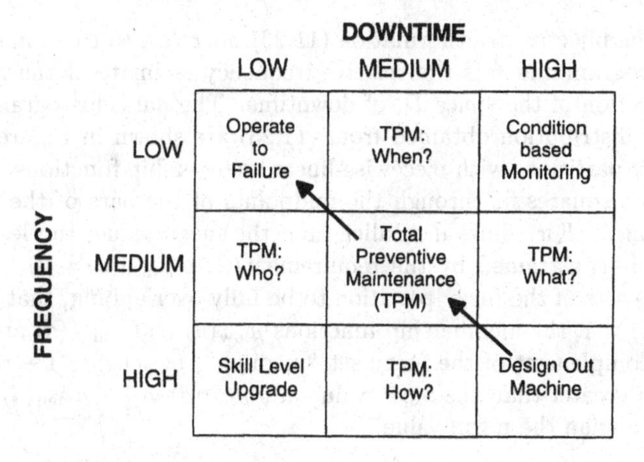

Fig. 11.4 Decision grid - fuzzy rules used in preventive maintenance decision making.

not occur too frequently. A successful 'control' of the maintenance problem is achieved if machines move, month by month from the lower right corner of the decision grid to the upper left corner. The problem is to define what is regarded as a "low downtime"? To obtain a fuzzy partition of the two spaces 'downtime' and 'frequency', the decision support system reviews monthly data of the worst ten machines. For example, consider the monthly evaluation of ten machines (out of 130) representing about 89% of the problems:

d	30	20	20	17	16	12	7	6	6	4
f	27	16	12	9	8	8	8	4	3	2

Since the downtime of machines presents the most important criteria, as it can readily be related to profit, loss or gain, we shall focus on downtime hereafter. The first step in obtaining a fuzzy partition is to plot a probability distribution obtained from the histogram. The average downtime, here 13.8, for that particular month is used to split the reference space into two halves. A downtime considerably larger than this value should be regarded as "high". It is also the point for which $\mu_{\text{med}}(d) = 1$. By definition, probability distributions summarize whole sets of machines; in preventive maintenance, however, decisions are made for individual machines. That is, we do not wish to consider *whether or not* we can (on average) expect a failure of any machine, but on the basis of the data collected, we wish to establish the ***degree of feasibility*** or ***possibility*** that a particular machine should be considered as having a "high downtime". We consequently require a transformation of the probabilistic information to a possibility distribution (or fuzzy restriction, which numerically is the same).

Using the bijective transformation (11.23), n_b refers to the number of bins of the histogram and p is the relative frequency estimate of the probability for that section of the space D, of downtime. The data, histogram, and the possibility distribution obtained from (11.23) are shown in Figure 11.5. Finally, fuzzy partitions with piecewise-linear membership functions is obtained from a least-squares fit through the midpoints of the bars of the possibility distribution π. For values d smaller than the mean value, the least-squares algorithm is constrained by the requirement that $\mu_{\text{med}}(d=0)=0$. If we require the sets of the fuzzy partition to be fully overlapping, that is, for any d, $\sum_i \mu_i(d)=1$, the membership functions $\mu_{\text{low}}(d)$ and $\mu_{\text{high}}(d)$ are obtained from the complement of the fuzzy set "medium", $\mu_{\text{high}}(d)=1-\mu_{\text{med}}(d)$ for values of d greater than the mean value and $\mu_{\text{low}}(d)=1-\mu_{\text{med}}(d)$ for values of d smaller than the mean value.

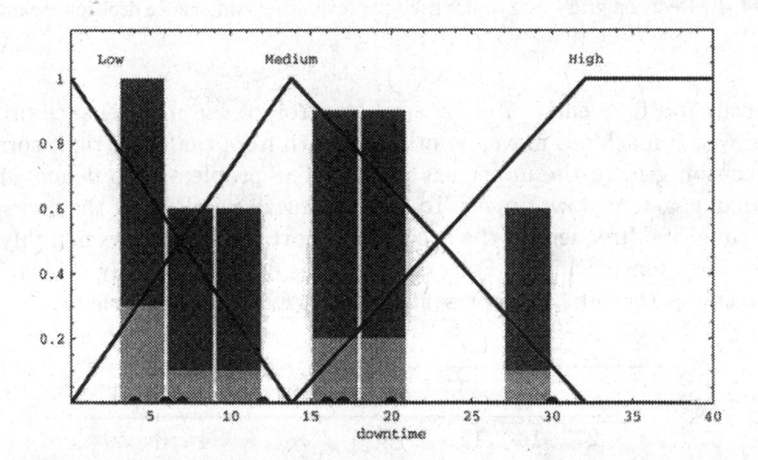

Fig. 11.5 Data, histogram, possibility distribution and fuzzy partition.

A (multi-stage fuzzy) control or optimization problem is introduced by considering time. Machines are reviewed on a monthly basis and, as indicated by the arrow in Figure 11.4. We aim to move (control) machines from the bottom-right to the upper-left corner of the decision grid.

11.7 EXAMPLE: EVALUATING STUDENT PERFORMANCES

In this example, we wish to evaluate student performances against set standards. At many universities, two thresholds are of particular importance. The student must obtain an overall result of, say $\geq 50\%$ to obtain the degree. Students with $\geq 70\%$ are awarded a degree with distinction. It is the neighborhood of these particular points we consider. Strict adherence to such hard thresholds is inevitably unfair in some cases where someone with, say 69.4%,

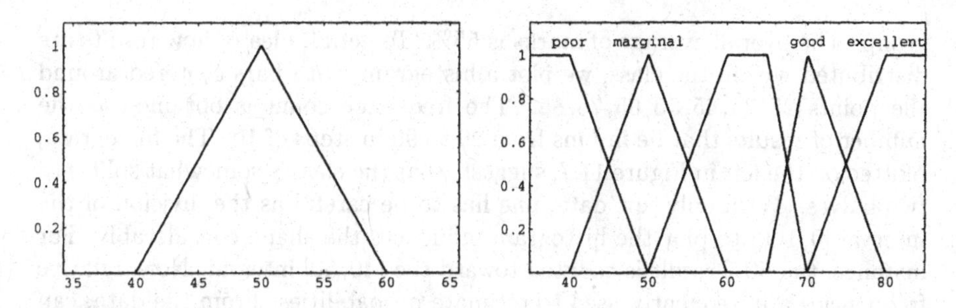

Fig. 11.6 Induced fuzzy set with $x_0 = 50$ and $\delta = 0.1$, and fuzzy partition for MSc degree evaluation.

may not achieve a distinction while a colleague with just 0.6% more would succeed. Let (X, d) be a metric space with $X = [0, 100]$ and the standard metric

$$
\begin{aligned}
d\colon\ \mathbb{R} \times \mathbb{R} &\to \mathbb{R}^+ \\
(x, x') &\mapsto |x - x'| \, .
\end{aligned}
$$

From (7.6), we know that the metric $d(\cdot)$ induces a similarity relation $\widetilde{E}\colon X \times X \to [0,1]$ on X by

$$
\widetilde{E} = 1 - \inf\left(d(x, x'), 1\right) , \tag{7.6}
$$

which formally is identical to a fuzzy restriction or fuzzy set. To identify 'border cases', that is, students which are close to the pass mark, denoted x_0, we introduce a function that depends on x_0 and a 'locality' parameter $\delta \in (0, \infty)$ used to scale the metric $d(\cdot)$ to become a proximity measure. As a result of these considerations, we obtain the triangular fuzzy set

$$
\mu_{x_0}(x) = 1 - \min\{|\delta \cdot x_0 - \delta \cdot x|, 1\} \, . \tag{11.24}
$$

Figure 11.6 (left plot), shows a fuzzy set induced by $\delta = 0.1$. A similar procedure can be applied to the 70% mark. Assuming fully overlapping sets ($\forall x \ \sum \mu(x) = 1$), the single parameter δ induces a complete fuzzy partition. Figure 11.6 (right plot) illustrates a general example of a fuzzy partition in performance evaluation.

Let us now consider the results of 23 students in a particular exam. Their results in % are:

47.33	34.5	71	84.17	76.5	39.67	62	55.83
35.67	81.67	65	78.83	60.33	43	40.5	40.67
37.33	64.33	62.67	50	56.67	46.83	26.67	

Analyzing the data, we wish to get a picture of how the class performed as a group compared to other classes and how results are distributed within a group. The overall average of marks is 55%. To get an idea of how results are distributed within the class, we plot a histogram, with bars centered around the points $25, 35, 45, 55, 65, 75, 85$. The frequency count is obtained as the number of results that lie in bins from 20 to 90 in steps of 10. The histogram, plotted on the left in Figure 11.7, suggests that the class is somewhat split into two halves. With only few data, one has to be careful as the division of the interval $[0, 100]$ to plot the histogram influences the shape considerably. For instance, the 50% result is counted toward the $(40, 50]$ interval. Now, relative frequencies are frequently used to estimate probabilities. From the data, can we determine what the probability was to score more than, say 50%? We would calculate such probability from the density function $p(x)$ as shown in the plot on the right in Figure 11.7 (obtained from Parzen's Gaussian kernel estimator with window width 5; cf. Section 3.2). Considering the definition of statistical laws, the meaning of such evaluation would be senseless, and the plot of the density function does not reveal any useful information.

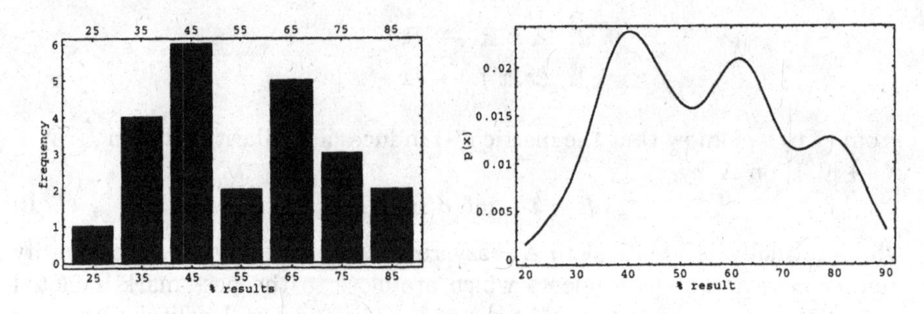

Fig. 11.7 Examination results: Histogram and density estimate.

12

Summary

data: known facts or things used as a basis for inference or reckoning.
system: a complex whole; a set of connected things or parts.
uncertainty: the fact or condition of being uncertain.

—*Oxford Dictionary*, 9th edition

In this section, we summarize some of the formalisms introduced to model natural systems. Further below, more philosophical ideas are discussed.

12.1 SYSTEM REPRESENTATIONS

Conceptually, the book presented three main topics connected by the following lines of reasoning:

Uncertainty:

Systems → Uncertainty → Expectation → Statistics: mean, (co-) variance, correlation.

Universal Function Approximation:

Systems → Modelling → Identification → Least-Squares Principle → Regression → Density Estimation → Basis Function Approximation.

Fuzzy Systems:

Uncertainty → Comparing & Reasoning → Equivalence Relations → (Fuzzy) Similarity Relations → Propositions as Subsets of a Data Space → (Fuzzy) Clustering → Fuzzy Rule-Based Systems → Random-Sets → Fuzzy-System Identification → Approximate Reasoning.

Virtually all the material is derived from, or directly related to, two fundamental concepts: The formulation of a system using some dependent variable(s), denoted $\mathbf{y}$, as a function of the independent variable(s) $\mathbf{x}$:

$$\boxed{\mathbf{y} = f(\mathbf{x}; \cdot)}$$

Second, the expectation operator forms the basis for describing (*measuring*) (un)certainty in systems and data:

$$\boxed{E[h(\cdot)] = \displaystyle\fint g(\cdot) \circ h(\cdot)}$$

In systems theory, a system $\mathfrak{S}$ is described by a formal model $\mathfrak{M}$ using a $\boldsymbol{\theta}$-parameterized mapping $f(\cdot)$ to describe the relationship among inputs and outputs as

$$y = f(\mathbf{u}; \boldsymbol{\theta}) \,, \tag{1.5}$$

which, in the regression formulation, was generalized to

$$y = f(\mathbf{x}; \boldsymbol{\theta}) \,. \tag{2.15}$$

Uncertainty in systems theory is due to

- ▷ Incomplete knowledge about the structure or nature of the system under consideration.
- ▷ Observations are imprecise, vague or fuzzy, as well as random.
- ▷ Preference; to have 'simple' models we choose to ignore aspects or accept incompleteness.

Consequently, sets of uncertain information need to be summarized; specifically *averaged* or *aggregated* using the expectation operator

$$E[h(\cdot)] \doteq \int_Y h(y) \cdot g(y) \; \mathrm{d}y \tag{2.1}$$

$$\doteq \fint_Y h(y) \circ g(y) \,. \tag{12.1}$$

Globally, uncertain knowledge about elements of some space is characterized by some distribution over this space which can also be described by measures of central tendency, $E[y] \doteq \eta$, and variability $E[(y-\eta)^2] \doteq \sigma^2$. Knowledge about events is quantified by the expectation leading to probability measures $Pr(\cdot)$ or possibility measures $\Pi(\cdot)$:

$$E[\zeta_A] = \int_Y \zeta_A(y)\; p(y)\; dy \doteq Pr(A),\; A \text{ crisp.} \tag{2.2}$$

$$= \sup_{y \in A}\; \pi(y) \doteq \Pi(A),\; A \text{ crisp.} \tag{2.14}$$

$$E[\mu_A] = \int_Y \mu_A(y)\; p(y)\; dy \doteq Pr(A),\; A \text{ fuzzy.} \tag{2.13}$$

$$= \underset{Y}{\fint} \mu_A(y) \circ \pi(y) = \sup_{y \in Y}\; \{\; \mu_A(y) \wedge \pi(y)\; \} \doteq \Pi(A),\; A \text{ fuzzy.} \tag{2.13}$$

Events are described by crisp and fuzzy membership functions, respectively:

$$\zeta : \; Y \to \{0, 1\} \qquad \text{and} \qquad \mu : Y \to [0, 1]\; .$$

The (un)certainty of a relationship between two variables is formalized using the covariance and the correlation ρ.

The least-squares principle is introduced as a criteria to quantify the quality of approximation of $\mathbf{y}$ by $f(\mathbf{x}; \boldsymbol{\theta})$. In regression, we thus try to find a function of the regressors in $\mathbf{x}$ such that the expected error is small. The examples 2.1.1 and 2.1.2 illustrated the least-squares principle applied to

- ▷ fitting functions to data,
- ▷ function approximation,

showing that the least-squares principle makes sense without a probabilistic framework. Modelling and identification of systems is, in general, based on the following concepts:

- ▷ A specific class of approximating functions $f(\mathbf{x}; \boldsymbol{\theta})$, parameterized by $\boldsymbol{\theta}$.
- ▷ The loss function $L(y, f(\mathbf{m}; \boldsymbol{\theta}))$ to quantify the quality of approximation using a set of training data samples $\mathbf{M} = \{\mathbf{m}_j\}$.
- ▷ The risk functional, quantifying the expected error:

$$E[L] = \int L(y, f(\mathbf{x}; \boldsymbol{\theta}))\; p(\mathbf{x}, y)\; \mathrm{d}\mathbf{x}\mathrm{d}y \tag{2.16}$$

$$\doteq R(\boldsymbol{\theta})\; .$$

For a stochastic framework, two closely related inductive principles play a fundamental role:

▷ **Maximum Likelihood (ML):** Determine the parameter(s) $\boldsymbol{\theta}$ for which the probability of observing the outcome (data) $\mathbf{M}$ is as high as possible. In other words, the likelihood (function) of the data

$$\ell(\boldsymbol{\theta}; \mathbf{m}_1, \mathbf{m}_2, \ldots, \mathbf{m}_d) = Pr(\mathbf{M}|\boldsymbol{\theta}) \tag{2.53}$$

is maximized to obtain a ML estimate:

$$\boldsymbol{\theta}_{\mathrm{ML}} = \arg\max_{\boldsymbol{\theta}} \ \ell(\boldsymbol{\theta}; \mathbf{M}) \ . \tag{2.54}$$

For convenience, the log-likelihood function $\mathcal{L}$ replaces ℓ in the actual calculation:

$$\mathcal{L}(\boldsymbol{\theta}; \mathbf{M}) \doteq \ln \ \ell(\boldsymbol{\theta}; \mathbf{M}) \ . \tag{2.55}$$

▷ **Empirical Risk Minimization (ERM):** In statistical learning theory, the training data $\mathbf{m} = (\mathbf{x}, y)$ are assumed to be independent identically distributed according to some (unknown) probability density function $p(\mathbf{m})$. The objective is to select a loss function from the class

$$Q(\mathbf{m}, \boldsymbol{\theta}) \doteq L\big(y, f(\mathbf{x}; \boldsymbol{\theta})\big) \tag{3.11}$$

such that it minimizes the risk functional

$$R(\boldsymbol{\theta}) = \int Q(\mathbf{m}, \boldsymbol{\theta}) p(\mathbf{m}) \ \mathrm{d}\mathbf{m} \ . \tag{3.10}$$

The ERM principle is then based on the empirical risk

$$R_{\mathrm{emp}}(\boldsymbol{\theta}) = \frac{1}{d} \sum_{j=1}^{d} Q(\mathbf{m}_j, \boldsymbol{\theta}) \ . \tag{3.12}$$

The class of approximating functions considered includes

▷ **Algebraic Polynomials:** For example, in linear regression

$$f(\mathbf{x}; \boldsymbol{\theta}) = \sum_{i=1}^{r} \ \theta_i \cdot x_i \ . \tag{3.2}$$

In a probabilistic model, $f(\mathbf{x}) = \int y \cdot p(y|\mathbf{x}) \mathrm{d}y$ and

$$L\big(y, f(\mathbf{x}; \boldsymbol{\theta})\big) = \big(y - f(\mathbf{x}; \boldsymbol{\theta})\big)^2 \ ,$$
$$R(\boldsymbol{\theta}) = \int \big(y - f(\mathbf{x}; \boldsymbol{\theta})\big)^2 p(\mathbf{x}, y) \ \mathrm{d}\mathbf{x}y \ .$$

▷ **Trigonometric Polynomials:** For example, in the Fourier series (2.52),

$$f(x; \boldsymbol{\theta}) = a_0 + \sum_{i=1}^{c} \ a_i \cdot \cos(i \ x) + \sum_{i=1}^{c} b_i \cdot \sin(i \ x) \ .$$

▷ **Local Basis Function Networks:** For example, in kernel density estimation, (3.17), or Radial Basis Function (RBF) neural networks,

$$f(\mathbf{x};\boldsymbol{\theta},\mathbf{w}) = \sum_{i=1}^{c} w_i \cdot K\left(\frac{\|\mathbf{x}-\boldsymbol{\theta}_i\|}{h}\right) .$$

In general, we use a linear combination of basis functions $\phi_i(\cdot)$:

$$f(\mathbf{x};\boldsymbol{\theta}) = \sum_{i=1}^{(c,n_R,r)} \theta_i \cdot \phi_i(\mathbf{x}) , \tag{3.22}$$

where, depending on the context, the upper limit of the sum is either c (number of clusters), or n_R (number of rules), or r (number of regressor variables). Or more generally, for example in mixture density estimation,

$$f(\mathbf{x};\boldsymbol{\theta},\mathbf{w}) = \sum_{i=1}^{(c,n_r,r)} w_i \cdot \phi_i(\mathbf{x},\boldsymbol{\theta}) . \tag{3.23}$$

In density estimation, the 'output' y becomes a density and hence $f(\mathbf{x};\boldsymbol{\theta})$ describes a class of density functions with

▷ Loss function: $L\big(y, f(\mathbf{x},\boldsymbol{\theta})\big)$ becomes $L\big(f(\mathbf{x},\boldsymbol{\theta})\big)$.

▷ Likelihood function: $\ell(\boldsymbol{\theta};\mathbf{M}) = \prod_{j=1}^{d} f(\mathbf{m}_j;\boldsymbol{\theta})$.

▷ Log-Likelihood function: $\mathcal{L}(\boldsymbol{\theta};\mathbf{M}) = \sum_{j=1}^{d} f(\mathbf{m}_j;\boldsymbol{\theta})$.

We noted that instead of maximizing $\mathcal{L}(\boldsymbol{\theta};\mathbf{M}) = \sum_{j=1}^{d} f(\mathbf{m}_j;\boldsymbol{\theta})$, we could equally minimize $-\sum_{j=1}^{d} \ln f(\mathbf{m}_j;\boldsymbol{\theta})$ implying that in density estimation, maximizing the log-likelihood function is equivalent to minimizing the ML empirical risk functional

$$R_{\text{emp}}(\boldsymbol{\theta}) = -\sum_{j=1}^{d} \ln f(\mathbf{m}_j;\boldsymbol{\theta})$$

in statistical learning theory (based on the ERM principle). Though the functional approximators given in (3.22) and (3.23) are universal - in principle they can approximate any nonlinear function, and their parameterizations also nonlinear and hence the maximization/minimization of an inductive principle poses a nonlinear optimization problem. The EM-algorithm was introduced as one possible technique to solve this optimization problem.

In Section 4 and 5, the demanded qualities of a model were further extended to allow for *interpretability* as well as precision. In other words, the identified model should allow an interpretation of local relationships in terms of (fuzzy) *if-then* rules. For this purpose, in Section 4, the data matrix **M** was partitioned according to (fuzzy) clusters identified in data space Ξ. Three partitional clustering algorithms were described: (1) the hard-c-means algorithm, (2) its fuzzy-c-means generalization, and (3) the Gustafson-Kessel algorithm, which is a further refinement to account for cluster shape and orientation. Subsequently clusters can be identified as rules. In partitional clustering the objective is to minimize the 'within-cluster-overall-variance', where the the within-cluster variance is described by some metric evaluating the distance of data points with respect to some prototypical point $\mathbf{c}^{(i)}$.

▷ **Hard-c-Means Clustering:** For all (i,j), $u_{ij} \in \{0,1\}$, the partition space is defined by:

$$M_{hc} = \left\{ \mathbf{U} \in V_{cd} \; : \; \sum_{i=1}^{c} u_{ij} = 1; \; 0 < \sum_{j=1}^{d} u_{ij} < d, \forall i \right\} . \tag{4.7}$$

Objective function:

$$J_{hc}(\mathbf{M}; \mathbf{U}, \mathbf{C}) = \sum_{i=1}^{c} \sum_{j=1}^{d} u_{ij} \cdot d_{\mathbf{A}}^2 \left(\mathbf{m}_j, \mathbf{c}^{(i)} \right) . \tag{4.8}$$

▷ **Fuzzy-c-Means Clustering:** For all (i,j), $u_{ij} \in [0,1]$, such that the fuzzy partition space is defined by:

$$M_{fc} = \left\{ \mathbf{U} \in V_{cd} \; : \; \sum_{i=1}^{c} u_{ij} = 1; \; 0 < \sum_{j=1}^{d} u_{ij} < d, \forall i \right\} . \tag{4.10}$$

Objective function:

$$J_{fc}(\mathbf{M}; \mathbf{U}, \mathbf{C}) = \sum_{i=1}^{c} \sum_{j=1}^{d} (u_{ij})^w \cdot d_{\mathbf{A}}^2 \left(\mathbf{m}_j, \mathbf{c}^{(i)} \right) . \tag{4.11}$$

▷ **Switching Regression Models:** In Section 5.7, switching regression models were introduced as an alternative to the EM-algorithm in mixture density estimation. The basic idea is to identify c-regression models

$$y = f_i(\mathbf{x}; \boldsymbol{\theta}_i) + \varepsilon_i \tag{5.32}$$

on the basis of an objective function from the family

$$E_w[\mathbf{U}, \{\boldsymbol{\theta}_i\}] = \sum_{i=1}^{c} \sum_{j=1}^{d} u_{ij}^w \cdot E_{ij}[\boldsymbol{\theta}_i] . \tag{5.36}$$

The outcome of a clustering algorithm is the partition matrix $\mathbf{U}$ and cluster centers $\mathbf{C}$:

$$(\mathbf{U}, \mathbf{C}) = \arg\min_{M} J(\mathbf{M}) \ .$$

The elements $u_{ij} \in U$ in the partition matrix are evaluations of an equivalence (proximity) relation. Section 7 illustrated the fact that a crisp equivalence relation leads to a paradox. Generalizing a transitive equivalence relation, any pseudo-metric (employed to quantify the proximity of a data point $\mathbf{m}_j$ to a cluster prototype $\mathbf{c}^{(i)}$) induces a fuzzy or similarity relation

$$\widetilde{E} : \Xi \times \Xi \ \rightarrow \ [0,1] \ , \tag{4.19}$$

for which transitivity is defined as

$$\widetilde{E}(\mathbf{o}_i, \mathbf{o}_k) \geq T\left(\widetilde{E}(\mathbf{o}_i, \mathbf{o}_j), \widetilde{E}(\mathbf{o}_j, \mathbf{o}_k)\right) \ . \tag{4.24}$$

▷ **Approximate Reasoning** is introduced via the *compositional rule of inference*, generalizing the 'crisp' rule

$$\text{IF } \mathbf{x} = a \text{ AND } \mathbf{y} = f(\mathbf{x}), \text{ THEN } y = f(a)$$

to be valid for fuzzy sets A' and B':

$$\mu_{B'}(y) = \sup_{x \in X} T\big(\mu_{A'}(x), \mu_R(x,y)\big) \ . \tag{7.15}$$

In other words, the fuzzy model describes a *fuzzy mapping*

$$\begin{aligned} \tilde{f} : \mathcal{F}(X) \ &\rightarrow \ \mathcal{F}(Y) \\ \mu_A(\mathbf{x}) \ &\mapsto \ \tilde{f}(A) \ . \end{aligned} \tag{11.8}$$

▷ The **Extension Principle** defines a general method to extend crisp mathematical concept to fuzzy versions:

$$\begin{aligned} \tilde{f} : \mathcal{F}(X) \ &\rightarrow \ \mathcal{F}(Y) \\ \big(\mu_{A_1}(x_1), \ldots, \mu_{A_r}(x_r)\big) \ &\mapsto \ \tilde{f}\big(\mu_{A_1}(x_1), \ldots, \mu_{A_r}(x_r)\big) \ , \end{aligned}$$

where

$$\mu_{\tilde{f}(A_1,\ldots,A_r)}(y) = \sup_{(x_1,\ldots,x_r) \in f^{-1}(y)} \left\{\mu_{A_1}(x_1) \wedge \cdots \wedge \mu_{A_r}(x_r)\right\} \ . \tag{11.11}$$

Four fuzzy rule-based system structures were discussed in detail. In general, the antecedent part of the rule calculates the degree of fulfillment of the rule via the degree of membership of $\mathbf{x}$ in A_i where A_i is a r-dimensional fuzzy set. Alternatively, in the conjunctive model, for each ith rule, the degree of fulfillment is defined as

$$\beta_i(\mathbf{x}) \doteq \mu_{A_{i1}}(x_1) \wedge \mu_{A_{i2}}(x_2) \wedge \cdots \wedge \mu_{A_{ir}}(x_r) \ , \tag{5.7}$$

where $i = 1, 2, \ldots, n_R$, the A_{ik} are one-dimensional fuzzy sets defined on subspaces x_k of Ξ, and $\wedge$ is calculated by a suitable t-norm operator.

▷ **Klawonn's Similarity-Based Models:** For a given set of points

$$\left\{ \left(x_{1_0}^{(i)}, \ldots, x_{r_0}^{(i)}, y_0^{(i)} \right) \right\} ,$$

the output of the fuzzy model, based on similarity relations, is the extensional hull M_0 of the points. For $i = 1, \ldots, n_R$, $k = 1, \ldots, r$:

$$\mu_{M_0}(x_1, .., x_r, y) = \max_i \left\{ \min \left\{ \min_k \left\{ \mu_{x_{k_0}^{(i)}}(x_k) \right\} , \ \mu_{y_0^{(i)}}(y) \right\} \right\} . \quad (4.32)$$

▷ **Random-Set Models:** Let Ω be a set of abstract states and Γ an observable multi-valued map

$$\begin{aligned} \Gamma : \Omega \quad &\rightarrow \quad \mathcal{U} \\ \omega \quad &\mapsto \quad \Gamma(\omega) \subset \Xi = X \times Y . \end{aligned} \quad (6.14)$$

The data set $\mathbf{M}$ generates a sample of Γ, denoted

$$\mathcal{C} = \{C_j\} \qquad \text{where} \quad C_j \subseteq X \times Y, \ j = 1, \ldots, d .$$

For the nonlinear autoregressive model

$$y(k+1) = f(\mathbf{x}) ,$$

where

$$\mathbf{x} = [y(k), \ldots, y(k - n_y + 1), u(k), \ldots, u(k - n_u + 1)]^T$$

such that each element in vector $\mathbf{x}$ defines an axis in the product space $X = X_1 \times \cdots \times X_r$. A prediction model is then build from the projections of random subsets $C_j \subset \Xi$ onto Y, the space of candidates for $y(k+1)$. Formally, for any $\mathbf{x} \in X$, extended into $X \times Y$, the projected random intervals on Y are defined by

$$C_i^{\mathbf{x}} = \{y : \ \mathbf{x}_{\text{ext}} \cap C_i \neq \emptyset\} . \quad (6.16)$$

From the conditional possibility distribution, for any $y' \in Y$, $\hat{\pi}(y'|\mathbf{x})$ determines the possibility of $y(k+1) = y'$:

$$\hat{\pi}(y|\mathbf{x}) = \frac{1}{n_p} \sum_{i=1}^{n_p} \zeta_{C_i^{\mathbf{x}}}(y) . \quad (6.17)$$

▷ **Linguistic Fuzzy Models:** Let A' be any fuzzy set in X. We then obtain the output of composition-based fuzzy inference engine as follows. For the n_R fuzzy if-then rules of the conjunctive linguistic model structure

$$R_i : \quad \text{IF } x_1 \text{ is } A_{i1} \text{ AND } x_2 \text{ is } A_{i2} \ldots \text{AND } x_r \text{ is } A_{ir}, \text{ THEN } y \text{ is } B_i . \quad (5.6)$$

▷ **Composition-based inference:**

▷ First determine the fuzzy set-membership functions

$$\mu_{A_{i1}\times\cdots\times A_{ir}}(x_1,\ldots,x_r) \doteq T\big(\mu_{A_{i1}}(x_1),\ldots,\mu_{A_{ir}}(x_r)\big)\ . \quad (8.3)$$

▷ Let (8.3) be viewed as the fuzzy set μ_A in the fuzzy implications (7.9)-(7.14) and $\mu_{R_i}(\mathbf{x},y)$, $i=1,\ldots,n_R$, is calculated according to any of the implications.

▷ Next, $\mu_R(\mathbf{x},y)$ is determined according to

▷ **Union Aggregation:** (8.1),

$$R \doteq \bigcup_{i=1}^{n_R} R_i = S\big(\mu_{R^1}(\mathbf{x},y),\ldots,\mu_{R^{n_R}}(\mathbf{x},y)\big)$$

▷ **Intersection Aggregation:** (8.2),

$$R \doteq \bigcap_{i=1}^{n_R} R_i = T\big(\mu_{R^1}(\mathbf{x},y),\ldots,\mu_{R^{n_R}}(\mathbf{x},y)\big)$$

▷ Finally, for an arbitrary input A', the output B' is obtained according to

$$\mu_{B'}(y) = \sup_{\mathbf{x}\in X} T\big(\mu_{A'}(\mathbf{x}),\mu_R(\mathbf{x},y)\big)\ .$$

▷ **Individual-rule-based inference:** The first two steps are identical to the composition-based inference:

▷ For a given input fuzzy set A' in X, determine the output fuzzy set B'_i in Y for each rule R_i according to the generalized modus ponens (7.15), that is,

$$\mu_{B'_i}(y) = \sup_{\mathbf{x}\in X} T\big(\mu_{A'}(\mathbf{x}),\mu_{R_i}(\mathbf{x},y)\big) \quad (8.6)$$

for $i=1,\ldots,n_R$.

▷ The output of the fuzzy inference engine is obtained from either the union

$$\mu_{B'}(y) = S\big(\mu_{B'_1}(y),\ldots,\mu_{B'_r}(y)\big) \quad (8.7)$$

or intersection

$$\mu_{B'}(y) = T\big(\mu_{B'_1}(y),\ldots,\mu_{B'_r}(y)\big) \quad (8.8)$$

of the individual output fuzzy sets $B'_1,\ldots,B'_r$.

Depending on the choice of operators used in calculating unions, intersections, implications, rule-aggregation and defuzzification of B', various inference engines are obtained.

▷ **Takagi-Sugeno Fuzzy Models:**

$$R_i : \quad \text{IF } \mathbf{x} \text{ is } A_i, \text{ THEN } y_i = f_i(\mathbf{x}) \,. \tag{5.8}$$

A common parameterization of y is the affine linear model

$$y_i = \mathbf{a}_i^T \mathbf{x} + b_i \,. \tag{5.10}$$

Takagi-Sugeno inference engine:

$$y = \frac{\sum_{i=1}^{n_R} \beta_i(\mathbf{x}) \cdot y_i}{\sum_{i=1}^{n_R} \beta_i(\mathbf{x})} \,. \tag{5.15}$$

Since $n_R = c$, introducing the normalized degree of membership

$$\phi_i(\mathbf{x}) \doteq \frac{\beta_i(\mathbf{x})}{\sum_{j=1}^{c} \beta_i(\mathbf{x})} \,, \tag{5.16}$$

and the singleton versions $y = b_i$ of the Linguistic and TS models, we notice that the fuzzy systems belong to the class of basis function approximators (3.22):

$$y = \sum_{i=1}^{n_R} \phi_i(\mathbf{x}) \cdot b_i \,. \tag{5.19}$$

Antecedent fuzzy sets are obtained by pointwise projection and consequent parameters were estimated by standard least-squares. In this book, we considered the Linguistic fuzzy model or equivalently the model based on similarity relations for approximate reasoning: qualitative knowledge processing, that is, capturing context-dependent, *ad hoc*, *a priori* expert knowledge. The other model structures, in particular the TS-model were introduced as models which can be identified from data using fuzzy clustering algorithms.

Analyzing data and systems has a purpose and inevitably decisions are made as the result of the analysis. Figure 1.2 outlined the process of systems analysis. In Figure 12.1, we can now consider a more detailed view of what the general term 'decision making' implies with respect to real-world problems.

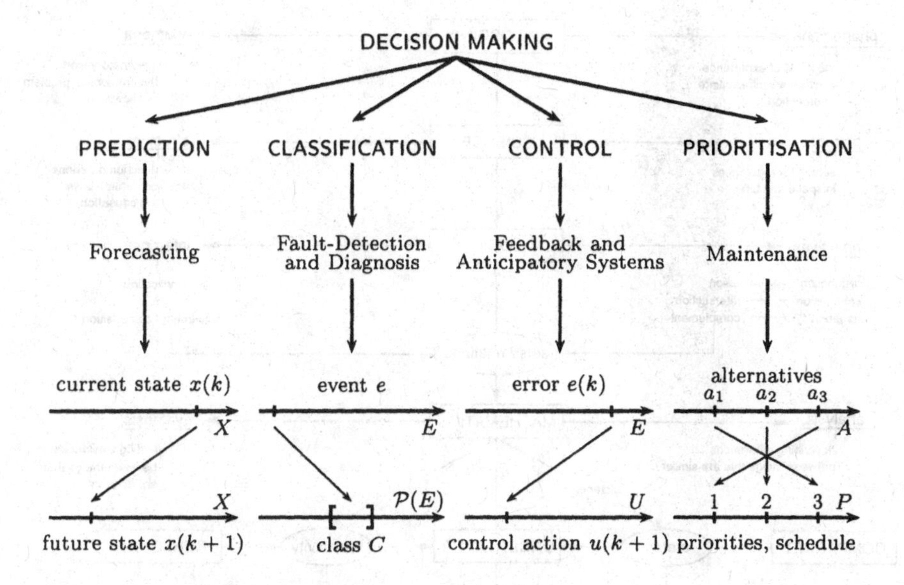

Fig. 12.1 Decision making concepts and applications.

12.2 MORE PHILOSOPHICAL IDEAS

> This harmony that human intelligence believes it discovers in nature - does it exist apart from that intelligence? No, without doubt, a reality completely independent of the spirit which conceives it, sees it or feels it, is an impossibility. A world so exterior as that, even if it existed, would be forever inaccessible to us. But what we call objective reality is, in the last analysis, that which is common to several thinking beings, and could be common to all; this common part, we will see, can be nothing but the harmony expressed by mathematical laws.
>
> —Henri Poincaré

The arguments put forward throughout the book reflect a similar position in epistemology (the philosophy of science – metaphysics). The view from which such "fuzzy logic of scientific discovery" is derived is outlined in Figure 12.2. We identified *'comparing'* and *'reasoning'* as the two fundamental concepts which take distinct forms in the areas of mathematics, science, logic, and philosophy[1].

[1] For those who would not consider reading the original work of Arthur Schopenhauer, the excellent book by Bryan Magee [Mag97] provides a comprehensive summary.

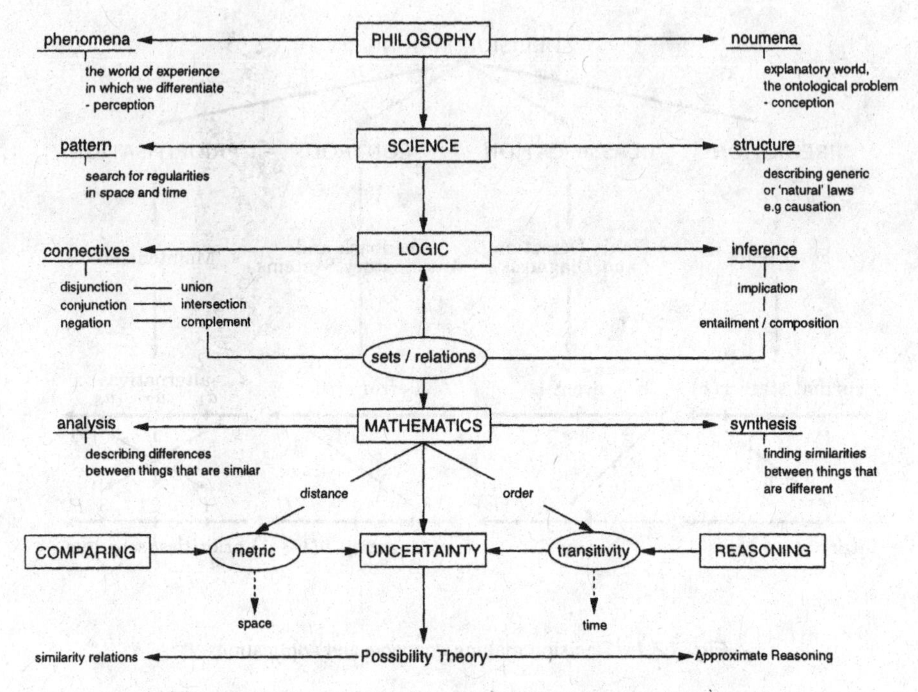

Fig. 12.2 "The fuzzy logic of scientific discovery" [Wol98].

The world as we know it is our interpretation of the observable facts in the light of theories that we ourselves invent. Reality is hidden but (transcendentally) real. This independent reality outside the world of all possible experience is Kant's world of the *noumenal*, the world of things as they are in themselves. Independent reality is something which human knowledge can approach only asymptotically. The world 'as we experience it' is dependent on the nature of our apparatus for experience, with the consequence that things as they appear to us, are not the same as they are in themselves. This world of experience is Kant's world of the *phenomena* – the empirical world. This epistemological view has its equivalent in modelling natural systems, the phenomenological perspective outlined in Section 1.

Since Hume demonstrated that 'scientific laws' are not empirically verifiable, Kant, Schopenhauer and Popper argued that the search for certainty is an error. Human knowledge does not reveal something objectively and timelessly true but remains 'our' belief, hypothesis, or theory – a creation of the human mind, and not embodied in the world waiting to be discovered. Instead of assuming a set of differential equations could describe the natural phenomena itself, we ought to focus on the *representation* of systems, which is limited and often subjective.

Our experience is made intelligible to us in terms of time, space and causality. Space and time are forms of sensibility whereas the subjective correlative is the 'understanding'. Schopenhauer introduced the notion of 'differentiation' tied to the concepts of space and time. Then, for anything to be different from anything else, either space or time has to be presupposed. That is, the concept of succession presupposes either spatial or temporal concepts, or both. Hence, causal connections exist only between objects in the phenomenal world. For mathematics, the concept of succession is required. This presupposes either spatial or temporal concepts, or both. This suggests that mathematics is created rather than being discovered and as rigorous it may be, every argument must have an absolute minimum of one premise and one rule of procedure (e.g., If A, Then B) before it begins, and therefore begins to be an argument at all. So, every argument has to rest on at least two undemonstrated assumptions, for no argument can establish either the truth of its own premises or the validity of the rules by which itself proceeds [Mag97].

Mathematical models of (perceived) reality are systems of *causal* entailments. To ask *"why"* is to assert a *"because"* and hence applies to the realm of causality only. In system theory, we study *organization* and *pattern* separate from their material embodiment. We shall now discuss this separation of a formal model from its real-world counterpart in more detail. In describing a probabilistic framework of processes we noted in previous sections of the book that it does not refer to an actual *measurement* but to an *observation*, in general. For instance, the *formal model* of the dice, based on the assumption of physical symmetry, leads to the probability of $1/6$ for the number 2 to occur. However, this gives us only limited help in predicting the outcome of rolling the particular dice I hold in my hand. $Pr(2) = 1/6$ is the result of a formal model – a *deduction*. On the other hand, *induction* is the process of matching measured data with the model. Induction, as a form of generalization, seeks to establish general (*i.e.* quantified) propositions on the basis of *instances*; deduction, conversely seeks to establish instances in terms of quantified or general propositions. Let X denote some set whose elements are called *instances*. For simplicity, let us assume we can enumerate the elements of X:

$$X = \{x_1, \ldots, x_i, \ldots\} .$$

A predicate or property f, possessed or manifested by an instance x_i is denoted $f(x_i)$. Let $\mathbf{x}$ be a variable taking values in X. Then, the expression $f(x_i)$ is an assertion about the specific instance x_i in X. The *universal quantifier* $\forall$ is used to make statements about the set or universe X. For example, the expression

$$\forall x \in X,\ f(\mathbf{x})$$

asserts "for all x in X, or for any x, x has the property f or $f(\mathbf{x})$ is true". This *general proposition* can be *interpreted* as a string of conjunctions

$$\forall x,\ f(\mathbf{x}) = f(x_1) \vee f(x_2) \vee \cdots \vee f(x_i) \vee \cdots$$

By deduction we can infer a *particular* $f(x_i)$ from the general proposition $\forall x, f(\mathbf{x})$. The problem of induction is the complementary attempt to establish a general proposition on the basis of particular instances (samples) $\{x_1, x_2, \ldots,\}$ in X for which it is known that $f(x_i)$ holds. *Empiricism* is then understood as the extrapolation from samples in X to a general truth about X.

Analogous to the definition of a general proposition above, in Section 1 we introduced the graph $F = \{(x, f(\mathbf{x}))\}$ as an alternative formulation of the function $\mathbf{y} = f(\mathbf{x})$ to model a system. We subsequently generalized the formulation of a graph $F \subset X \times Y$ by means of fuzzy if-then rules and fuzzy graphs $\widetilde{F}$. A system can therefore be defined as a relation in the set-theoretic sense. We define a system in terms of relationships between observed features rather than what these features actually are (physical, biological, ...). The general notion of a system as a subset of the space Ξ can be made more specific by any of the methodologies presented in this book. Most of the time we specified a system in terms of equations defined on appropriate variables. Each variable is associated with a system object X_i or Y which represents the range of the respective variable. In particular, we considered the construction of such relations by means of measured data and in terms of a set of verbal statements. A system is understood as a particular kind of set, that is, a relation, capturing the collection of appearances of the object in consideration rather than the object of study itself. To develop a kind of theory starting from the definition of $\mathfrak{S}$ as a graph $F \subset X \times Y$, more structure needs to be introduced a) into the elements of the system objects X_i as a set itself with additional appropriate structure or b) by introducing the structure in the object sets X_i and Y themselves. The first approach leads to the concept of (abstract) time systems and the second to the concept of an algebraic system. The former was introduced in Section 1 with the elements of objects being time functions.

Mathematics and set-theoretic formulations in particular are our chosen *language* to describe aspects of the real-world. As with every natural language, we have to distinguish between *syntactic* aspects (referring to the language itself) and *semantic* aspects pertaining to the interpretation of statements produced by the language. From a syntactical point of view, divorced from any external referents, *propositions* in the language are in general not *about* anything. *Production rules* are themselves propositions but refer to other propositions in the language. The syntactical production rules of a language are its internal vehicles for what Robert Rosen [Ros91] calls *inferential entailment*. The rules allow us to say that propositions imply each other. By talking of a 'formal model', we already assumed that we choose mathematics as the *formalism*. Said another way, a formalism is a finite list of production rules. A formal system is therefore distinguished from a *natural system*, an aspect of, or process in, the real world we wish to study. An algorithm is the application of production rules to specific propositions. By addressing a *"why"* or *"what"* to a proposition, we assume that it *entails* a *"because"* or *"this"*. Inferential entailment or implication describes a relation between propositions in a formalism; p *entails* c, or p implies c.

The study of entailment relations at the level of phenomena; the entailment of one phenomenon by another, is usually described as *causality*. As indicated in Section 7.2, propositional calculus founded in bivalent logic is entirely inappropriate in modelling causation. The two-valued logical correlates of a causal connections is that of material implication, associating antecedent p with a cause and the consequent c with an effect, an absent cause may have an effect; things may be self-caused. To model causal links, relations between p and c are required to be non-reflexive (nothing is self-caused) and asymmetrical (irreversibility). Based on this requirement, conventional logic is inadequate for modelling cause-effect links. However, recent studies of fuzzy relations offer a relational approach that provide appropriate mappings that are irreflexive, transitive (chain-like) and asymmetrical. Sections 4.6 and 4.7 together with the discussion in Section 7 form the background for the development of a formal framework which allows us to analyze causal or dynamic systems in a measure-free, that is, (fuzzy) logical or (fuzzy) relational framework. In all this we need to separate *inferential entailment* from *causal entailment*. Formal or pure propositional calculus founded in bivalent logic is an example of the former, where new propositions are generated by inferential rules. The relation of inferential entailment to the entirely different *causal entailment* between phenomena is embodied in the concept of a *natural law*. In establishing (hypothesizing) natural laws we make the following assertions:

- ▷ The succession of events or phenomena that we perceive are not completely arbitrary; there are relations (e.g., causal relations) manifested in the world of phenomena.
- ▷ The relations between phenomena are, at least in parts, observable, that is, can be perceived and grasped by the human mind.

12.2.1 Data Engineering

What we can observe is not nature itself, but nature exposed to our method of questioning.

—Werner Heisenberg

Data engineering is the combination of pattern recognition techniques with system theory (Figure 12.3). To approach challenges, provided by the natural sciences and engineering, we require knowledge of more than one methodology and how the different theoretical frameworks can be related. Various theoretical concepts – often described as distinct or unrelated – can shown to be closely related if not formally equivalent. Starting the problem-solving process not from a theory but from the data *and* context-dependent knowledge available, the book introduced a number of concepts for system modelling and data analysis with the aim of decision making: prediction and classification, prioritization and control.

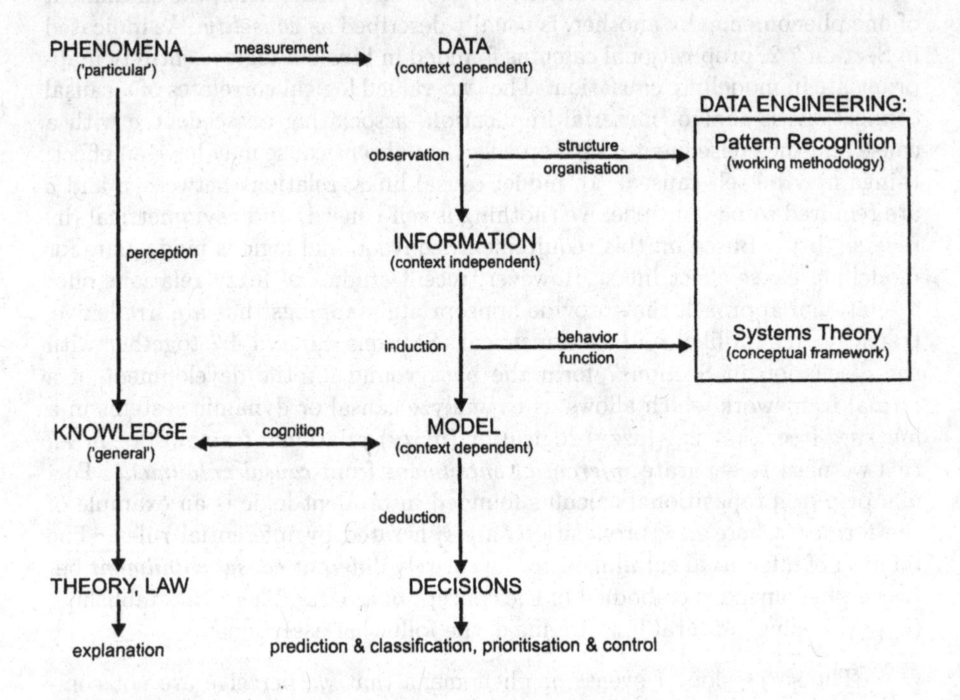

Fig. 12.3 Data engineering is the combination of pattern recognition techniques with system theory.

In modern systems theory and its applications, the physical structure of the system under consideration is of secondary interest. Instead, a model is intended to appropriately represent the *behavior* of the system under consideration. This requires methodologies and paradigms which provide *interpretability* in addition to accuracy. Accuracy is thereby understood not as the ability to provide a numerical value which minimizes (for example) the mean-square-error, but rather as the ability to be precise and honest about uncertainty. The forecast of a conventional time-series model, based on various simplifying assumptions, is not necessarily more accurate than the forecast of a rule-based fuzzy system generating a possibility distribution. A fuzzy set may be more precise than a real number!

Data mining, extends classical statistics by emphasizing *retrospective* analysis of data; primarily interested in understandability rather than accuracy or predictability. The focus is on relatively simple interpretable models involving rules, trees, graphs, and so forth with current practice being pattern-focused rather than model-focused. We therefore call 'data engineering' the practice of matching data with models or, more generally, the art of *turning data into information*. This requires the combination of concepts from various disciplines such as

▷ Applied Statistics (multi-variate, descriptive statistics)

▷ Systems Theory (modelling and identification of dynamic processes)

▷ Decision Theory (multi-criteria decision making, knowledge-based system).

13

Appendices

13.1 SETS, RELATIONS, MAPPINGS

A *set* X is a collection of (mathematical) objects such as points in the plane, real numbers, functions, and so on, called the *elements* of X. If x is an element of X, we write $x \in X$ and if X and X' are two sets for which

$$x \in X \quad \text{implies that} \quad x \in X' ,$$

X is *contained* in X', X is a *subset* of X' and we write $X \subseteq X'$. The subset of X which contains no elements is called the *empty set*, denoted $\emptyset$. If A and B are two sets, their *union* $A \cup B$ is defined to be the set of elements which belong to at least one of the sets A and B; their *intersection* $A \cap B$ is defined as the set of elements which belong to both A and B; the *difference* $A - B$ is defined to be the set of elements which belong to A and not to B. If $A \cap B = \emptyset$, we say that the sets are *disjoint*; if $A \cap B \neq \emptyset$, we say that the sets *intersect*. If $A \subseteq X$, the difference $X - A$ is called the *complement* of A relative to X and is denoted by A^c. If $A \subseteq X$, the *characteristic function* of A relative to X is a map ζ_A, defined on the elements of X by

$$\zeta_A(x) = \begin{cases} 1 & \text{if } x \in A \\ 0 & \text{if } x \in X - A. \end{cases}$$

The following relations hold:

$$\zeta_{A\cup B}(x) = \max\{\zeta_A(x), \zeta_B(x)\} = \zeta_A(x) + \zeta_B(x) - \zeta_A(x)\zeta_B(x) ,$$
$$\zeta_{A\cap B}(x) = \min\{\zeta_A(x), \zeta_B(x)\} = \zeta_A(x) + \zeta_B(x) ,$$
$$\zeta_{A^c}(x) = 1 - \zeta_A(x) .$$

The set consisting of the elements $x_1, x_2, \ldots, x_r$ is denoted by $\{x_1, x_2, \ldots, x_r\}$. For example, the set of positive integers $\{0, 1, 2, \ldots\}$ is denoted by $\mathbb{N}$ and the set of real numbers is denoted by $\mathbb{R}$ and often called the Euclidean or *real line*. The following sets are called *intervals*:

$$[a, b] = \{x : x \in \mathbb{R}, x \geq a, x \leq b\} \quad \text{"closed interval"},$$
$$(a, b) = \{x : x \in \mathbb{R}, x > a, x < b\} \quad \text{"open interval"},$$
$$(a, b] = \{x : x \in \mathbb{R}, x > a, x \leq b\} \quad \text{"left-open interval"},$$
$$[a, b) = \{x : x \in \mathbb{R}, x \geq a, x < b\} \quad \text{"right-open interval"}.$$

A set X is *finite* iff there is an $n \in \mathbb{N}$ such that there exists a one-to-one (onto) correspondence between all elements in X and the set $\{1, 2, \ldots, n\}$. A set X is said to be *infinite* if it is not finite. A set equivalent to $\mathbb{N} = \{1, 2, \ldots\}$ is called *countably infinite* (or *denumerable*). A set is *countable* if it is either finite or countably infinite. Alternatively, a set is countable if its elements can be arranged as the terms of a sequence.

A set I and a correspondence $i \mapsto a_i$ associating each $i \in I$ to an element $a_i \in A$ is called a *family of elements* in A and is denoted by $\langle a_i : i \in I\rangle$; I is called the *index set*. If $I = \{1, 2, \ldots, n\}$, we have a set called an n-tuple, and if $n = 2$, this tuple is called a *pair*. If I is the set of strictly positive integers, we obtain a *sequence*

$$\langle a_1, a_2, a_3, \ldots\rangle \doteq \langle a_n\rangle .$$

A set I and a correspondence associating each element $i \in I$ a subset A_i of A is called a *family of sets* in A and is denoted by

$$\mathcal{A} \doteq \langle A_i : i \in I\rangle .$$

A family $\mathcal{A} = \langle A_i : i \in I\rangle$ is called a (crisp or hard) *partition* of the set A if

$$(1) \quad \forall\, i : A_i \neq \emptyset, A_i \subseteq A,$$
$$(2) \quad i \neq j \Rightarrow A_i \cap A_j = \emptyset,$$
$$(3) \quad \textstyle\bigcup_{i\in I} A_i = A .$$

In other words, every point of A belongs to one and only one A_i. A family $\mathcal{A} = \langle A_i : i \in I\rangle$ is called a *covering* of A if

$$(1) \quad \forall\, i : A_i \neq \emptyset,$$
$$(2) \quad \textstyle\bigcup_{i\in I} A_i = A .$$

Alternatively, each element of A belongs to at least one A_i. The *order* of a covering $\mathcal{A}$ is the greatest integer m for which there exist $m+1$ sets of $\mathcal{A}$ having a non-empty intersection; a partition is a covering of order 0. A family $\mathcal{A} = \langle A_i : i \in I \rangle$ is called a *filter base* (or just base) if

$$(1) \quad \forall\, i : A_i \neq \emptyset,$$
$$(2) \quad \forall\, i \quad \forall\, j \quad \exists\, k : A_k \subseteq A_i \cap A_j .$$

We say that a family $\mathcal{A}$ is a *lattice* with respect to the operations $\cup$ and $\cap$ if

$$A \in \mathcal{A},\ B \in \mathcal{A} \ \Rightarrow\ A \cup B \in \mathcal{A},\ A \cap B \in \mathcal{A} .$$

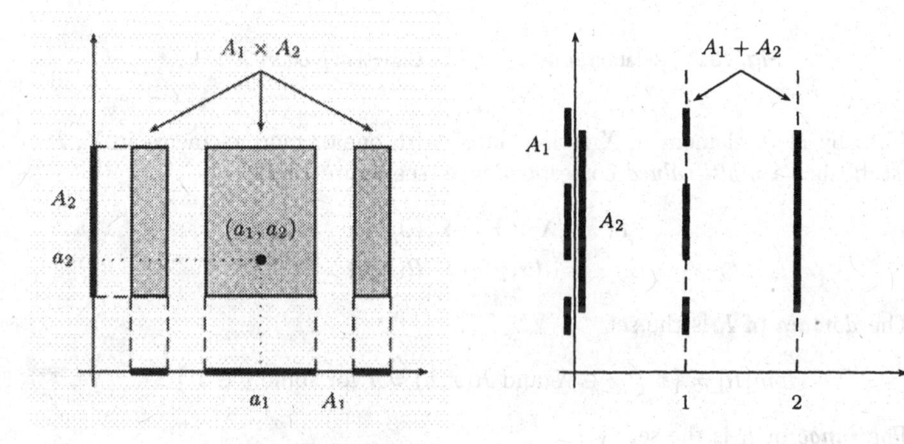

Fig. 13.1 Cartesian product (left) and Cartesian sum (right) of two sets A_1 and A_2.

The *Cartesian sum* $A_1 + A_2$ of two sets A_1 and A_2 is the set formed by the pairs $(1, a_1)$, $(2, a_2)$, where $a_1 \in A_1$ and $a_2 \in A_2$ as shown in Figure 13.1. Suppose that two elements, first $x \in X$, followed by $y \in Y$, are chosen. Then, this choice is denoted by the pair (x, y) and is called an *ordered pair*. The set of all such ordered pairs is called the *Cartesian product* (or *product set*) of X and Y,

$$X \times Y = \{(x, y) : x \in X,\ y \in Y\} .$$

Note that, in general, $X \times Y \neq Y \times X$. If furnished with some structure, the set $X \times Y$ leads to a *Cartesian product space* (see Figure 13.1).

Any subset $R \subseteq X \times Y$ of $X \times Y$ defines a binary relation between the elements of X and the elements of Y. A *relation* is therefore a set of ordered pairs, denoted

$$R = \{(x, y) \in X \times Y : R(x, y) \text{ holds}\} .$$

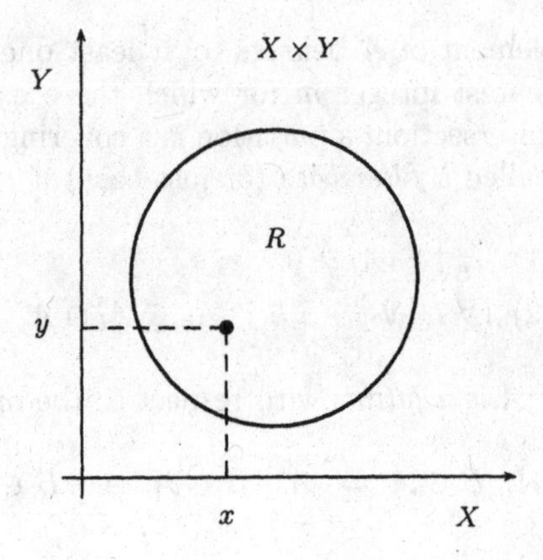

Fig. 13.2 Relation R defined on the Cartesian product $X \times Y$.

Since by R an element in X is associated with one or more elements in Y, R establishes a *multi-valued* correspondence (see figure 13.2):

$$\begin{aligned} R: \quad X \times Y &\rightarrow \{0,1\} \\ (x,y) &\mapsto R(x,y) \, . \end{aligned}$$

The *domain* of R is the set

$$dom[R] = \{x : x \in X \text{ and } R(x,y) = 1 \text{ for some } y \in Y\} \, .$$

The *range* of R is the set

$$ran[R] = \{y : y \in Y \text{ and } R(x,y) = 1 \text{ for some } x \in X\} \, .$$

The relation $R \subseteq X \times X$ establishes a relation among the elements of X. An important example of such relations are *equivalence relations* on X possessing the following properties:

$$\begin{aligned} R(x,x) \qquad & \forall\, x \in X \qquad && \text{(reflexivity)} \\ R(x,x') \;\Rightarrow\; & R(x',x) && \text{(symmetry)} \\ R(x,x') \wedge R(x',x'') \;\Rightarrow\; & R(x,x'') && \text{(transitivity)} \, . \end{aligned}$$

Intuitively, an equivalence relation is a generalization of equality (which itself is an equivalence relation). If R is an equivalence relation on a set X, and if $a \in X$ is any element of X, then we can form a subset of X defined

$$[a]_R = \{x : R(a,x) = 1\} \, .$$

If b is another element of X, we can likewise form the subset

$$[b]_R = \{x : R(b,x) = 1\} \, .$$

We then have the following conditions:

$$[a]_R = [b]_R \quad \text{or} \quad [a]_R \cap [b]_R = \emptyset \,.$$

The subset $[a]_R$ defined above is called *equivalence class* of a (under relation R). Since any two such classes are either identical or disjoint, the relation R can be regarded as partitioning the set X into a family of subsets. The set of equivalence classes of X under an equivalence relation R is called *quotient set* of X modulo R and denoted

$$X/R \,.$$

A *function* (or mapping[1]) from a set X into a set Y, denoted $f\colon X \to Y$, is a 'single-valued' relation $f \subset X \times Y$ such that for every $x \in X$ there exists a *unique* $y \in Y$ written as

$$y = f(x) \,.$$

If $f\colon X \to Y$, X, the set of arguments of f is called the *domain* of f, denoted $dom[f]$ and Y, the set of values of $f(\cdot)$, is called the *codomain* of f. The element $y \in Y$ in $f(x) = y$ is called the *image* of $x \in X$, or *the value of the function* at x. The *range* of a function $f\colon X \to Y$, denoted $ran[f]$, is the set of elements in Y that are images of elements in X, that is, $ran[f]$ is the set of all images of f:

$$ran[f] = \{f(x) \,:\, x \in X\} \,.$$

$ran[f]$ is sometimes referred to as the *image set*. The *graph* of function $f\colon X \to Y$ is the set (relation)

$$\{(x, f(x)) \,:\, x \in X\} \;=\; \{(x,y) \,:\, x \in X,\; y = f(x)\} \,.$$

The set

$$f(A) = \{f(a) \,:\, a \in A \subseteq X\}$$

is called the *direct image* of A, $f(A) \subset ran[f]$. Suppose $f\colon X \to Y$ and $B \subset Y$. Then the set

$$f^{-1}(B) = \{x \,:\, f(x) \in B\}$$

is called the *inverse image* of B *under* f, $f^{-1}(B) \subset X$, that is, $f^{-1}(B)$ is a subset of the domain of f. Let $f\colon X \to Y$ and $g\colon Y \to Z$. Then the *composite* or *product* of the functions f and g is the function $g \circ f$ from X into Z:

$$g \circ f \,:\, X \to Z \,.$$

The composite $g \circ f$ is defined for every $x \in X$

$$(g \circ f)(x) = g(f(x)) \,.$$

[1] It is customary to use the terms *function*, *mapping* and *transformation* synonymously.

There exists an important relation between mappings and equivalence relations. Let $f\colon X \to Y$ be a mapping. For each $x \in X$, let us define

$$[x]_f = \{x' : f(x') = f(x)\} .$$

These sets $[x]_f$ are all subsets of X and can be regarded as the equivalence classes of a corresponding equivalence relation R_f on X. The relation R_f specifies the extent to which the elements of X can be distinguished, or resolved, by f. Thus, if we regard the mapping f as associating with an element $x \in X$ and element $y \in Y$, the equivalence relation R_f specifies those distinct elements of X which are assigned the same value by f.

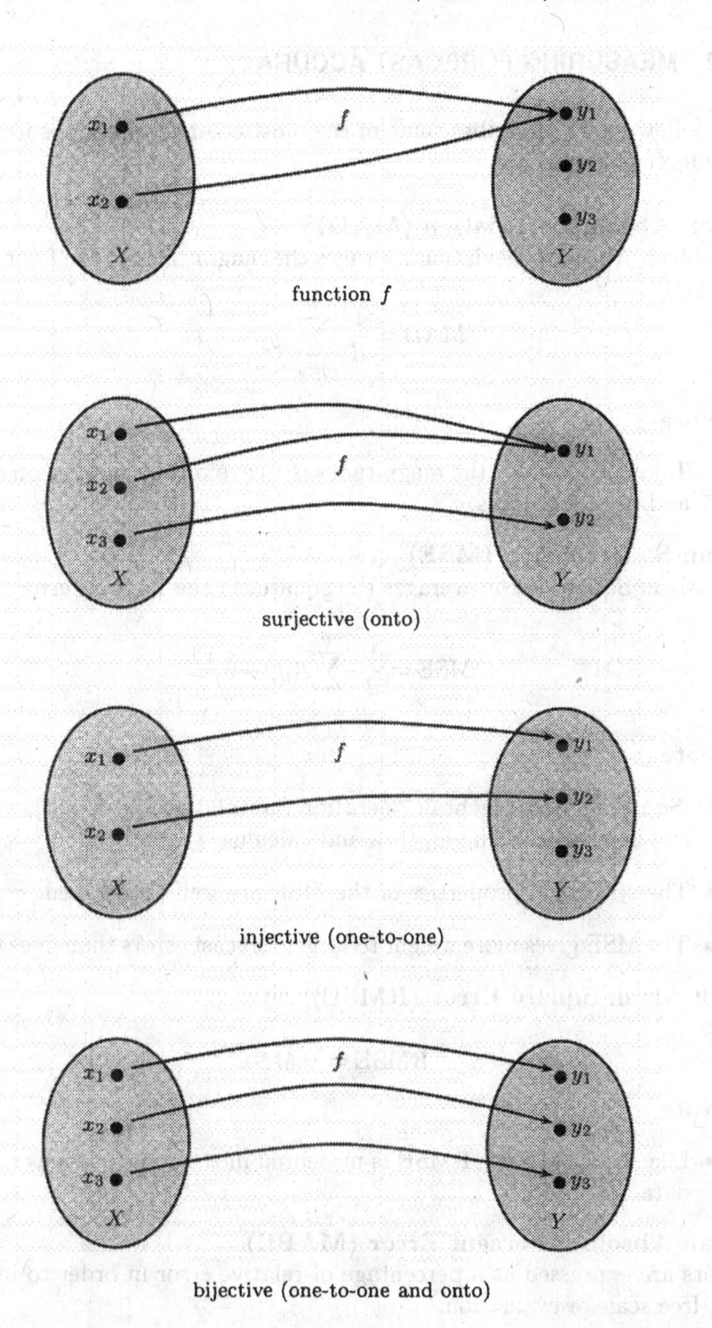

Fig. 13.3 Classification of mappings.

13.2 MEASURING FORECAST ACCURACY

The following list describes some of the most common measures used to evaluate forecasting accuracy.

Mean Absolute Deviation (MAD)
The Mean Absolute Deviation averages the magnitudes of the forecast errors:

$$\text{MAD} = \frac{1}{d} \cdot \sum_{j=1}^{n} |y_j - \hat{y}_j| \, .$$

Note

- Taking squares or the magnitudes is to avoid the cancellation of positive and negative terms.

Mean Square Error (MSE)
The Mean Square Error averages the squares of the forecast errors:

$$\text{MSE} = \frac{1}{d} \cdot \sum_{j=1}^{d} (y_j - \hat{y}_j)^2 \, .$$

Note

- 'Squaring' is an algebraic operation ('absolute value' is not), so the MSE can be studied using algebra and calculus.
- The statistical properties of the MSE are well established.
- The MSE gives more weight to larger forecast errors than does the MAD.

Root Mean Square Error (RMSE):

$$\text{RMSE} = \sqrt{MSE} \, .$$

Note

- Like the MAD, the RMSE is measured in the same units as the original data.

Mean Absolute Percent Error (MAPE)
Errors are expressed as a percentage of relative error in order to introduce a unit-free scale of evaluation:

$$\text{MAPE} = \frac{1}{d} \cdot \sum_{j=1}^{d} \left| \frac{y_j - \hat{y}_j}{y} \right| \cdot 100\% \, .$$

Note

- Because it is dimensionless, or unit-free, the MAPE can be used to compare the accuracy of the same or different models on two entirely different series.

Multiple Correlation Coefficient (MCC):

$$\text{MCC} = \frac{\sum_{j=1}^{d} \hat{y}_j}{\sum_{j=1}^{d} y_j^2}$$

Note:

- The ratio MCC measures the proportion of the total variation of y that is explained by the regression.
- MCC is often expressed in percent.
- Sometimes the mean value of y is subtracted from $\hat{y}$ and y before calculating MCC.

13.3 (HIERARCHICAL) CLUSTERING

A hierarchical clustering is a sequence of nested partitions. Instead of working on an object data set $\mathbf{M}$ directly, this method processes sets of d^2 numerical *relationships* between pairs of objects represented by the data. It is convenient to array the relational values as a $d \times d$ proximity or *relation matrix* whose elements describe a binary relation. For example, let $\mathbf{M} = \{\mathbf{m}_1, \mathbf{m}_2, \ldots, \mathbf{m}_d\} \subset \mathbb{R}^n$ be a set of d *feature vectors* in n-space, then every metric or distance measure d produces a (dis-)similarity relation matrix whose elements are $d(\mathbf{m}_i, \mathbf{m}_j) \doteq d_{ij} : \mathbb{R}^n \times \mathbb{R}^n \to \mathbb{R}^+$; for instance, object data are converted into relational data using $d_{ij} = \|\mathbf{m}_i - \mathbf{m}_j\| = \sqrt{\mathbf{m}_i^T \mathbf{m}_j}$, $1 \leq i, j \leq d$.

An *agglomerative* algorithm starts with each object describing an individual cluster. On the basis of a proximity matrix, disjoint clusters are merged sequentially according to some local connectivity criterion. The latter is formulated by any positive semi-definite, symmetric set-distance function on $\mathcal{P}(\mathbb{R}^n) \times \mathcal{P}(\mathbb{R}^n)$. Different linkage algorithms correspond to different choices of this distance $d(M, M')$ between two subsets of $\mathbf{M}$. A *dendrogram* provides a convenient illustration of hierarchical clustering. At each step, the clusterings are assigned sequence numbers $0, 1, \ldots, (d-1)$ and $\delta(l)$ is the level of the lth clustering. More specifically, the steps involved are as follows:

1. Start with disjoint clusters at level $\delta(0) = 0$ and sequence number $l = 0$.

2. Find the least dissimilar pair of clusters in current clustering, $d(M, M') = \min_{i \neq j}\{d(M_i, M_j)\}$, where the minimum is over all pairs of clusters in the current clustering.

3. Increment the sequence number from l to $l + 1$. Merge clusters M and M' into a single cluster M'' to form the next clustering l. Set the level of this clustering to $\delta(l) = d(M, M')$.

4. Update the proximity matrix by deleting rows and columns corresponding to M and M', and adding a row and column corresponding to M''. The proximity between the new cluster and the remaining ones is defined for *single linkage clustering* as

$$d(M'', M''') = \min\{d(M''', M), d(M''', M')\} .$$

5. Repeat steps 2 – 4 until all objects are in one cluster.

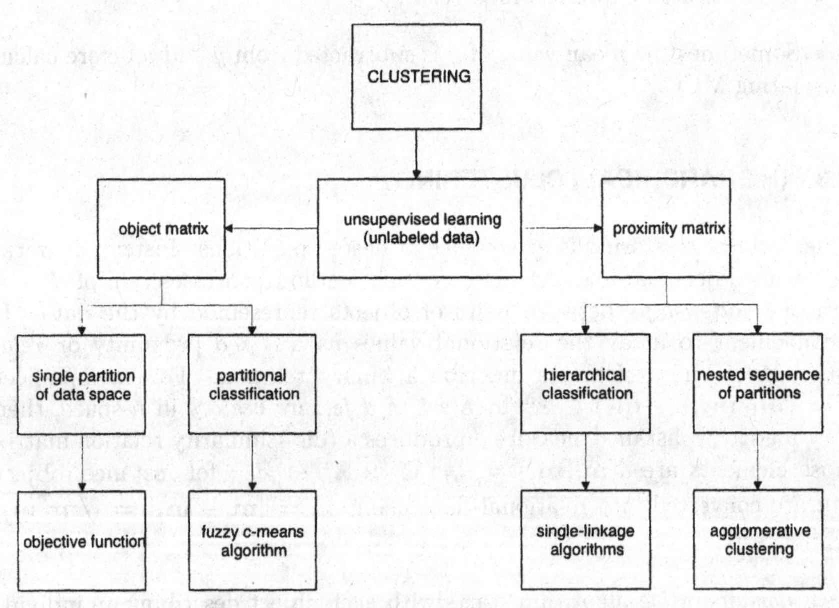

Fig. 13.4 Overview of clustering techniques.

13.4 MEASURE SPACES AND INTEGRALS

Uncertainty techniques use the basic notion of an *event* and its occurrence or non-occurrence to model a phenomenon. To every event A there exists a contrary event "not A", denoted by $\neg A$, and $\neg A$ occurs only if A does not occur. "A implies B", that is, when A occurs, B occurs necessarily, is denoted by

$A \subseteq B$. Logical combinations of events create new events. The uncertainty of the process under consideration can take various distinct forms (randomness, fuzziness, vagueness, ambiguity, imprecision) which require the definition of an appropriate *measure* to quantify (un)certainty [Wol98]. Formally, events are represented as sets of *elementary outcomes*. The sets of events form a *field* (or *algebra*) of sets.

Let Ω be an arbitrary non-empty set. A collection $\mathcal{A}$ of subsets of Ω is a *σ-field*, if

(a) $\Omega \in \mathcal{A}$.

(b) If $A \in \mathcal{A}$, then $A^c \in \mathcal{A}$.

(c) If $A_i \in \mathcal{A}$, $i = 1, 2, \ldots \in I$, a countable index set, then $\bigcup_{i=1}^{\infty} A_i \in \mathcal{A}$.

For a finite number of A_i, $\mathcal{A}$ is called a *field* or an *algebra*. The pair $(\Omega, \mathcal{A})$ is called a *measurable space*. In probability theory, Ω denotes the sample space of elementary outcomes while $\mathcal{A}$ is the set of events (measurable sets). Ω refers to the certain event while the empty set $\emptyset$ refers to the impossible event. The power set $\mathcal{P}(\Omega)$ of Ω is a σ-algebra. It is the largest σ-algebra of subsets of Ω, and when $\Omega = \mathbb{R}$, the σ-algebra generated by the collection of all open subsets of $\mathbb{R}$ is called the *Borel* σ-algebra of $\mathbb{R}$ and denoted $\mathcal{B}$. The empty set $\emptyset$ is contained in σ_Ω since it is the complement of Ω.

Let $(\Omega, \mathcal{A})$ be a measurable space. A function $g\colon \mathcal{A} \to [0, \infty]$ is a *measure* if $g(\emptyset) = 0$ and if $A_i, i = 1, 2, \ldots$ is a sequence of pairwise disjoint elements of $\mathcal{A}$, then the condition

$$g\left(\bigcup_{i=1}^{\infty} A_i = \sum_{i=1}^{\infty} g(a_i)\right)$$

is called *σ-additivity*. If it is required for only finitely many A_i, it is called *additivity*. The triple $(\Omega, \mathcal{A}, g)$ is called *measure space*. If (X, σ_X) is another measurable space and $\Gamma\colon \Omega \to X$ with

$$\forall\, A \in \sigma_X,\ \Gamma^{-1}(A) = \{\omega\colon \Gamma(\omega) \in A\} \in \mathcal{A}\,,$$

then $g(\Gamma^{-1})\colon \sigma_X \to [0, 1]$ is a measure on (X, σ_X), referred to as the *measure image* of g by X. The measure $g\big((a, b)\big) = b - a$ on $(\mathbb{R}, \mathcal{B})$ is called the *Lebesgue measure* on $\mathbb{R}$. The measure space $(\Omega, \mathcal{A}, g)$ is a *probability space* if $g(\Omega) = 1$ and the mapping Γ is called a *random variable* and $g(\Gamma^{-1})$ is the *probability law* of Γ on X. On $(\mathbb{R}, \mathcal{B})$, probability measures can be defined in terms of *probability density functions*. These are functions $p\colon \mathbb{R} \to [0, \infty)$ such that $\int_{-\infty}^{\infty} p(x)\mathrm{d}x = 1$. The associated probability measure is given by $g_\Gamma(A) = \int_A p(x)\mathrm{d}x$ for $A \in \mathcal{B}$. The measure g for a probability space is then denoted by Pr.

A probability space is a set (such as the set of possible outcomes of an experiment) associated with a function $Pr(\cdot)$ that measures subsets (events) of that space. A discrete probability space has a finite or infinite number of possible outcomes. For example, the set of d integers $X = \{1, 2, 3, \ldots, d\}$ or the set of non-negative numbers $X = \{0, 1, 2, \ldots\}$. A problem occurs with the latter example. If each outcome is to be 'equally likely', that is, each element has the same probability, their sum does not add up to one, $\sum_j^\infty p(x_j) \neq 1$, a basic requirement (axiom) in probability theory. If, on the other hand, the number of outcomes is countable, or the space is finite, we calculate the probability of an event $A \subseteq X$ as the sum of the probabilities of the individual outcomes, $Pr(A) = \sum_{x_j \in A} p(x_j)$.

A sample space that has as elements all the points in an interval, or all the points in a union of intervals, on the real line $\mathbb{R}$ is called *continuous*, for example, $X = \{x \colon x > 0\}$. These spaces have always an infinite number of elements. Considering the interval between two points in X, we find that there are 'more' points in any continuous interval than there are integers. Thus a continuous sample space is said to contain an 'uncountable' or 'non-denumerable' number of points.

When the probability spaces are not just infinite but are also continuous, the probability of events cannot be described in terms of a finite sum but leads to the introduction of integrals. Let $A = [a, b]$ be an interval in $X = \mathbb{R}$ from a to b, describing an event in question. Then the integral $Pr(A) = \int_a^b p(x)\, dx$ calculates the probability of A as the area under the graph $p(x)$ (probability density function) restricted by $[a, b]$ (cf. Fig. 2.1). The (mathematical) problem which occurs with continuous spaces is the fact that the probability of getting the exact outcome, say $Pr(a)$, is zero. That is, we cannot 'sum up' the probabilities of the single-element events (points) that are subsets of A. The solution leads to the concepts of σ-*algebra* and the *Lebesgue integral* (measures).

To allow probabilities of events to be calculated by accumulating the likelihood of individual outcomes constituting the event, we do not allow 'all' possible subsets of X to be events: Let all intervals that are subsets of X, and any finite unions and intersections of such intervals be events. In other words, if A and B are events, then $A \cap B$ and $A \cup B$ are possible events. Similarly, if A is an outcome, then A^c is a candidate as are the empty set $\emptyset$ and X.

To introduce the Lebesgue and then fuzzy integral, we first look at the principle idea behind the standard Riemann integral. The idea of the Riemann integral is to measure the area under the graph $f(x)$ by cutting it into narrower and narrower columns and approximating each column with two rectangles; one rectangle just above the graph, the other just below (Figure 13.5). If, as the columns get narrower, the sum of the areas of the smaller rectangles

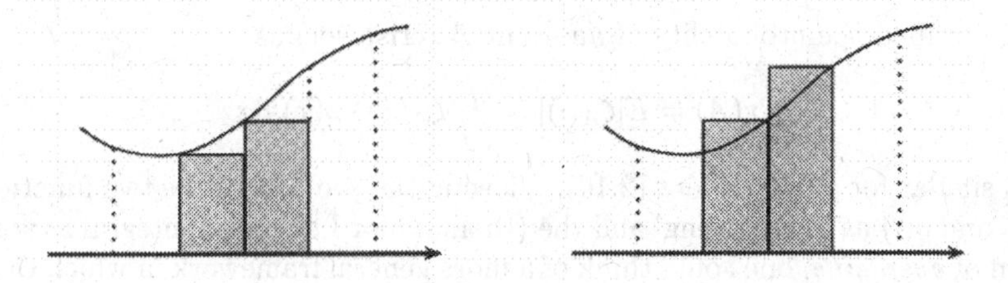

Fig. 13.5 Riemann integration.

approaches that of the larger rectangles, the function is called *Riemann integrable*, and the common limit is the integral. The problem with Riemann integrals is that they are only applicable to continuous functions. This problem can be solved by taking a different view of the integral. The Lebesgue integral takes a somewhat opposite direction than the Riemann integral,. It first looks at the values of $f(x)$ and then at the values of x that produced y (Figure 13.6). As with the Riemann integral, we cut the function $f(x)$ into narrower and narrower intervals but this time into horizontal slices, dividing the function's domain.

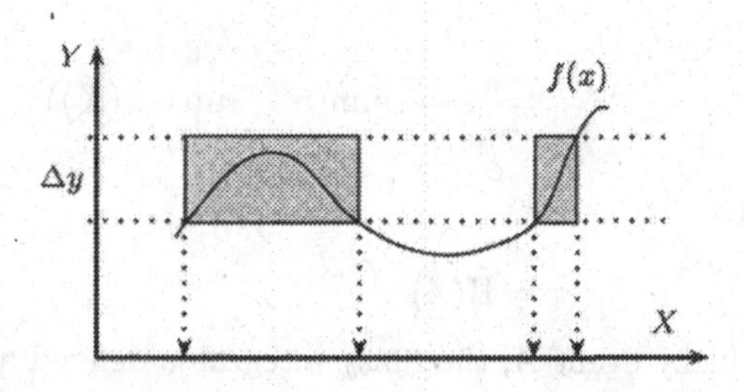

Fig. 13.6 Lebesgue integration.

The calculation of the integral becomes a question of *measuring* the length of the subsets of X. Thought of in higher dimensions we would measure the area, the volume and so forth. It is for this reason that Lebesgue integration is synonymous with measure theory. It was A. Kolmogorov who, by realizing this relationship, based probability theory on measure theory.

Fuzzy or Sugeno integrals are nonlinear generalized Lebesgue-Stieltjes integrals which we use for modelling expectations using a fuzzy measure $g(\cdot)$. Fuzzy integrals are developed analogous to Lebesgue integrals. From the definition of an expectation (2.1) of a function $f(\cdot)$ on X with respect to some function $g(\cdot)$,

$$E[f(\cdot)] = \int_X f(x) \cdot g(x) \, \mathrm{d}x \,,$$

we described the probability of an event A (crisp set) as

$$Pr(A) = E[\zeta_A(\cdot)] = \int_X \zeta_A(x) \cdot p(x)\, dx \tag{2.2}$$

and similar for a fuzzy event (2.10). Viewing the product of the two functions $f(\cdot)$ and $g(\cdot)$ as a 'weighting' and the (Riemann or Lebesgue) integral as some form of *averaging*, one could think of a more general framework in which those *composition* and *accumulation* operations are replaced. For instance, let A be a crisp set and f a function from X to $[0,1]$. Then the fuzzy integral of a function f over A with respect to a fuzzy measure $g(\cdot)$ is defined as

$$⨍_A f(x) \circ g(\cdot) = \sup_{\alpha \in [0,1]} \min\left(\alpha, g(A \cap F_\alpha)\right) ,$$

where $F_\alpha = \{x,\ f(x) \geq \alpha\}$. In terms of the expectation of a characteristic function of a crisp set A we write

$$\begin{aligned} E[\zeta_A(\cdot)] &= ⨍_X \zeta_A(x) \circ g(x) \\ &= \sup_\alpha \min\{\alpha, g(A_\alpha)\} \end{aligned}$$

with $g(\cdot) = \Pi(\cdot)$,

$$= \sup_{\alpha \in [0,1]} \min(\alpha, \sup_{x \in A \cap F_\alpha} \pi(X))$$

and for $f(\cdot) = 1$,

$$= \Pi(A) . \tag{2.14}$$

Similarly, for a fuzzy event A, the fuzzy integral is defined as

$$⨍_A f(x) \circ g(\cdot) = ⨍_X \min(\mu_A(x), f(x)) \circ g(\cdot)$$

with $f(\cdot) = 1$,

$$= ⨍_X \mu_A(x) \circ g(\cdot) .$$

As with probabilities, the expectation of the fuzzy set membership function describes the possibility of fuzzy event A:

$$\begin{aligned} E[\mu_A(\cdot)] &= ⨍_X \mu_A(x) \circ \pi(x) \qquad (2.13) \\ &= \sup_x \min(\mu_A(x), \pi(x)) \\ &= \Pi(A) . \end{aligned}$$

We can interpreted the expectation of a fuzzy event as the degree of consistency of the fuzzy event with the possibility distribution π. In that sense, possibility relates to the degree of feasibility and need not to be associated with a degree of likelihood or frequency. See also Figure 2.1.

Fuzzy measures occur if we replace σ-additivity by requiring the mapping g to be monotonic with respect to set-inclusion only. More specifically, let $\mathcal{A}$ be a family of subsets of a set Ω, with $\emptyset \in \mathcal{A}$. The mapping $g\colon \mathcal{A} \to [0, \infty)$ is called a *fuzzy measure* if

(a) $g(\emptyset) = 0$.

(b) If $A, B \in \mathcal{A}$ and $A \subseteq B$, then $g(A) \leq g(B)$.

The triple $(\Omega, \mathcal{A}, g)$ is a *fuzzy measure space.*

13.5 UNBIASEDNESS OF ESTIMATORS

Let x_j, $j = 1, 2, \ldots, d$ be a random sample of a random variable $\mathbf{x}$ associated with a probability law that has a mean $\eta_{\mathbf{x}}$ and variance $\sigma^2_{\mathbf{x}}$. The *unbiasedness* of the estimator of the variance

$$\hat{\sigma}^2_{\mathbf{x}} = \frac{1}{d-1} \sum_{j=1}^{d} (x_j - \hat{\eta}_{\mathbf{x}})^2 \tag{2.9}$$

is proved by showing that $E[\hat{\sigma}^2_{\mathbf{x}}] = \sigma^2_{\mathbf{x}}$:

$$\begin{aligned}
E\left[\sum_{j=1}^{d} (x_j - \eta_{\mathbf{x}})^2\right] &= E\left[\sum_{j=1}^{d} (x_j^2 - 2x_j\hat{\eta}_{\mathbf{x}} + \hat{\eta}^2_{\mathbf{x}})\right] \\
&= E\left[\sum_{j=1}^{d} x_j^2 - 2\sum_{j=1}^{d} x_j\hat{\eta}_{\mathbf{x}} + \sum_{j=1}^{d} \hat{\eta}^2_{\mathbf{x}}\right] \\
&= E\left[\sum_{j=1}^{d} x_j^2 - 2\sum_{j=1}^{d} x_j\hat{\eta}_{\mathbf{x}} + d\hat{\eta}^2_{\mathbf{x}}\right] \\
&= E\left[\sum_{j=1}^{d} x_j^2 - 2\hat{\eta}_{\mathbf{x}} d\hat{\eta}_{\mathbf{x}} + d\hat{\eta}^2_{\mathbf{x}}\right] \\
&= E\left[\sum_{j=1}^{d} x_j^2 - d\hat{\eta}^2_{\mathbf{x}}\right].
\end{aligned}$$

All x_j are assumed to follow the same probability law. Therefore, $E[x_j^2] = E[x_1 \cdot x_2] = E[\mathbf{x}^2]$. Hence

$$E\left[\sum_{j=1}^{d}(x_j-\hat{\eta}_{\mathbf{x}})^2\right] = \sum_{j=1}^{d} E\left[x_j^2\right] - dE\left[\hat{\eta}_{\mathbf{x}}^2\right]$$
$$= \underbrace{dE\left[\mathbf{x}^2\right]}_{**} - \underbrace{dE\left[\hat{\eta}_{\mathbf{x}}^2\right]}_{**}$$
$$= d\left(\sigma_{\mathbf{x}}^2+\eta_{\mathbf{x}}^2\right) - d\left(\eta_{\mathbf{x}}^2+\frac{\sigma_{\mathbf{x}}^2}{d}\right)$$
$$= d\sigma_{\mathbf{x}}^2 + d\eta_{\mathbf{x}}^2 - d\eta_{\mathbf{x}}^2 - \sigma_{\mathbf{x}}^2 = (d-1)\sigma_{\mathbf{x}}^2 \, .$$

If we define the statistic as (2.9), it follows that

$$E\left[\hat{\sigma}_{\mathbf{x}}^2\right] = \frac{1}{d-1}\, E\left[\sum_{j=1}^{d}(x_j-\hat{\eta}_{\mathbf{x}})^2\right] = \frac{(d-1)\sigma_{\mathbf{x}}^2}{d-1} = \sigma_{\mathbf{x}}^2 \qquad \text{q.e.d.}$$

Notes

* We have

$$\sigma_{\mathbf{x}}^2 = E\left[(\mathbf{x}-\eta_{\mathbf{x}})^2\right] - E\left[\mathbf{x}^2 - 2\mathbf{x}\eta_{\mathbf{x}} + \eta_{\mathbf{x}}^2\right]$$
$$= E\left[\mathbf{x}^2\right] - 2\eta_{\mathbf{x}}E\left[\mathbf{x}\right] + E\left[\eta_{\mathbf{x}}^2\right]$$
$$= E\left[\mathbf{x}^2\right] - \eta_{\mathbf{x}}^2 \qquad \text{or} \qquad E\left[\mathbf{x}^2\right] = \eta_{\mathbf{x}}^2 + \sigma_{\mathbf{x}}^2 \, .$$

** Since $\sigma_{\hat{\eta}}^2 = \frac{\sigma_{\mathbf{x}}^2}{d}$ and $\eta_{\hat{\eta}} = \eta_{\mathbf{x}}$, we have $E\left[\hat{\eta}_{\mathbf{x}}^2\right] = \eta_{\mathbf{x}}^2 + \frac{\sigma_{\mathbf{x}}^2}{d}$.

13.6 STATISTICAL REASONING

In general, we can distinguish the following modes of statistical reasoning:

1. Tests of significance.
2. Statistical generalization (randomization for causal inference and induction).
3. Statistical causality (regression models and time-series analysis).
4. Subjective inference (such as Bayesian theory).

Tests of significance, rooted in the Neyman-Pearson framework of statistical hypothesis testing, are often considered as irrelevant for the empirical sciences suggesting that it is a theory not a method [Wan92]. Throughout the book we used *descriptive* or deductive techniques analyzing a given sample without drawing any conclusions or inferences about the population. Apart from *explorative data analysis* (EDA) and *descriptive statistics*[2], inferential (or inductive) statistics plays a prominent role in statistical theory. Given a representative sample of a *population*, we aim to infer conclusions about the population. The uncertainty of such analysis is quantified using probabilities. Here we briefly list some of the key issues and terms.

Sampling Theory: Methods for obtaining a representative sample.

Estimation Theory

- point estimates vs. interval estimates.
- distribution or 'quality' of the estimators of $\hat{\eta}$ and $\hat{\sigma}^2$.

Confidence Interval: For example, we are confident of finding the statistic (unknown parameter) θ in the interval $\theta \pm 2\sigma_\theta$ (confidence limits), that is, in 95.45% (confidence level) of the time.

Statistical Decision Theory: Decision making about the population:

Tests of Significance: Testing whether an outcome is due to chance or something else. The key idea is that if an observed value is too many standard errors away from its expected value, it is unlikely to be due to chance. Note that chances are assumed to be in the measuring procedure, not the thing being measured.

Hypotheses: Statements about the probability distribution of the population.

H_0: null hypothesis, expressing the idea that an observed difference is due to chance, for example, the probability of no difference is $p = 0.5$.

H_1: alternative hypothesis, for example, $p > 0.5$, $p \neq 0.5$, $p = 0.7$.

Hypothesis Testing: Procedures to decide whether to accept or reject a hypothesis or to determine whether the observed samples differ significantly from expected results.

Type I error: hypothesis rejected when it should be accepted.

Type II error: hypothesis accepted when it should be rejected.

Test Statistics: used to quantify the difference between the data and what is expected on the null hypothesis. For example, the z-statistic

$$z = \frac{\text{observed} - \text{expected}}{\text{standard error}}$$

[2] A good introduction, from a 'descriptive perspective' can be found in [Fre91].

says how many standard errors away an observed value is from its expected value, where the expected value is calculated using the null hypothesis.

Standard Error: describes how big the chance error is.

Significance Level: The maximum probability with which we would be willing to risk a type I error. For example, if there is a 0.05 or 5% level of significance in designing a test of a hypothesis, then there are about 5 chances in 100 that we would reject the hypothesis when it should be accepted, that is, we are about 95% confident that we have made the right decision. We can say that the hypothesis has been rejected at a 0.05% level of significance, which means that we could be wrong with probability 0.05.

***p*-value:** The *p*-value of a test is the chance of getting a big statistic assuming the null hypothesis to be right.

13.7 FREQUENCY ANALYSIS

In this book the focus has been on the analysis of data in the time domain if they were obtained from a time-series. The analysis of signals in the frequency domain is another very powerful framework which we briefly introduce in the present section. The focus will be on the transition from the time domain to the frequency domain by means of integral transformations.

Though many natural or physical systems may, in principle, operate in continuous-time, we assume the situation in which we have a set of observations such as the following sampled sequence:

k	0	1	2	3	4	$\cdots$
$x(k)$	0	0.25	0.5	0.75	1	$\cdots$

In the time domain, we would model such data by means of linear difference equations and the aim in this section is to develop a model of the data in terms of frequency components. Before going any further, we need to make some basic assumptions:

1. **Period Sampling:** We assume the values are sampled periodically every T_s seconds, that is, we have

$$x(kT_s) \doteq x(k), \quad k = 0, 1, \ldots .$$

For most engineering applications this may be no problem as the hardware collecting data is sampling periodically anyway. However, consider medical data collected from patients on a "monthly basis". The number of days between samples usually vary considerably.

2. **Reactive Paradigm:** We assume that the system or process which generated the data can be described sufficiently by means of historical data alone. In other words, the value of x at time k is a function of past values only:
$$x(k) = f\left(x(0), x(1), \ldots, x(k-1)\right) .$$

3. **Linearity:** We assume a linear relationship for $f(\cdot)$:
$$x(k) = a_0 + a_1 x(k-1) + a_2 x(k-2) + \cdots .$$

These assumptions lead to a *linear recurrence difference equation* in correspondence to ordinary differential equations in continuous-time. We develop frequency domain models of sampled-data systems from harmonic analysis in continuous time.

In Section 2.1.2, the Fourier series was introduced as an approximation of a signal $f(t)$ by means of periodic functions. One can also say that the Fourier series decomposes a signal into periodic components:

$$f_p(t) = \frac{a_0}{2} + \sum_{i=1}^{\infty} \left(a_i \cdot \cos(i\omega_0 t) + b_i \cdot \sin(i\omega_0 t)\right) . \tag{2.52}$$

An alternative equivalent formulation is the complex Fourier series

$$f_p(t) = \sum_{i=-\infty}^{\infty} c_i \cdot e^{\mathrm{j} i\omega_0 t},$$

where the complex Fourier coefficients c_i are obtained as

$$c_i = \frac{\omega_0}{2\pi} \int_{-\pi/\omega_0}^{\pi/\omega_0} f_p(t) e^{\mathrm{j} i\omega_0 t} \mathrm{d}t .$$

Here, the period of $f_p(t)$ equals $2\pi/\omega_0$. Each c_i is the value of the complex frequency component of $f_p(t)$ at $i\omega_0$. As shown in Figure 13.7, the amplitude spectrum of $f_p(t)$ can be visualized. Apart from the visualization of frequency components in the signal, the transformation into the frequency domain has several advantages: For many signals and systems, simple parametric expressions can be found analytically; solutions of differential/difference equations become algebraic in nature, allowing simple manipulation and analysis of complex systems. Hence a transformation into the frequency domain is not only useful for harmonic analysis but for systems analysis in general.

A generalization of the Fourier series is obtained if we wish to consider more and more frequencies, that is, shrinking $\Delta\omega$ in the complex Fourier series. To achieve this, $\omega_0 \to 0$, the fundamental period $2\pi/\omega_0$ must increase infinitely.

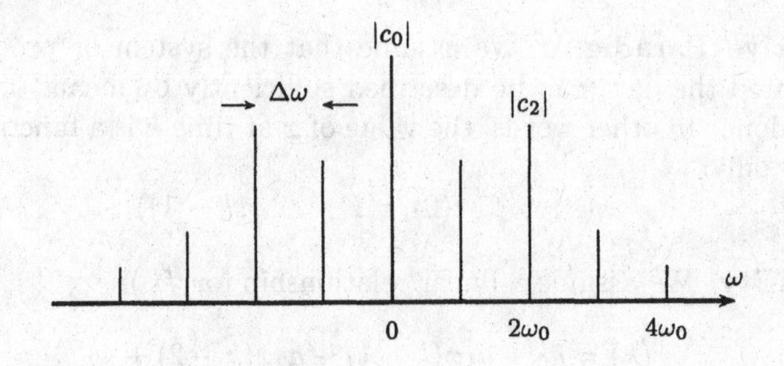

Fig. 13.7 Discrete amplitude spectrum obtained from Fourier analysis.

Replacing ω_0 by $\Delta\omega$ to emphasize the vanishing frequency interval, we obtain the following expression:

$$
\begin{aligned}
f_p(t) &= \sum_{i=-\infty}^{+\infty} c_n \cdot e^{\mathrm{j}i\Delta\omega t} \\
&= \sum_{i=-\infty}^{+\infty} \left(\frac{\Delta\omega}{2\pi}\right) \int_{-\pi/\Delta\omega}^{\pi/\Delta\omega} f(\tau) e^{-\mathrm{j}i\Delta\omega\tau} \mathrm{d}\tau \; e^{\mathrm{j}i\Delta\omega t} \\
&= \frac{1}{2\pi} \sum_{i=-\infty}^{+\infty} \left[\int_{-\pi/\Delta\omega}^{\pi/\Delta\omega} f(\tau) e^{-\mathrm{j}i\Delta\omega\tau} \mathrm{d}\tau \right] e^{\mathrm{j}i\Delta\omega t} \Delta\omega \,.
\end{aligned}
$$

As $\Delta\omega$ approaches zero, the product $i\omega_0$ always equals ω. As a result, the fundamental period of $f_p(t)$ increases to infinity such that $f_p(t)$ approaches $f(t)$:

$$
\begin{aligned}
f(t) &= \frac{1}{2\pi} \lim_{\Delta\omega\to 0} \sum_{i=-\infty}^{+\infty} \left[\int_{-\pi/\Delta\omega}^{\pi/\Delta\omega} f(\tau) e^{-\mathrm{j}i\Delta\omega\tau} \mathrm{d}\tau \right] e^{\mathrm{j}i\Delta\omega\tau} \Delta\omega \\
&= \frac{1}{2\pi} \int_{-\infty}^{\infty} \left[\int_{-\infty}^{\infty} f(\tau) e^{-\mathrm{j}\omega\tau} \mathrm{d}\tau \right] e^{\mathrm{j}i\omega t} \mathrm{d}\omega \\
f(t) &\doteq \frac{1}{2\pi} \int_{-\infty}^{\infty} F(\mathrm{j}\omega) e^{\mathrm{j}\omega t} \mathrm{d}\omega \,.
\end{aligned}
$$

Here

$$
F(\mathrm{j}\omega) = \int_{-\infty}^{\infty} f(t) e^{-\mathrm{j}\omega t} \mathrm{d}t
$$

is called the *Fourier transform* or *spectrum* of $f(t)$. For the integral to converge we have the condition that $f(t) \to 0$ for $t \to \infty$ which often may not be case. To ensure convergence, we use the following trick. Multiplication of $f(t)$ by a factor $e^{-|\sigma|t}$ ensures that the product $f(t)e^{-|\sigma|t}$ diminishes for increasing t. Let us also assume, without loss of generality, that $f(t)$ starts at

$t = 0$ with $f(t) = 0$ for $t < 0$. Using the notation $s \doteq \sigma + \mathrm{j}\omega$, we obtain the so-called *Laplace transform* of $f(t)$ as

$$F(s) = \int_0^{\sigma+\mathrm{j}\omega} f(t)e^{-st}\mathrm{d}t$$

and vice versa from $F(s)$ we obtain $f(t)$ as

$$f(t) = \frac{1}{2\pi\mathrm{j}} \int_{\sigma-\mathrm{j}\omega}^{\sigma+\mathrm{j}\omega} F(s)e^{st}\mathrm{d}s \; .$$

Next, we derive the Laplace transform of a sampled, that is, discrete-time, signal. The Laplace transform assumes a continuous-time signal while here we assume we are given the sequence $x(kT_s)$. Once again we use a trick to develop a mathematical model of the sampling process by introducing the Dirac impulse $\delta(t)$ with the property $\int \delta(t) = 1$. The sampled signal is then described by

$$\begin{aligned} x^*(t) &= x(t) \cdot \sum_{i=0}^{\infty} \delta(t - kT_s) \\ &= \sum_{k=0}^{\infty} x(kT_s)\delta(t - kT_s) \; . \end{aligned}$$

The Laplace transform of $x^*(t)$ is then obtained as

$$\begin{aligned} F^*(s) &= \int_0^{\infty} \sum_{i=-\infty}^{\infty} x(t)\delta(t - kT_s)e^{-st}\mathrm{d}t \\ &= \sum_{k=-\infty}^{\infty} \int_0^{\infty} x(t)e^{-st}\delta(t - kT_s)\mathrm{d}t \\ &= \sum_{k=0}^{\infty} x(ksT_s) \qquad \text{where } t = kT_s \; . \end{aligned}$$

If we denote e^{sT_s} by z, we obtain what is called the *z-transform* of a discrete-time signal:

$$F^*(s) = \sum_{k=0}^{\infty} x(k)z^{-k} \; .$$

Bibliography

Bab98. R. Babuska. *Fuzzy Modelling for Control.* Kluwer, Dodrecht, The Netherlands, 1998.

Bez81. J. C. Bezdek. *Pattern Recognition with Fuzzy Objective Function Algorithms.* Plenum Press, New York, 1981.

Boo58. G. Boole. *An Investigation of the Laws of Thought on which are Founded the Mathematical Theories of Logic and Probabilities.* Dover, New York, 1858 (1958).

Bun98. M. Bunge. *The Philosophy of Science.* Transaction Publishers, New Brunswick, USA, 1998.

Cas92. J. L. Casti. *Reality Rules: Picturing the World in Mathematics.* Wiley, New York, 1992.

CM98. V. Cherkassky and F. Mulier. *Learning from Data.* Wiley, New York, 1998.

DP83. D. Dubois and H. Prade. Unfair coins and necessity measures: Towards a possibilistic interpretation of histograms. *Fuzzy Sets and Systems*, 10:15–20, 1983.

FPP97. D. Freedman, R. Pisani, and R. Purves. *Statistics.* Norton and Company, 1997.

FR94. J. F. Fodor and M. Roubens. *Fuzzy Preference Modelling and Multicriteria Decision Support.* Kluwer, Dodrecht, The Netherlands, 1994.

Fre91. D. Freedman. *Statistics.* W. W. Norton, 1991.

GMN97. I. R. Goodman, R. P. S. Mahler, and H. T. Nguyen. *Mathematics of Data Fusion.* Kluwer, Dodrecht, The Netherlands, 1997.

HB93. R. J. Hathaway and J. C. Bezdek. Switching regression models and fuzzy clustering. *IEEE Transactions on Fuzzy Systems*, 3:195–204, 1993.

HKKR99. F. Höppner, F. Klawonn, R. Kruse, and T. Runkler. *Fuzzy Cluster Analysis.* Wiley, Chichester, England, 1999.

Höh88. U. Höhle. Quotients with respect to similarity relations. *Fuzzy Sets and Systems*, 27:31–44, 1988.

JW98. R. A. Johnson and D. W. Wichern. *Applied Multivariate Statistical Analysis.* Prentice-Hall, New Jersey, fourth edition, 1998.

Kal60. R. E. Kalman. A new approach to linear filtering and prediction problems. *Journal of Basic Engineering*, pages 35–45, March 1960.

KGK94. R. Kruse, J. Gebhardt, and F. Klawonn. *Foundations of Fuzzy Systems.* Wiley, Chichester, England, 1994.

KK97. F. Klawonn and R. Kruse. Constructing a fuzzy controller from data. *Fuzzy Sets and Systems*, 85:177–193, 1997.

Kun96. L. I. Kuncheva. On the equivalence between fuzzy and statistical classifiers. *International Journal of Uncertainty, Fuzziness and Knowledge-Based Systems*, 4(3):245–253, 1996.

Lju87. L. Ljung. *System Identification.* Prentice-Hall, New Jersey, 1987.

LS97. F. W. Lawvere and S. H. Schanuel. *Conceptual Mathematics.* Cambridge University Press, 1997.

Mag97. B. Magee. *The Philosophy of Schopenhauer.* Clarendon Press, 1997.

MHL72. T. J. McAvoy, E. Hsu, and S. Lowenthal. Dynamics of ph in controlled stirred tank reactor. *Ind. Eng. Chem. Process Des. Develop.*, 11(1):68–70, 1972.

NW97. H. T. Nguyen and E. A. Walker. *A First Course in Fuzzy Logic.* CRC Press, Boca Raton, Florida, 1997.

Pap91. A. Papoulis. *Probability, Random Variables, and Stochastic Processes.* McGraw-Hill, New York, third edition, 1991.

Par62. E. Parzen. On estimation of a probability density function and mode. *Annals of Math. Stat.*, 33:1065–1076, 1962.

Ped93. W. Pedrycz. *Fuzzy Control and Fuzzy Systems.* RSP - Research Studies Press, Taunton, Somerset, England, second edition, 1993.

PY98. K. M. Passino and S. Yurkovich. *Fuzzy Control.* Addison Wesley Longman, Menlo Park, California, 1998.

Ros85. R. Rosen. *Anticipatory Systems.* Pergamon Press, 1985.

Ros91. R. Rosen. *Life Itself.* Columbia University Press, New York, 1991.

Sil86. B. W. Silverman. *Density Estimation.* Chapman and Hall, London, 1986.

SS83. B. Schweizer and A. Sklar. *Probabilistic Metric Spaces.* North-Holland, New York, Oxford, 1983.

SY89. W. Siler and H. Ying. Fuzzy control theory: The linear case. *Fuzzy Sets and Systems*, 33:275–290, 1989.

TS85. T. Takagi and M. Sugeno. Fuzzy identification of systems and its application to modeling and control. *IEEE Transactions on Systems, Man, and Cybernetics*, SMC-15:116–132, 1985.

Vap98. V. N. Vapnik. *Statistical Learning Theory.* Wiley, New York, 1998.

Wan92. C. Wang. *Sense and Nonsense of Statistical Inference.* Marcel Dekker, New York, 1992.

Wan97. L. X. Wang. *A Course in Fuzzy Systems and Control.* Prentice-Hall, New Jersey, USA, 1997.

Wol98. O. Wolkenhauer. *Possibility Theory with Applications to Data Analysis.* RSP - Research Studies Press, Taunton, Somerset, England, 1998.

Wol01. O. Wolkenhauer. Random-sets: Theory and applications. In W. Pedrycz, editor, *Granular Computing: An Emerging Paradigm.* Springer Verlag, Berlin, Germany, 2001.

YSB90. H. Ying, W. Siler, and J. J. Buckley. Fuzzy control theory: A nonlinear case. *Automatica*, 26(3):513–520, 1990.

Zad71. L. A. Zadeh. Similarity relations and fuzzy orderings. *Information Sciences*, 3:177–200, 1971.

Index

9 780471 416562